Computing Supplementum 7

G. Tinhofer, E. Mayr, H. Noltemeier,
M. M. Syslo (eds.)
in cooperation with R. Albrecht

Computational Graph Theory

Springer-Verlag Wien New York

Prof. Dr. G. Tinhofer
Institut für Mathematik
Technische Universität München
Federal Republic of Germany

Prof. Dr. E. Mayr
Department of Computer Science
Stanford University
Calif., U.S.A.

Prof. Dr. H. Noltemeier
Institut für Informatik
Universität Würzburg
Federal Republic of Germany

Prof. Dr. M. M. Syslo
Institute of Computer Science
University of Wrocław
Poland

Prof. Dr. Rudolf Albrecht
Institut für Informatik
Universität Innsbruck
Austria

With 68 Figures

ISSN 0344-8029
ISBN-13:978-3-211-82177-0   e-ISBN-13:978-3-7091-9076-0
DOI: 10.1007/978-3-7091-9076-0

Printed on acid-free paper

Preface

One of the most important aspects in research fields where mathematics is applied is the construction of a formal model of a real system. As for structural relations, graphs have turned out to provide the most appropriate tool for setting up the mathematical model. This is certainly one of the reasons for the rapid expansion in graph theory during the last decades. Furthermore, in recent years it also became clear that the two disciplines of graph theory and computer science have very much in common, and that each one has been capable of assisting significantly in the development of the other. On one hand, graph theorists have found that many of their problems can be solved by the use of computing techniques, and on the other hand, computer scientists have realized that many of their concepts, with which they have to deal, may be conveniently expressed in the language of graph theory, and that standard results in graph theory are often very relevant to the solution of problems concerning them. As a consequence, a tremendous number of publications has appeared, dealing with graphtheoretical problems from a computational point of view or treating computational problems using graph theoretical concepts. Due to these facts, graph theory and computer science have become so strongly connected that it seems no overstatement to say that in our days a good deal of modern graph theory is part of computer science or, on the other hand again, to say that computer science is at least partially based on graph theory.

The purpose of this supplementary volume is to draw attention to problems and applications which represent the strong connection between the two disciplines. It contains a collection of invited papers, each devoted to a particular class of graphtheoretical problems and their solution by computational techniques. Although the papers are all written on an expert and not on a teaching level they are supposed to be suitable for a first contact with the contents. The collected papers cover a broad spectrum of graphtheoretical topics. Techniques for hard graph problems, problems on planar graphs, path problems, coloring problems, graphs and orders and several other topics are discussed, as well as computational aspects like data structures, the probabilistic behaviour of algorithms, the design and analysis of parallel algorithms and VLSI-structures. No claim for completeness is made, however, we believe that the collection is representative in the sense that all the main topics of „computational graph theory" are included.

February 1990   The Editors

Contents

Computing Suppl. 7, 1–15 (1990)

Computing

© by Springer-Verlag 1990

Efficient Computations in Tree-Like Graphs

Andrzej Proskurowski*, Eugene, Oregon, and **Maciej M. Sysło†**, Berlin

Abstract — Zusammerfassung

Efficient Computations in Tree-Like Graphs. Many discrete optimization problems are both very difficult and important in a range of applications in engineering, computer science and operations research. In recent years, a generally accepted measure of a problem's difficulty became a worst-case, asymptotic growth complexity characterization. Because of the anticipated at least exponential complexity of any solution algorithm for members in the class of $\mathcal{NP}$-hard problems, restricted domains of problems' instances are being studied, with hopes that some such modified problems would admit efficient (polynomially bounded) solution algorithms. We survey investigations of the complexity behavior of $\mathcal{NP}$-hard discrete optimization problems on graphs restricted to different generalizations of trees (cycle-free, connected graphs.) The scope of this survey includes definitions and algorithmic characterization of families of graphs with tree-like structures that may guide the development of efficient solution algorithms for difficult optimization problems and the development of such solution algorithms.

AMS Subject Classifications: 05C; 68B, C, E, F; 90B, C.

Key words: tree-like graphs, decomposable graphs, k-trees, tree-width

Effiziente Algorithmen für Graphen mit Baum-ähnlichen Graphen. Viele diskrete Optimierungsprobleme sind einerseits schwer zu lösen, haben andererseits aber viele Anwendungen in den Ingenieurwissen-schaften, in der Informatik oder in Operations Research. Ein allgemein akzeptiertes Maß für die Schwierigkeit eines Problems ist die asymptotische worst-case Komplexität. Für NP-schwere Probleme ergibt sich danach für jeden Algorithmus eine exponentielle Laufzeit. Schränkt man sich jedoch auf Teil-klassen der zugrunde liegenden Strukturen ein, so lassen diese oft effiziente (polynomial beschränkte) Algorithmen zu. Wir geben einen Überblick über die Komplexität NP-schwerer Probleme auf Graphen mit verallgemeinerter Baumstruktur (z.B. zykelfrei, zusammenhängend). Es werden Definitionen und algorithmische Charakterisierungen von Familien solcher Graphen gegeben, die bei der Entwicklung von effizienten Lösungsalgorithmen hilfreich sein können.

1. Motivation

The framework in which we are interested in tree-like graphs consists of finding restricted classes of graphs for which many generally difficult decision and optimization problems are efficiently solvable (in the worst case). These graphs often exhibit

---

* Department of Computer and Information Science, University of Oregon, Eugene, Oregon 97403, USA. Research supported in part by the Office of Naval Research Contract N00014-86-K-0419.
† FB 3-Mathematik, TU Berlin. On leave from the Institute of Computer Science, University of Wrocław, Przesmyckiego 20, 51151 Wrocław, Poland. Research supported by the grant RP.I.09 from the Institute of Informatics, University of Warsaw and by a grant from the Alexander von Humboldt-Stiftung.

some decomposability properties. Our own research has concentrated recently on algorithmic aspects of graph representations of orders and on partial k-trees, also known as graphs with tree-width k. Those graphs are all partial subgraphs of chordal graphs with the maximum clique size bounded by k. We will discuss alternative views of these and other families of graphs, classes of problems efficiently solvable on these graphs, and the relevant algorithm design paradigms.

Application domain problems often translate into optimization problems on the graphs representing the application; these *combinatorial problems* are often very difficult. A measure of complexity of a combinatorial problem is the *worst-case, asymptotic* behavior of the time to compute a solution as a function of the problem size. A class of problems notorious for their difficulty is that of $\mathcal{NP}$-*complete* problems.

Let us first state our vocabulary for discussing *discrete optimization problems* defined on *combinatorial graphs*. A (combinatorial) graph $G = (V, E)$ consists of the set V of *vertices* and the set E of *edges*, each edge *incident* with its two *end-vertices* (which are thus *adjacent*). A *subgraph* of a given graph $G = (V, E)$ *induced* by a subset of vertices $V' \subset V$ consists of all edges from E that are not adjacent to any vertex from $V - V'$. A *partial subgraph* on the same set of vertices involves a subset of edges. A sequence of different vertices $v_0, v_1, \ldots, v_n$ such that v_{i-1} and v_i $(0 < i \leq n)$ are adjacent is called a *path* of length n; if v_0 and v_n are identical $(n \geq 2)$, we have a *cycle*. A graph with n vertices and edges between all pairs of distinct vertices is called *complete* and denoted by K_n. A graph in which there exists a path between any two of its vertices is said to be *connected*; a set $S \subset V$ such that the subgraph of a graph $G = (V, E)$ induced by $V - S$ is not connected is called a *separator* of G. A *tree* is a connected graph without any cycles (it is easy to see that any minimal separator in a tree consists of exactly one vertex).

2. Definitions of Some Tree-Like Graphs

We first present definitions of several classes of graphs by their *generative* descriptions. Often they also have *analytical* descriptions based on a process of *decomposition*. Such a description gives a *parse tree* in which each node corresponds to a subgraph of the original graph.

2.1. Generative Definitions of Classes of Graphs

Many families of graphs admitting recursive descriptions that invoke some kind of decomposability can be described by their *iterative construction*, often expressible by a recursive (hierarchical) construction rules as well. The former consists of primitive graphs and composition rules, the latter takes often a formal linguistic form.

The grammatical approach to defining families of graphs is exemplified by *context-free hyper edge replacement* grammars [76], [41], [39], [40], [49], [50], [51]. A grammar consists of a finite set of non-terminal labels N with a distinguished start label $s \in N$ and a finite set of rules, each having the *left-hand side*, a hyperedge with a label $a \in N$, and the *right-hand side*, a labeled hypergraph H. During an application of the rule, a hyperedge labeled a is replaced by the hypergraph H in such a way that some distinguished nodes of H ('terminals', 'sources') are identified with the corresponding nodes of the replaced hyperedge. The right-hand side of a *terminal rule* has only unlabeled two-edges, so that the language of such grammar contains only combinatorial graphs. Below, we list some other formalisms following this approach in defining tree-like classes of graphs.

k-trees are defined recursively as either K_{k+1}, the completely connected graph on $k + 1$ vertices, or two k-trees 'glued' along K_k subgraphs. Iterative definition includes K_{k+1} as the primitive graph and defines an $n + 1$-vertex k-tree ($n > k$) as an n-vertex k-tree T augmented by an extra vertex adjacent to all vertices of a K_k subgraph of T (see [11]).

hook-up graphs generalize k-trees by allowing any base graph A and any subgraph B of a Hook-up (A, B) graph to which a new vertex is made adjacent. (Thus, k-trees are Hook-up (K_{k+1}, K_k) graphs, see [47].)

k-terminal graphs are closed with respect to a finite number of composition operations, where two graphs G_1 and G_2 with terminal label sets $\{1 \ldots k\}$ each are composed either by identifying terminals in G_1 and G_2 that induce isomorphic subgraphs, or by adding edges between terminals in G_1 and terminals in G_2. The composition is completed by determination of terminal vertices in the new graph. A k-terminal family contains also basis graphs with all vertices terminal [82], [83], [84].

k-terminal recursive family involves recursive operations on graphs with at most k *terminal* vertices labeled $1 \ldots |T|$. An operation is determined by a *connection matrix* that indicates which of the input graphs' terminal vertices are identified to create a vertex of the resulting graph; some of these vertices are labeled terminals of the new graph. There is also a set of k-terminal base graphs with no non-terminal vertices, and a set of operations [18].

Any graph family defined by a context-free hyperedge replacement grammar (or an equivalent formalism) has bounded tree-width. For any given k, there is a context-free hyperedge replacement grammar generating all partial k-trees. Specifically, such a grammar would have $N = \{a\}$, the only label of a hyperedge with k vertices and the replacement rule set that consists of (terminal) rules substituting each of the $2^{\binom{k}{2}}$ edge combinations for the hyperedge and of the rule replacing such a hyperedge with a hypergraph consisting of $k + 1$ hyperedges spanned on $k + 1$ vertices (including the k original vertices). Similarly, all graphs generated by any k-terminal recursive family of graphs have bounded tree-width. It is also possible to generate all partial k-trees using that formalism but we will not show it here for the large size of rules. Actually, the formalisms of hyperedge replacement and that of k-terminal recursive family are equivalent.

2.2. Decomposability

2.2.1. Undirected Graphs

k-trees are alternatively defined as connected graphs with no K_{k+2} subgraphs and with all minimal separators inducing K_k [68]. This property of the existence of small (constantly bounded) separators is inherited by their subgraphs, partial k-trees. Namely, partial subgraphs of k-trees are exactly the k-*decomposable* graphs: a graph G is k-decomposable if and only if it either has at most $k + 1$ vertices, or has a separator S with at most k vertices ($G - S$ has $m \geq 2$ connected components C_1, $C_2, \ldots, C_m$) such that each graph G_i obtained from the component C_i ($1 \leq i \leq m$) extended by S with its vertices completely connected is k-decomposable (Arnborg and Proskurowski [5]). This motivates a closer look at partial k-trees, their recognition and embedding algorithms, as well as efficient algorithms solving $\mathcal{NP}$-hard optimization problems restricted to partial k-trees (for fixed k).

The *tree-width* of a graph G ([67]) is defined as one less than the maximum size of vertex sets $V_1, V_2, \ldots, V_m$ into which one can pack vertices of G (one vertex belonging possibly to more than one set) such that

- $V_1 \cup \ldots \cup V_m = V(G)$,
- edges are only between vertices in the same sets,
- G is representable by a tree T which has V_i as nodes and, for $V_i, V_j, V_l \in V(T)$, if there is a path in T between V_i and V_l containing V_j then $V_i \cap V_l \subseteq V_j$.

It is not too difficult to see that partial k-trees are exactly graphs with tree-width k (assuming that the definition calls for the minimum value of the parameter k): In one direction, given a partial k-tree, take an embedding k-tree and its $(k + 1)$-cliques as V_i's. In the other direction, since V_i's intersect on at most k vertices, complete each V_i by adding edges between all pairs of its vertices, and then increase the cliques' sizes to $k + 1$ by adding edges between some vertices of neighboring cliques.

By a similar construction, one can see that graphs generated by the other formalisms mentioned above have bounded tree-width, as well.

Hochberg and Reischuk, [43], define (k, μ)-decomposable graphs for which any decomposition into k-connected components yields μ as the maximum size of a component. It is again easy to see that these graphs have constantly bounded tree-width.

Lauteman [49, 50] defines s-decomposition trees similarly to the parse trees of the above k-terminal graphs. He gives a finite set of *rewrite rules*, where left-hand-side graphs have distinguished terminal vertices to be identified with the corresponding terminal vertices of the right-hand-side graphs. The constant bound s is equal to the maximal size of the right-hand-side graphs of the rewrite rules.

2.2.2. Directed decomposable graphs

In the past decade, there has been very active research in the area of a general theory of set decomposition and in particular directed graph (digraph) decomposition.

Cunningham and Edmonds, [29], discuss general decomposition of sets, followed by [28] who discusses digraph decomposition. A *composition* of two diagraphs G_1 and G_2 is defined as the digraph with the vertex set equal to union of their vertices (with the exception of a special vertex v repeated in both). The arcs of the resulting digraph reflect transitivity of arcs through v (i.e., $(u_1, u_2) \in E$ if and only if $((u_1, v) \in E_1 \wedge (v, u_2) \in E_2) \vee ((u_1, v) \in E_2 \wedge (v, u_2) \in E_1)$). Each strongly connected digraph has a unique minimal decomposition into components that can be 'prime' (non-decomposable), 'brittle' (every partition is a *split*), or 'semi-brittle' (*circular splits*). Furthermore, the decomposition of semibrittle digraphs can proceed into distars and '*circles of transitive tournaments*'.

A special case of the composition operation is the *substitution*, when $G_2 = vG_2'$ is a *pointed* graph, with vertex v adjacent to all the remaining vertices of G_2. Graphs obtained by substitution can be described using the notion of *autonomous sets*: A subgraph G_2 of G is autonomous if and only if for every vertex u in $V(G) - V(G_2)$ either $\forall v \in V(G_2) : (u, v) \in E(G)$ or $\forall v \in V(G_2) : (u, v) \notin E(G)$. This treatment allows to include in the consideration graphs with disconnected components. There is also a unique decomposition theorem for autonomous sets: Each graph has the composition tree whose nodes are blocks of partitions into autonomous sets of the graph, each denoted by D (degenerate, disconnected, 'parallel'), C (complete, 'series'), or P (prime, decomposable arbitrarily into C and D components). Fast algorithms for finding decomposition of directed graphs into such subgraphs are presented in [19].

Decomposition often allows efficient solution algorithms for some discrete optimization problems. However, the range of those problems is severely limited in the split case requiring strong connectivity. Decomposition by substitution allows using the *divide and conquer* algorithms parallelling a natural factorization of objective functions in many discrete optimization problems. An excellent survey is given in [62]. A subsequent work ([42]) treat problems on posets and uses decomposition with prime elements of bounded size.

2.3. Other Combinatorial Structures with Parse Trees

Chordal graphs can be defined by a similar recursive construction (or decomposition) description as the graph families from section 2.1: Starting with a single vertex, any chordal graph (and only such graphs) can be constructed by adding a new vertex adjacent to all vertices of any complete subgraph of a chordal graph ([30, [36], [75]]). Here, the generic definition of an *infinite set* of primitive graphs (K_k for *any* value of k) makes the major difference. Nevertheless, chordal graphs can be represented by their parse trees (clique trees) with help of which some algorithmic problems can be solved efficiently. Chordal graphs constitute an example of graphs decomposable by clique separators, however of unbounded size ([37, 80]).

Chordal graphs can be interpreted as intersection graphs of subtrees in trees, see for instance Golumbic [38]. An important subfamily of chordal graphs consists of interval graphs, the intersection graphs of intervals of a line. On the other hand, a class of intersection graphs not properly contined in chordal graphs is the class of circular-arc graphs, the intersection graphs of intervals of a circle.

The existence of the unique parse tree (corresponding to the constructive definition of this class of graphs) contributes to the design of many efficient algorithms for *complement reducible* graphs. These are the graphs that can be reduced to single vertices by recursively complementing all connected subgraphs (Corneil *et al.* [23]).

3. Complexity of Parsing of Tree-Like Graphs

The problem of recognition and embedding of a partial k-tree for a fixed value of k is polynomially solvable, see Arnborg, Corneil and Proskurowski [2]. The algorithm recognizing a partial k-tree with n vertices has complexity $\mathcal{O}(n^{k+2})$; any lower bounds result on the complexity of the problem might help to explain difficulties with finding a system of confluent rewrite rules recognizing partial k-trees for $k > 3$. A related—and very important from the applications point of view—problem is that of finding the minimum value of k for which a given graph is k-decomposable (or, equivalently, is a partial k-tree). This problem is $\mathcal{NP}$-hard, as shown by Arnborg *et al.* [2].

Any sequence of applications of rewrite rules that reduce a given partial 2-tree to the empty graph determines also an embedding of the graph in a full 2-tree. This is so, because the reduction 'reverses' a feasible generation process of the full 2-tree. An application of a reduction rule can be thought of as 'pruning' of a 2-*leaf* (vertex of degree 2) which is deleted, leaving as a trace an edge connecting its two neighbors. A similar pruning of 3-leaves (completion of a triangle spanned on neighbors of a vertex of degree 3, in a 'star-triangle substitution' process) in recognition of partial 3-trees must be done with care, since not all vertices of degree 3 in a partial 3-tree can be 3-leaves of an embedding in a full 3-tree, and an indiscriminate pruning may lead to irreducible graphs (Arnborg and Proskurowski [8]).

The system of confluent rewrite rules reducing any partial 3-tree (and only a graph from this class) to the empty graph allows for a linear recognition of a partial 3-tree and construction of its embedding in a full 3-tree (Matousek and Thomas [57]). One could describe those rules reducing vertices of degree 3 as based on a combination of 'strength' of their neighborhood (existing edges between their neighbors), and of 'relation' to other vertices of degree 3 (the nature of shared neighborhoods with those vertices). The rewrite rules are given in Arnborg and Proskurowski [8]. Thus, one could suspect that for a safe reduction of vertex v of degree k in a partial k-tree G, there seems to be required certain trade-off between the amount of mutual connection among the k neighbors of v, the number of other vertices of degree k sharing their neighborhood with v, and the strength of this sharing. For some general rules see Arnborg and Proskurowski [6].

Attempts to generalize this approach to higher values of k have not brought any success, so far. A reason might be that while the two abovementioned rules of thumb are straightforward enough for $k = 3$ the sheer number of combinations to consider for $k > 3$ is difficult to handle. Another reason might be that such a complete system of confluent rewrite rules does not exist for higher values of k.

4. Problems with Efficient Solution Algorithms on Tree-Like Graphs

Discrete optimization problems that do not involve counting and that are defined on graphs, can be viewed simply as graph properties that a given graph has or does not have. Typical examples are 2-colorability ('Is a given graph 2-colorable?') and Hamiltonicity ('Does a given graph have a Hamilton cycle?'). These properties can be expressed as well-formed formulea in some formalism utilizing variable symbols, relational symbols (over some domains), logical connectives, and quantifiers. Depending on the restrictions on the use of these symbols, one defines languages of varying descriptive power. For instance, one could restrict relations to a single domain or use many-sorted structures, allow only existential quantification, restrict the domains of quantifiers, and so on. It is important to find formalisms that balance their power of expression and the ease of analysis (the complexity of property recognition).

In [24], Courcelle presents an excellent survey of the interaction between logic languages and graph properties, defining and analyzing the power of First Order Logic, Second Order Logic, Monadic Second Order Logic, and their extensions.

First Order Logic: The domain: graph elements (vertices and edges).
Basic relations: $V(x)$, $E(x)$, $R(x, y, z)$ denoting vertex set, edge set, and edge with incident vertices, respectively.
Quantification: over domain variables.
Examples: A given graph labeling is a proper coloration. All vertices have degree bounded by a given integer.

Second Order Logic: Variables: graph elements, relations over graph elements.
Quantification: over binary relation variables (and, consequently, over relational variables of any arity).
Example: Two given graphs are isomorphic.

Monadic Second Order Logic: Restriction: relational variables denoting sets only (relations on one variable).
Examples: A given graph is Hamiltonian. A given graph is m-colorable.

Although First Order Logic is a rather weak formalism as far as the expressive power is concerned, it is in general undecidable whether a general graph has a property described in this language. Thus, an interesting avenue of investigations is to consider the status of problems defined in these formalisms but restricted to some narrower classes of graphs. For instance, when applied to context-free hyper-edge replacement graphs, even Monadic Second Order Logic ($MSOL$) is decidable. When the class of graphs is restricted to confluent Node Label Controled graphs, (NLC [70]), Monadic Second Order Logic with quantification only over vertex sets

($MSOL_v$) is decidable. Thus, it makes sense to inquire about the computational complexity of such problems on those graphs. An important connection between investigations of decidability of logical theories and the tree-like graphs is established by the following statement: 'For a property described by a Monadic Second Order Logic expression, one can decide in polynomial time whether a given partial k-tree has this property.' ([27], [3].) Similarly, if the property is expressed in $MSOL_v$ and the graph belongs to the class of confluent NLC languages, then there exist efficient decision algorithms, as well.

To be able to deal with discrete problems optimizing over some objective functions, the $MSOL$ formalism has been extended by Courcelle [26] and by Arnborg et al. [3] allowing counting set cardinalities, and evaluating sums of functions of sets, respectively. Thus, the properties in the above statement have to be extended to those described by $CMSOL$ and $EMSOL$ expression, respectively.

The importance of the bounded tree-width is shown by the following theorem of Seese [73]: Any class of graphs that has a decidable $MSOL$ property has bounded tree-width.

Arnborg et al. [3] present a detailed description of applications of $MSOL$ to partial k-trees. The authors' main result is the efficient solvability of a number of problems $\mathcal{NP}$-hard for general graphs. They prove it constructively by reducing decidability of an $EMSOL$ property for a partial k-tree G to the problem of deciding the corresponding, linearly definable property of a binary tree representing parsing of G (its 'tree decomposition'). For the latter problem, they construct a tree automaton that computes a solution in linear time. This tree automaton is found using results about Decision Problems in SOL, obtained in the 1960's (Doner [31], Thatcher and Wright [81]). Since the transformation itself (the derived property) is linear and the parse tree is assumed to be given with the input graph, a linear time solution algorithm for the original problem is obtained.

An interesting exception to the spirit of the recent results on efficient algorithms for problems on partial k-trees is the polynomial-time algorithm for the graph isomorphism problem (Bodlaender [16]), since that problem is *not* expressible by the proposed extensions to $MSOL$.

We should mention other recent attempts to characterize problems solvable efficiently on partial k-trees, notably Bodlaender's [17] and Scheffler's [71]. Each of those authors defines languages for some 'locally verifiable' properties, extends them by conjuctions with some 'non-local' statements (designed mainly to deal with the notion of connectivity), and designs a paradigm for constructing a solution algorithm for a given property and a given bound k on the tree-width of the problem instance.

Scheffler [71] considers optimization problems that can be described by formulea involving predicates expressing properties of a bounded neighborhood of a vertex. These are existentially quantified over a fixed set of subgraphs and universally quantified over all vertices of a graph. (The author follows the approach introduced by Seese [73].) She extends the class of properties by allowing conjunction with connectedness and acyclicity, and presents algorithm paradigms for deciding the above properties for a partial k-tree given together with an ordering of vertices corresponding to a perfect elimination of an embedding chordal graph. These algorithms use the given ordering of vertices and combine the properties of individual vertices (expressed by values of the objective function) into the global answer. Assuming an additive objective function, the corresponding optimization problem is solved following the general dynamic programming strategy.

The time complexity of these algorithms, while linear in the size of the input graph, depends exponentially on the problem (the number of subgraphs defining the property) and on the parameter k defining the class of input graphs.

Borie et al. [18] define regular properties of graphs based on the existence of a homomorphism between members of a given k-terminal, recursive family of graphs and some finite set. These properties are preserved under the homomorphism and the integrity of composition operators is maintained. (Their definition follows that of Bern et al. [12].) They prove constructively that the recognition, optimization, and enumeration of solutions for a given regular property are linearly solvable on recursively constructed graph families.

Monien et al. [60] use the notion of tree-width to investigate completeness for the class of languages that are acceptable by non-deterministic auxiliary push-down automata in polynomial time and logarithmic space (equal to $LOGCFL$ complexity class). They define the tree-width of a conjunctive form of a propositional formula as the tree-width of the corresponding hypergraph and show that many algorithms reducing 3-SAT with bounded tree-width preserve this bound for the instances of problems to which 3-SAT is reduced. This allows them to show these problems to be $LOGCFL$-complete when restricted to instances with tree-width bounded by $\log n$.

5. Algorithm Design Paradigm

Already 19th century physicists knew that certain difficult problems, hopeless in general can be solved in some 'tree-like graphs': the series-parallel reduction computing the equivalent resistance of a ladder circuit, or the star-triangle replacement in other electrical networks. However, the theory of these operations had to wait until 1980's. Slisenko [76] observed that the Hamilton cycle problem can be efficiently (in time polynomial in the size of the graph) solved on graphs obtained by the context-free replacement of hyperedges by hypergraphs, with terminal replacements of a hyperedges by edges between some of its vertices. Takamizawa et al. [79] developed a methodology for solving many such hard problems ($\mathcal{NP}$-hard) in linear time on series-parallel graphs. Intuitively, this 'good' algorithmic behavior of partial 2-trees can be explained by their bounded decomposability property that follows from a separation property of ('full') 2-trees: every minimal separator consists of both end-vertices of an edge.

The approach taken by Arnborg and Proskurowski in [5] to attack hard discrete optimization problems restricted to partial k-trees given with their embedding follows the general dynamic programming strategy. In a k-decomposable graph, the decomposability structure (an embedding in a full k-tree) is followed in solving pertinent subproblems. Solutions to these subproblems mutually interact only through the bounded interface of a minimal separator. Assuming that many instances of discrete optimization problems of interest are partial k-trees for relatively small values of k (say, about 10), the following algorithm paradigm for solving optimization problems on partial k-trees is of practical interest (Arnborg and Proskurowski [5]):

Depending on the problem being solved for partial k-tree G, each minimal separator S of a full k-tree embedding G is assigned a number of 'states'. Each such state represents constraints on a subproblem of optimization on the graph G_i (cf. the definition of decomposability), where feasible solutions agree on the subgraph induced in G by S. A solution to the problem corresponding to a state of S associates with the state the optimal value of the objective function. The algorithm requires successive 'pruning' of the k-leaves of the embedding k-tree (and of the resulting k-trees). In each pruning step, it solves the corresponding subproblems and updates the values of states of the corresponding minimal separator. When pruning a k-leaf v, this state update of the separator S (the remaining neighbors of v) involves combination of solutions to k subproblems (represented by the k separators of G consisting of v and $k - 1$ vertices of S). To find a solution to the overall problem, the eventual 'root optimization' is necessary, whereby the states of up to $k + 1$ minimal separators constituting the definitional K_{k+1} root of the embedding full k-tree are combined to yield the solution. If the problem being solved admits constant time pruning steps and a constant time 'root optimization', the resulting algorithm is linear in the size of the input graph. (So do for instance, *Independent Set, Vertex Cover, Chromatic Number, Graph Reliability*, cf. Arnborg and Proskurowski [5].) This follows from the fact that the number of states is independent of the size of the graph (although it can grow quite rapidly with k), and the number of minimal separators to consider is only linear with the size of the graph. It is important to realize, that the low order polynomial time complexity of the algorithm is achieved when the input consists of a suitable embedding of the given graph in a full k-tree. Otherwise, the complexity of the exact optimization algorithm is likely to be dominated by the complexity of an embedding algorithm.

A similar idea of combining states of components of a k-terminal graph according to its parse tree has been expressed by Wimer *et al.* [84] who list a score of families of k-terminal graphs and several dozens of problems to which their methodology applies.

Although the approach of [5] was the first attempt to describe efficient algorithms on partial k-trees by a common paradigm, it did not address the question of mechanical derivation of an efficient algorithm solving a difficult problem on those graphs from the problem description. It took several more years for some of those problems to be identified.

The important results of Arnborg *et al.* [3] have been mentioned in the preceding section. The efficient algorithm solving a given *EMSOL* problem is constructed as a tree automaton following the formal description of the corresponding property.

Borie *et al.* [18] describe an algorithm design paradigm based on their definition of *k*-terminal, recursive family of graphs and of regular properties. Following the decomposition tree of the graph in the problem's instance and using the "states" indicated by the homomorphism classes (by definition, there is only a finite number of those), the dynamic programming technique is used to compute a solution to the problem in linear time.

6. Graph Minors and Existence of Polynomial Time Algorithms

Major progress has been made possible by the results of Robertson and Seymour [67]. Their results on minor containment gave rise to a new non-constructive tools for establishing polynomial-time solvability [67] and a new interest in *forbidden substructures* characterization of classes of graphs [4], [33].

A graph *H* is a *minor* of a graph *G* if it can be obtained from a subgraph of *G* by contracting edges (*contracting* an edge introduces a new vertex replacing the two end vertices of the contracted edge and inheriting their adjacencies). Robertson and Seymour proved that every class of graphs closed under minor-taking has a finite number of *minimal forbidden minors* (graphs not in the class with all minors belonging to the class). Because every such class of graphs has constantly bounded tree-width, the membership of a graph in the class can be decided in time growing at most with the cube of the graph size, but with astronomical multiplicative constants. Similarly, many problems are now known to be decidable in low-degree polynomial time, based on the knowledge of the finite set of forbidden minors for a given class of graphs. However, but for a very few exceptions, there is no indication of how those graphs can be efficiently found, and even if they are known, the complexity of solution algorithms exhibit multiplicative constants of astronomical magnitude.

The class of graphs with path-width 2 has been characterized in [33] by 110 minimal forbidden minors. The class of partial 3-trees has a small set of minimal forbidden minors characterizing it [4]. The completeness of this set was proved using the knowledge of a small complete set of confluent reduction rules for this class of graphs. For higher values of *k*, this approach will not yield results as long as such rules are not known.

7. Parallel computation

Recent research on parallel algorithms shows that trees are amenable to the combination of the dynamic programming techniques (*pruning* of tree leaves) and the standard parallel techniques of *contraction* of long branches resulting in efficient parallel algorithms [58]. This discovery seems to generalize to graphs of tree-like structure, prime example of which are the partial k-trees. Bodlaender [15] uses it arguing the existence of poly-log algorithms for partial k-trees. Very recently, first efficient parallel algorithms for chordal graphs have been designed (Chandra-sekhran and Iyengar [22], Naor *et al.* [63], Kleim [48]). Chandrasekhran and Hedetniemi [20] describe an efficient parallel algorithm for the partial k-tree embedding problem.

References

[1] S. Arnborg, Efficient algorithms for combinatorial problems on graphs with bounded decomposa-bility—a survey, BIT 25 (1985), 2–23.

[2] S. Arnborg, D. G. Corneil, and A. Proskurowski, Complexity of finding embeddings in k-trees, SIAM Journal of Algebraic and Discrete Methods 8 (1987), 277–284.

[3] S. Arnborg, J. Lagergren, and D. Seese, Problems easy for decomposable graphs, Proceedings of ICALP 88, Springer-Verlag Lecture Notes in Computer Science 317 (1988), 38–51.

[4] S. Arnborg, A. Proskurowski, and D. G. Corneil, Forbidden minors characterization of partial 3-trees, UO-CIS-TR-86-07, University of Oregon (1986), to appear in Discrete Mathematics (1990).

[5] S. Arnborg and A. Proskurowski, Linear time algorithms for NP-hard problems restricted to partial k-trees, TRITA-NA-8404, The Royal Institute of Technology (1984), Discrete Applied Mathematics 23 (1989), 11–24.

[6] S. Arnborg and A. Proskurowski, Recognition of partial k-trees, Proceedings of the 16th South-Eastern International Conference on Combinatorics, Graph Theory and Computing, Utilitas Mathematica, Winnipeg, Congressus Numerantium 47 (1985), 69–75.

[7] S. Arnborg and A. Proskurowski, Problems on graphs with bounded decomposability, Bull. EATCS (1985).

[8] S. Arnborg and A. Proskurowski, Characterization and recognition of partial 3-trees, SIAM Journal of Algebraic and Discrete Methods 7 (1986), 305–314.

[9] B. Baker, Approximation algorithms for NP-complete problems on planar graphs, Proceedings FOCS 24 (1983), 105–118.

[10] M. Bauderon and B. Courcelle, Graph expressions and graph rewritings, Mathematical Systems Theory 20 (1987), 83–127.

[11] L. W. Bineke and R. E. Pippert, Properties and characterizations of k-trees, Mathematica 18 (1971), 141–151.

[12] M. W. Bern, E. L. Lawler, and A. L. Wong, Linear Time Computation of Optimal Subgraphs of Decomposable Graphs, Journal of Algorithms 8 (1987), 216–235.

[13] H. L. Bodlaender, Classes of graphs with bounded tree-width, RUU-CS-86-22 (1986).

[14] H. L. Bodlaender, Planar graphs with bounded tree-width, Bull. EATCS (1988).

[15] H. L. Bodlaender, NC-algorithms for graphs with small tree-width, Proceedings of the Workshop on Graph-Theoretic Concepts in Computer Science WG-88, Springer-Verlag Lecture Notes in Computer Science 344 (1988), 1–10.

[16] H. L. Bodlaender, Polynomial algorithms for graph isomorphism and chromatic index on partial k-trees, Proceedings of the Scandinavian Workshop on Algorithm Theory, Spring-Verlag, Lecture Notes in Computer Science 318 (1988), 223–232.

[17] H. L. Bodlaender, Dynamic programming on graphs with bounded tree-width, RUU-CS-87-22, Proceedings of ICALP'88, Springer-Verlag Lecture Notes in Computer Science 317 (1988), 105–118.

[18] R. B. Borie, R. G. Parker, and C. A. Tovey, Automatic generation of linear algorithms from predicate calculus descriptions of problems on recursively constructed graph families, manuscript (July 1988).

[19] H. Buer and R. H. Möhring, A fast algorithm for decomposition of graphs and posets, Math. Oper. Res. 8 (1983), 170–184.

[20] N. Chandrasekharan and S. T. Hedetniemi, Fast parallel algorithms for tree decomposing and parsing partial k-trees, Proceedings of 26 Annual Allerton Conference in Communications, Con-trol, and Computing (1988).

[21] N. Chandrasekharan, S. T. Hedetniemi, and T. V. Wimer, A method for obtaining difference equations for the number of vertex subsets having a given property in restricted k-terminal families of graphs, manuscript (October 1988).

[22] N. Chandrasekharan and S. S. Iyengar, NC algorithms fr recognizing chordal graphs and k-trees, IEEE Trans. on Computers 37 (1988), 1170–1183.

[23] D. G. Corneil, H. Lerchs, and L. Stewart Burlingham, Complement reducible graphs, Discrete Applied Mathematics 3 (1981), 163–174.

[24] B. Courcelle, Some applications of logic of universal algebra, and of category theory to the theory of graph transformations, Bull EATCS (1988), 161–213.

[25] B. Courcelle, Recognizabilty and second-order definabilty for sets of finite graphs, I-8634, Université de Bordeaus, (1987).

[26] B. Courcelle, The monadic second order logic of graphs I: recognizable sets of finite graphs, I-8837, Université de Bordeaux (1988).

[27] B. Courcelle, The monadic second order logic of graphs III: tree-width, forbidden minors and complexity issues, I-8852, Université de Bordeaux (1988).

[28] W. M. Cunningham, Decomposition of directed graphs, SIAM Journal of Algebraic and Discrete Methods 3 (1982), 214–228.

[29] W. M. Cunningham and J. Edmonds, A combinatorial decomposition theory, Canadian J. Mathematics 32 (1980), 734–765.

[30] G. A. Dirac, On rigid circuit graphs, Abh. Math. Sem. Univ. Hamburg 25 (1961), 71–76.

[31] J. E. Doner, Decidability of the weak second-order theory of two successors, Notices of the American Mathematical Society 12 (1966), 513.

[32] E. S. ElMallah, Decomposition and Embedding Problems for Restricted Networks, PhD. Thesis, University of Waterloo (1987).

[33] M. R. Fellows, N. G. Kinnersley, and M. A. Langston, Finite-basis theorems and a computational-integrated approach to obstruction set isolation, Proceedings of Computers and Mathematics Conference (1989).

[34] M. R. Fellows and M. A. Langston, Non-constructive advances in polynomial-time complexity, Information Processing Letters 26 (1987), 157–162.

[35] M. R. Fellows and M. A. Langston, Non-constructive tools for proving polynomial-time decidability, Journal of the ACM 35 (1988), 727–739.

[36] D. R. Fulkerson and O. A. Gross, Incidence matrices and interval graphs, Pacific Journal of Mathematics 15 (1965), 835–855.

[37] F. Gavril, Algorithms on clique separable graphs, Discrete Math. 19 (1977), 159–165.

[38] M. C. Golumbic, Algorithmic graph theory and perfect graphs, Academic Press (1980).

[39] A. Habel, Graph-theoretic properties compatible with graph derivations, Proceedings of WG-88, Springer-Verlag Lecture Notes in Computer Science 344 (1988).

[40] A. Habel, Hyperedge replacement: grammars and languages, Phd. Dissertation, Bremen 1989.

[41] A. Habel and H. J. Kreowski, May we introduce to you: hyperedge replacement, Proceedings of the 3rd International Workshop on Graph Grammars and their Applications to Computer Science, Springer-Verlag Lecture Notes in Computer Science 291 (1987), 15–26.

[42] M. Habib and R. H. Möhring, On some complexity properties of N-free posets and posets with bounded decomposition diameter, Discrete Mathematics 63 (1987), 157–182.

[43] W. Hochberg and R. Reischuk, Decomposition graphs—a uniform approach for the design of fast sequential and parallel algorithms on graphs, manuscript (1989).

[44] D. S. Johnson, The NP-Completeness column: an ongoing guide, Journal of Algorithms 6 (1985), 434–451.

[45] D. S. Johnson, The NP-completeness column: an ongoing guide; Journal of Algorithms 8 (1987), 285–303.

[46] Y. Kajitani, A. Ishizuka, and S. Ueno, A Characterization of the Partial k-tree in Terms of Certain Structures, Proceedings of ISCAS'85 (1985), 1179–1182.

[47] M. Klawe, D. G. Corneil, and A. Proskurowski, Isomorphism testing in hook-up graphs, SIAM Journal of Algebraic and Discrete Methods 3 (1982), 260–274.

[48] P. N. Klein, Efficient parallel algorithms for chordal graphs, Proceedings of the 29th Symposium on FoCS (1988), 150–161.

[49] C. Lauteman, Decomposition trees: structured graph representation and efficient algorithms, Proceedings of CAAP'88, Springer-Verlag Lecture Notes in Computer Science 299 (1988), 217–244.

[50] C. Lauteman, Efficient algorithms on context-free graph languages, Proceedings of ICALP'88, Springer-Verlag Lecture Notes in Computer Science 317 (1988), 362–378.

[51] T. Lengauer and E. Wanke, Efficient analysis of graph properties on context free graph languages, Proceeding of ICALP'88, Springer-Verlag Lecture Notes in Computer Science 317 (1988), 379–393.

[52] A. Lingas and A. Proskurowski, Fast parallel algorithms for the subgraph homeomorphism and the subgraph isomorphism problems for classes of planar graphs, Theoretical Computer Science 68, 2 (1989), 155–174.

[53] A. Lingas and M. M. Sysło, A polynomial-time algorithm for subgraph isomorphism of two-connected series-parallel graphs, Proceedings of ICALP'88, Springer-Verlag Lecture Notes in Computer Science *317* (1988), 394–409.

[54] G. S. Lueker and K. S. Booth, A linear time algorithm for deciding interval graphs isomorphism, Journal of the ACM *26* (1979), 183–195.

[55] S. Mahajan and J. G. Peters, Algorithms for regular properties in recursive graphs, Proceedings of 25 Annual Allerton Conference in Communications, Control, and Computing (1987), 14–23.

[56] J. Matousek and R. Thomas, On the complexity of finding iso- and other morphisms for partial k-trees, manuscript, (May 1988).

[57] J. Matousek and R. Thomas, Algorithms finding tree decompositions of graphs, manuscript (May 1988).

[58] G. L. Miller and J. H. Reif, Parallel Tree Contraction and its Applications, Proceedings of the 26th FoCS (1985), 478–489.

[59] B. Monien and I. H. Sudborough, Bandwidth constrained NP-complete problems, Proceedings of the 13th STOC (1981), 207–217.

[60] B. Monien, I. H. Sudborough, and M. Wiegers, Complexity results for graphs with treewidth $O(\log n)$, manuscript (1989).

[61] J. H. Muller and J. Spinrad, Incremental modular decomposition, Journal of the ACM *36* (1989), 1–19.

[62] R. H. Möhring and F. J. Radermacher, Substitution decomposition for discrete structures and connections with combinatorial optimization, Annals of Discrete Mathematics *19* (1984), 257–356.

[63] J. Naor, M. Naor, and A. A. Schäffer, Fast parallel algorithms for chordal graphs, SIAM Journal on Computing *18* (1989), 327–349.

[64] T. Politof, Δ-Y Reducible Graphs, Concordia University, Montreal, manuscript (1985).

[23] A. Proskurowski, Centers of 2-Trees, Proceeding of the 2nd Combinatorial Conference France-Canada, Annals of Discrete Mathematics *9* (1980), 1–5.

[66] A. Proskurowski, Separating subgraphs in k-trees: cables and caterpillars, Discrete Mathematics *49* (1984), 275–285.

[67] N. Robertson and P. D. Seymour, Graph Minors, (series of 23 papers in varying stages of editorial process, 1983–1988).

[68] D. J. Rose, On simple characterization of k-trees, Discrete Mathematics *7* (1974), 317–322.

[69] A. Rosenthal and J. A. Pino, A generalized algorithm for centrality problems on trees, Journal of the ACM *36* (1989), 349–361.

[70] G. Rozenberg and E.. Welzl, Boundary NLC grammars: basic definitions, normal forms, and complexity, Information and Control *69* (1986), 136–167.

[71] P. Scheffler, Linear time algorithms for NP-complete problems for partial k-trees, R-MATH-03/87 (1987).

[72] P. Scheffler and D. Seese, Tree-width and polynomial-time solvable graph problems, manuscript (1986).

[73] D. Seese, The structure of the models of decidable monadic theories of graphs, to appear in Journal of Pure and Applied Logic.

[74] D. Seese, Tree-partite graphs and the complexity of algorithms, Proceedings of FCT-85, Springer-Verlag Lecture Notes in Computer Science *199* (1985), 412–421.

[75] Y. Shibata, On the tree representation of chordal graphs, Journal of Graph Theory *12* (1988), 421–428.

[76] A. O. Slisenko, Context-free grammars as a tool for describing polynomial subclasses of hard problems, Information Processing Letters *14* (1982), 52–56.

[77] M. M. Sysło, NP-complete problems on some tree-structured graphs: a review, Proceedings of WG-83, Trauer Verlag (1984), 342–353.

[78] M. M. Sysło, A graph-theoretic approach to the jump number problem, in: I. Rival(ed.), Graphs and Order, Reidel, Dodrecht 1985, 185–215.

[79] K. Takamizawa, T. Nishizeki, and N. Saito, Linear-time Computability of Combinatorial Problems on Series-parallel Graphs, Journal of the ACM *29* (1982), 623–641.

[80] R. E. Tarjan, Decomposition by clique separators, Discrete Mathematics *55* (1985), 221–232.

[81] J. W. Thatcher and J. B. Wright, Generalized finite automata theory with an application to a decision problem in second-order logic, Mathematical Systems Theory *2* (1968), 57–81.

[82] T. V. Wimer, Linear algorithms on k-terminal graphs, PhD. Dissertation, Clemson University (August 1988).

[83] T. V. Wimer and S. T. Hedetniemi, k-terminal recursive families of graphs, Proceedings of 25th Annual Conference on Graph Theory, Utilitas Ma thematica, Winnipeg, Congressus Numerantium *63* (1988), 161–176.

[84] T. V. Wimer, S. T. Hedetniemi, and R. Laskar, A methodology for constructing linear graph algorithms, Clemson University, TR-85-SEP-11, (September 1985).

[85] P. Winter, Steiner problem in networks: a survey, Networks *17* (1987), 129–167.

Andrzej Proskurowski
Department of Computer
and Information Science
University of Oregon
Eugene, Oregon 97403,
U.S.A.

Computing Suppl. 7, 17–51 (1990)

Computing
© by Springer-Verlag 1990

Graph Problems Related to Gate Matrix Layout and PLA Folding

Rolf H. Möhring*, Berlin

Abstract — Zusammerfassung

Graph Problems Related to Gate Matrix Layout and PLA Folding. This paper gives a survey on graph problems occuring in linear VLSI layout architectures such as gate matrix layout, folding of programmable logic arrays, and Weinberger arrays. These include a variety of mostly independently investigated graph problems such as augmentation of a given graph to an interval graph with small clique size, node search of graphs, matching problems with side constraints, and other. We discuss implications of graph theoretic results for the VLSI layout problems and survey new research directions. New results presented include NP-hardness of gate matrix layout on chordal graphs, efficient algorithms for trees, cographs, and certain chordal graphs, Lagrangean relaxation and approximation algorithms based on on-line interval graph augmentation.

Key words: linear VLSI layout architectures, gate matrix layout, PLA folding, interval graph augmentation, graph search, path width, vertex separation, complexity, approximation algorithms, matching with side constraints.

Graphentheoretische Probleme beim Gate Matrix Layout und PLA Folding. Der Artikel behandelt graphentheoretische Probeme, die bei linearen VLSI-Layout Architekturen wie Gate-Matrix-Layout, Programmierbaren Logischen Arrays und Weinberger Arrays auftreten. Zu diesen gehören u.a. die Einbettung von Graphen in Intervallgraphen mit kleiner Cliquengröße, Suchspiele auf Graphen, und Zuordnungsprobleme mit Nebenbedingungen. Wir diskutieren Folgerungen aus graphentheoretischen Ergebnissen für das VLSI Layout und geben eine Übersicht über neue Forschungsergebnisse. Hierzu gehören u.a.: die NP-Vollständigkeit des Gate-Matrix-Layout auf chordalen Graphen, effiziente Algorithmen für Bäume, Cographen sowie gewisse chordale Graphen, Langrange Relaxation und approximierende Algorithmen, die auf der On-line Konstruktion von Intervallgrapheinbettungen basieren.

1. Introduction

We study a class of graph-theoretic problems that arise in certain linear VLSI layout problems such as Weinberger arrays, gate matrix layout and PLA-folding.

The term *linear* refers to the fact that the most important degrees of freedom in the underlying VLSI architecture consist of linear (i.e. one-dimensional) arrangements of the relevant physical objects, the *gates*.

---

* Technical University of Berlin, Fachbereich Mathematik, Straße des 17. Juni 136, 1000 Berlin 12. This work was supported by the Deutsche Forschungsgemeinschaft (DFG).

In the most general form, an instance of such a linear layout problem consists of a $m \times n$ 1-0 matrix $M = (m_{ij})$ (the *net-gate matrix*), whose rows and columns represent the nets $N_1, \ldots, N_m$ and the gates $G_1, \ldots, G_n$ of the circuit, respectively.

The gates may be thought of as the basic electronic devices that are arranged linearly in a row, and the nets as realizing connections between them (details are given in Section 2). Net N_i must connect all gates G_j with $m_{ij} = 1$. Connections are realized rowwise by reserving for a given permutation of the gates (columns) for every net N_i the part of the row from the leftmost to the rightmost gate to which a connection must be established.

This can be expressed more formally by considering for a permutation π of the gates the *augmented net-gate matrix* $M_\pi = (\bar{m}_{ij})$ with

$$\bar{m}_{ij} := \left\{ \begin{array}{ll} 1 & \text{if there are Gates } G_r, G_s \text{ with} \\ & \pi(r) \leq j \leq \pi(s) \text{ and } m_{ir} = m_{is} = 1 \\ 0 & \text{otherwise} \end{array} \right\}$$

Nets of the augmented net-gate matrix may share the same row (called *track*) if they have no gate in common. An assignment of augmented nets to tracks preserving this property is called a *feasible track assignment*.

The additional ones in M (with respect to the same column permutation of M) are called *fill-ins*. They are represented by a "*" to distinguish them from the given ones in M (see Figure 1.1 below).

The result of a permutation of the gates (*gate arrangement*) and an associated feasible track assignment is called a *layout*. Its area is proportional to (#gates) · (#tracks) = $n \cdot$ (#tracks).

So constructing an area-minimal layout (*optimal layout*) is equivalent to finding a gate arrangement and an associated feasible track assignment such that the number of tracks is minimum.

In matrix terminology, this leads to the following *matrix permutation problem* (MPP):

Given: A 0-1 matrix (net-gate matrix) M.

Problem: Find a permutation of the columns and an assignment of the augmented rows (nets) to tracks such that the number of tracks is minimum.

An example is given in Figure 1.1. We denote the minimum number of tracks by $t(M)$ and call a layout with $t(M)$ tracks an *optimal layout*. Due to the mentioned origin of this problem in VLSI-applications, there is an enormous body of papers on it. This article gives an overview on available results and new developments.

Section 2 deals with the VLSI background and models the linear VLSI layout technologies "Weinberger arrays", "gate matrix layout" and "PLA folding" as instances of the general MPP. Some of these applications lead to restricted MPP's in the sense that either the permutation of the gates (e.g. fixed first and last gate) or the assignment of nets to tracks (e.g. at most two per track) are restricted.

	G₁	G₂	G₃	G₄
N₁	1	1		
N₂	1		1	
N₃		1		1
N₄			1	1
N₅		1		

net - gate matrix

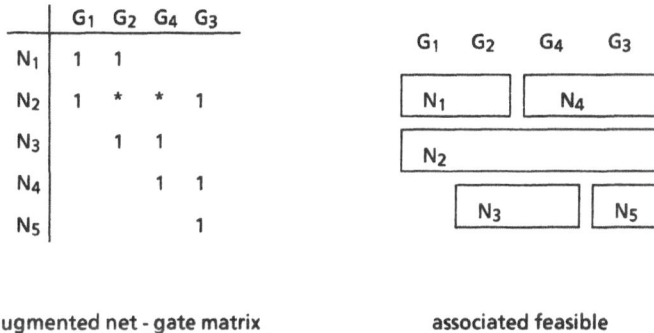

	G₁	G₂	G₄	G₃
N₁	1	1		
N₂	1	*	*	1
N₃		1	1	
N₄			1	1
N₅				1

augmented net - gate matrix   associated feasible
for π = (1, 2, 4, 3)   track assignment

Figure 1.1. An example MPP

Section 3 gives several equivalent graph theoretic problems with different and partly independent background. Among them are: augmentation of a graph to an interval graph with small clique size, a node search problem, determining the path-width of a graph, matching problems with side constraints and others.

Section 4 is devoted to the complexity of the problem, and to reductions between some of the specialized versions. The general problem is already NP-hard on chordal graphs, but solvable in polynomial time on trees and cographs and if the number of tracks is fixed. Sharper version of NP-completeness results are also obtained for certain variants of PLA folding.

Finally, Section 5 deals with exact and approximation algorithms for solving MPP's. Due to the practical relevance of the problem, many heuristics have been proposed in the literature. Nevertheless, the problem of the existence of an approximation algorithm with constant relative performance bound has remained open. We present a class of on-line algorithms based on incremental interval graph generation that may be promising in this respect.

Our graph-theoretic notation is usually standard. For all notions and definitions not explicitly stated here, we refer to [Go80] or [Ev79].

2. The VLSI Background

The matrix permutation problem is typical for a number of "regular" layout styles for the generation of random logic modules in VLSI.

Such a random logic module may be seen as an irregular structure of basic components such as transistors, gates, flip-flops etc. It is given by some input description, e.g., by a transistor scheme, a logic scheme, or a set of Boolean functions. From this description, a concrete layout (a physical module) must be constructed according to some layout architecture style. A *regular* layout style is a style in which basic topological relationships between the physical components on the chip area are known in advance (e.g. restricted placement, predefined locations for the arrangements of gates etc.).

Typical such layout styles are *Weinberger arrays, gate matrix layout* and *programmable logic arrays* discussed below, see also [GL88] and [BMHS84].

Weinberger Array. Weinberger arrays were introduced in [Wei67] as a layout architecture for Boolean functions that are given by a circuit consisting only of NOR-gates (see Figure 2.1). Each NOR-gate is converted into an nMOS gate (i.e. a gate in the nMOS VLSI technology, see e.g. [MC80] for details); and these gates are arranged in a linear array that constitute the columns of an associated MPP (see Figure 2.2).

Each column of this MPP consists of two vertical wires. One wire is connected to the pull-up transistor and serves as the output port, while the other wire is connected with the ground power line. (Usually, two neighboring gates share a common ground wire.) The input signals to the gates are obtained from horizontal polysilicon intervals on a row. A transistor is formed by the intersection of an extension of such a polysilicon interval with a diffusion segment between the output the ground lines. For example, transistors a and b are formed by connecting row 1 and 2 to diffusion segments in gate A. The output of gate A is connected to the last row which serves as input to gate F, etc.

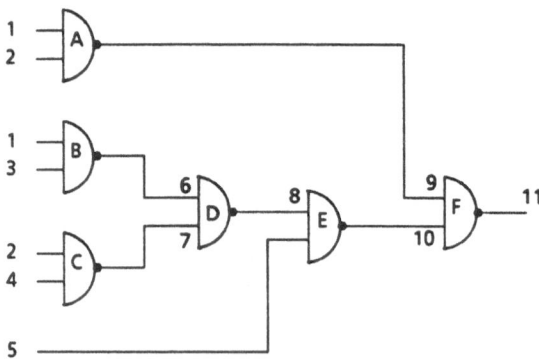

Figure 2.1. A circuit of NOR gates

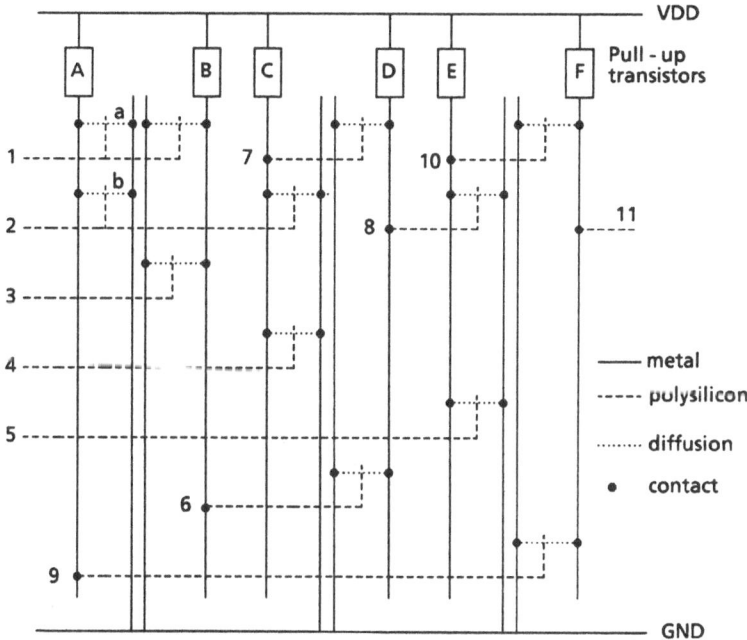

Figure 2.2. A Weinberger array layout of the circuit of Figure 2.1

Note that connections (i.e. polysilicon intervals) may be placed on the same line if they do not overlap. Since the number of gates is fixed, minimization of the layout area is equivalent to reducing the number of rows, i.e. by finding a suitable permutation of the gates (which defines the length of the polysilicon intervals) and an associated row assignment of the intervals such that the number of rows is minimum.

So we obtain the following *Weinberger MPP* (WMPP):

Given: —A collection $G_0, G_1, \ldots, G_n, G_{n+1}$ of gates, where $G_1, \ldots, G_n$ represent the NOR gates of the circuit, and where G_0 and G_{n+1} represent the input (on the left) and output signals (on the right) of the circuit, respectively.
—A collection of nets $N_1, \ldots, N_m$. Each net N_i consists of those gates G_j, to which it is output or input.
—The net-gate matrix M.

Problem: Find a permutation of $G_1, \ldots, G_n$ (i.e. the positions of G_0 and G_{n+1} are fixed) and an associated feasible track assignment (layout) such that the number of tracks is minimum.

Figure 2.3 shows the matrix M, a layout corresponding to Figure 2.2 and an optimal layout for the above example.

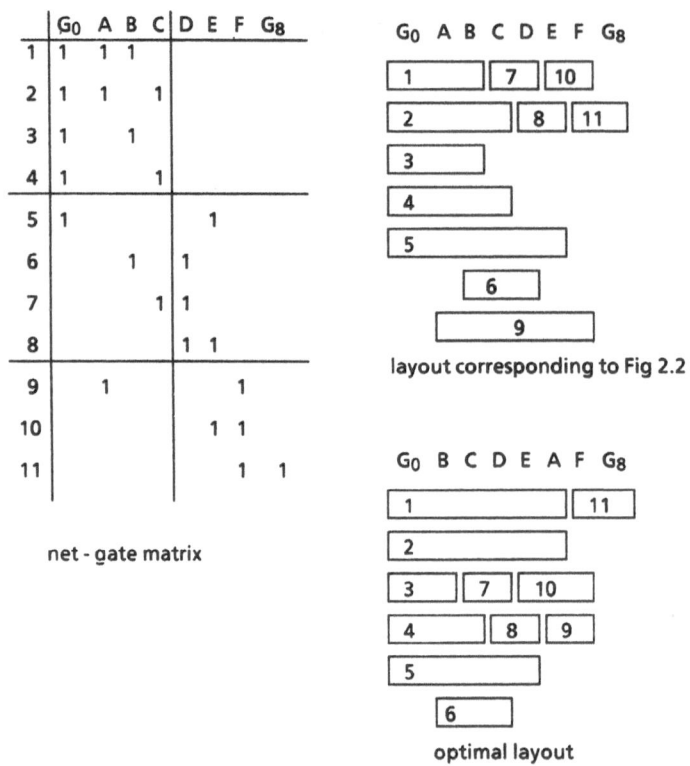

Figure 2.3. Net-gate matrix and two layouts for the WMPP of Figure 2.1

Gate Matrix Layout. This architecture was introduced by [LLa80] as a regular layout style for large scale transistor circuits in the CMOS technology. In such a layout, a vertical polysilicon wire corresponding to an input, internal or output signal is placed in every column (columns A, B, C, D, E, F, G and Z in Figure 2.4). All transistors using the same signal are constructed along the same column (e.g. transistors 1 and 7 in column A of Figure 2.4). Connections among transistors are made by horizontal metal lines, while connections to Vdd/Vss are in a second metal layer (and irrelevant to the underlying MPP). They are indicated by up and downward arrows in Figure 2.4.

A *net* is a collection of metal lines and transistors to which it must be connected. Net N_1 in row A of Figure 2.4 spans from column A to G and is connected to three transistors (1, 2, and 3) and one metal line (G). The (slightly simplified) assumption about the realization of nets is that the series-parallel transistor circuit of each net and its output signals can be realized in a row regardless the permutation of the metal lines and transistors.

As with Weinberger arrays, minimizing the layout area leads to the following *gate matrix permutation problem* (GMPP).

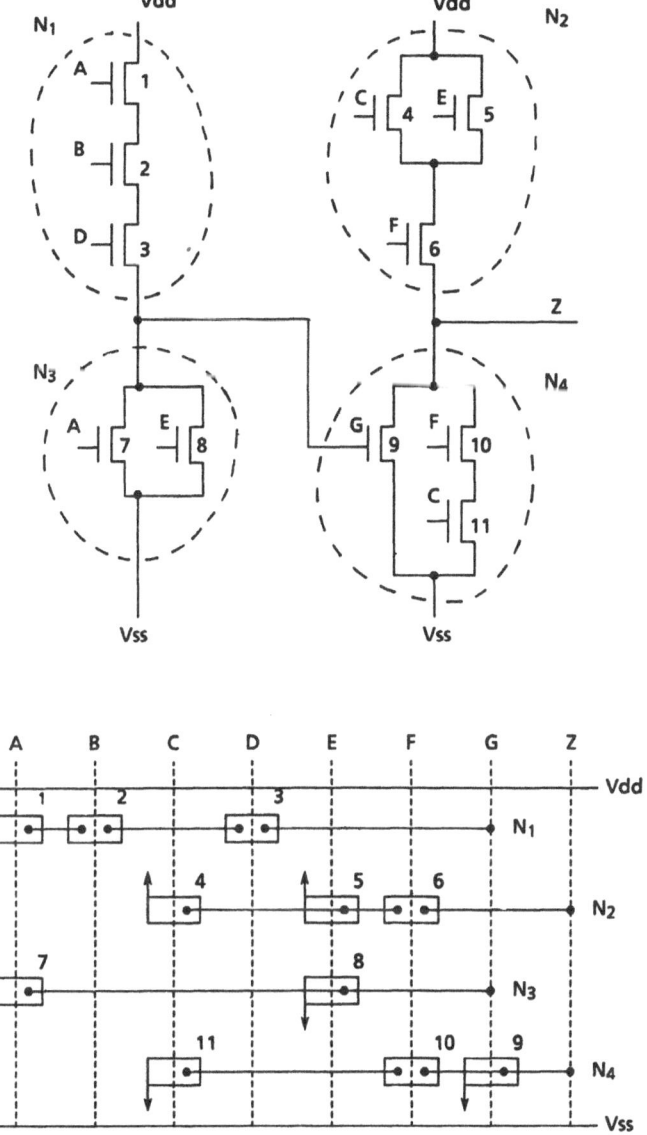

A...Z : signals   N₁...N₄ : nets
1...11 : transistors   a...d   : rows

Figure 2.4. A transistor circuit and an associated gate matrix layout

Given: —A collection $G_1, G_2, \ldots, G_n$ of gates representing metal lines or tran-
          sistors of a gate matrix.
          —A collection $N_1, \ldots, N_m$ of nets, where each N_i is a subset of $\{G_1, \ldots, G_n\}$.
          —The net-gate matrix M.

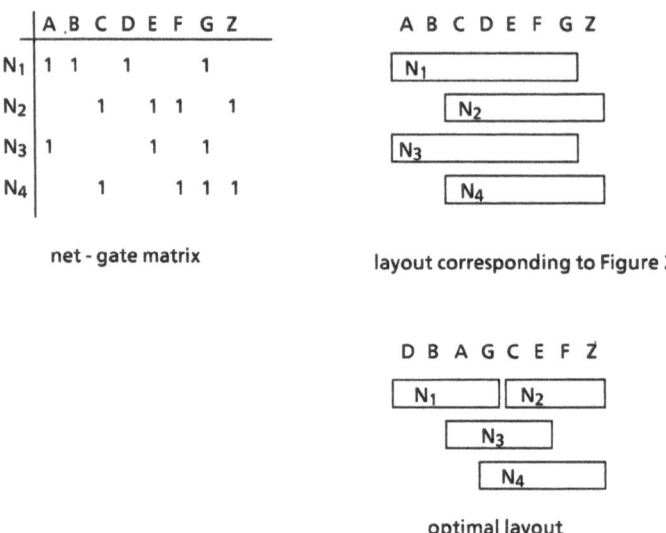

net - gate matrix

layout corresponding to Figure 2.4

optimal layout

Figure 2.5. Net-gate matrix and two layouts for the GAMP of Figure 2.4

Problem: Find a permutation of $G_1, \ldots, G_n$ and an associated feasible track assign-
            ment (layout) such that the number is tracks is minimum.

Figure 2.5 displays the net-gate matrix and two layouts for the above example.

The GMMP is the combinatorial core of the problem to construct an area minimal
gate matrix layout. Additional (and usually neglected) features are 1) that one may
distinguish two collections of rows (p-devices and n-devices) that permit independent column permutations, and 2) that a net may require more than one row
depending on the permutation of its gates. While this second feature can be modeled
within the *MPP* formulation by appropriate net splitting, incorporation of the first
feature may lead to better layouts [NFKY86]. For further technical information
we refer to [SM83], [WHW85].

Programmable Logic Array. A programmable logic array (PLA) realizes a collection of Boolean functions given in disjunctive form (two-level sum of product form)
on a two-dimensional array (see Figure 2.6).

This array consists of an AND-plane and an OR-plane. For every variable x_i of the
Boolean functions, there is an input signal to the AND-plane (in fact, both inputs
x_i and $\bar{x}_i$ are generated). Each row of the AND-plane produces a term that is an
input to the OR-plane. The columns of the OR-plane correspond to the different
Boolean functions and combine the appropriate product terms by an OR operation.
By adding storage elements and simple feedback connections, a PLA can very easily
be used to implement a sequential circuit. This application of PLA's is popular
in the design of microcontrollers. For further technical information, we refer to
[FM75].

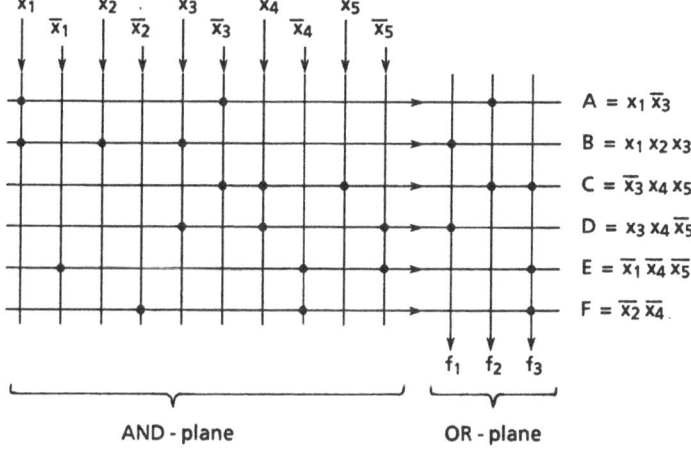

$$f_1 = B + D, \, f_2 = A + C, \, f_3 = C + E + F$$

Figure 2.6. A PLA layout of 3 Boolean functions

For reducing the area of a PLA, two techniques can be applied. *Logic minimization* for reducing the number of rows (= product terms) and *PLA folding* for reducing the number of columns. The first technique is the same as finding the minimum number of prime implicants for a set of Boolean functions (see e.g. [BMHS84]). It is usually applied before the folding.

The folding allows two (sometimes also more) signals to share a row (in the AND-plane) or a column (in the OR-plane). This leads to the same class of MPP's in both the AND and the OR-plane.

PLAMPP

Given: —A collection of gates $G_1, \ldots, G_n$ that correspond to the signals in one plane of a PLA.
             —A collection of nets $N_1, \ldots, N_m$, where each net is a set of signals that have to be combined by AND (in the AND-plane) or OR (in the OR-plane).
             —The net-gate matrix M.

Problem: Find a permutation of the gates $G_1, \ldots, G_n$ and a feasible assignment of at most two nets to a track (*PLA layout*) such that the number of tracks is minimum.

Figure 2.7 displays the net-gate matrix and several layouts for the example of Figure 2.6.

There are several more restrictive versions of the PLA folding problem. If gates occuring in the second net of a track may not occur in the first net of a track with two nets, one speaks of *block folding*. In that case, any folding defines a partition

Rolf H. Möhring

		A	B	C	D	E	F
1	x_1	1	1				
2	$\bar{x}_1$					1	
3	x_2		1				
4	$\bar{x}_2$						1
5	x_3		1		1		
6	$\bar{x}_3$	1		1			
7	x_4			1	1		
8	$\bar{x}_4$					1	1
9	x_5			1			
10	$\bar{x}_5$				1	1	

net - gate matrix

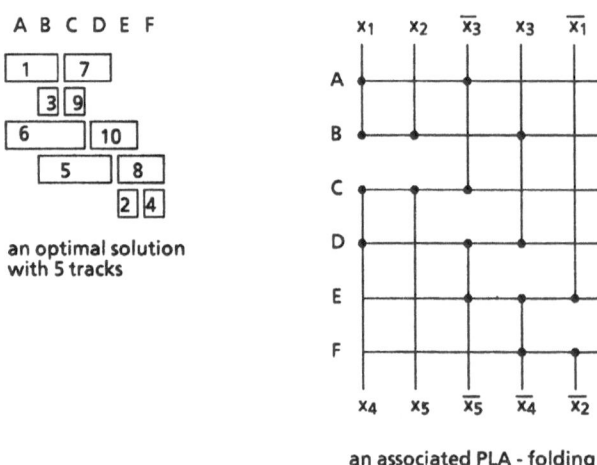

an optimal solution with 5 tracks

an associated PLA - folding

Figure 2.7. The net-gate matrix and an optimal layout for the AND-plane of Figure 2.6

G_1, G_2 of the gates such that if net N_i is before N_j in the same track, then $N_i \subseteq G_1$ and $N_j \subseteq G_2$.

If the nets are pre-assigned to the two sides of the layout, one speaks of *constrained folding*. In that case, a partition N_1, N_2 of the nets is given as input to the problem, and any restricted folding may only assign net N_i before N_j on the same track if $N_i \in N_1$ and $N_j \in N_2$.

The combination of block folding and restricted folding is called *constrained block folding*. Figure 2.8 shows optimal layouts for these different folding problems for the example of Figure 2.7.

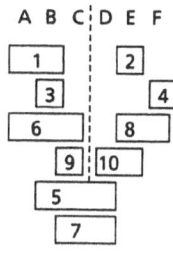 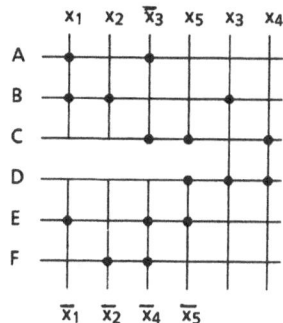

a) optimal block folding and associated PLA with 6 tracks

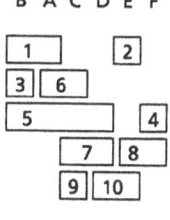 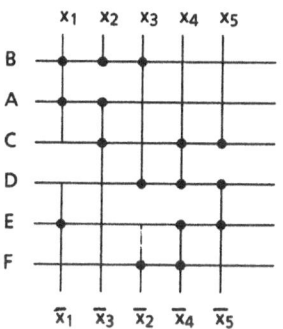

b) optimal constrained folding and associated PLA for the net
partition {1,3,5,7,9}, {2,4,6,8,10} with 5 tracks

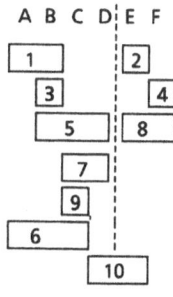 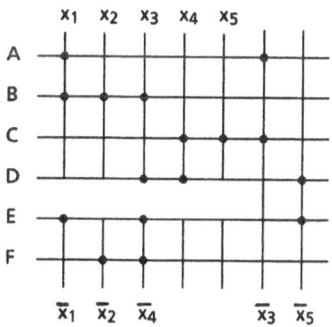

c) optimal constrained block folding and associated PLA for the
partition from b) with 7 tracks

Figure 2.8. Restricted PLA-foldings

3. Graph-Theoretic Formulations and Related Problems

We will now consider several graph theoretic problems that are equivalent to the VLSI layout problems discussed in the previous sections. These problems have their own graph theoretic background and have to a large extent been investigated inependently of each other.

Interval graph augmentation. This formulation occurs already in the first papers on gate matrix layout [Win82], [Win83]. For Weinberger arrays, similar considerations are made in [OMKK79].

Let V be a finite set and $(I_v)_{v \in V}$ be a collection of (not necessarily distinct) intervals I_v of a linearly ordered universe (such as the real line or a permutation of the gates). Such a collection of intervals $(I_v)_{v \in V}$ defines a partial order $P = (V, <)$ on V by putting

(3.1) $u < u \Leftrightarrow I_u$ is entirely to the left of I_v.

It also defines an undirected graph $G = (V, E)$ on V by putting

(3.2) $(u, v) \in E \Leftrightarrow I_u$ and I_v intersect (i.e. $I_u \cap I_v \neq \emptyset$)

A partial order P and a graph G obtained in that way are called an *interval order* and an *interval graph*, respectively, and an associated collection of intervals $(I_v)_{v \in V}$ is called an *interval representation* of P or G. An example is given in Figure 3.1.

Interval orders and interval graphs model the sequential and intersection structure of a set of intervals of the real line. This is why they have many applications dealing with intersection and consecutiveness such as the gene structure in molecular biology, seriation in archeology, preference and indifference relations in measurement theory, and consecutive retrieval, VLSI channel routing, and gate matrix layout in computer science. For more information on these applications, see [Go80], [Go85], [Mö85], [Mö89].

Note that different interval representations with the same intersection behavior define the same interval graph but possibly different interval orders. We call all interval orders related in this way to a fixed interval graph G the interval orders *associated* with G.

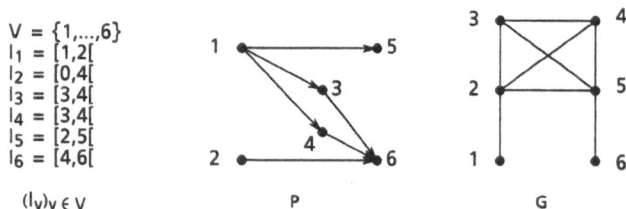

Figure 3.1. A collecting of intervals with associated interval order P (as transitively reduced directed acyclic graph) and interval graph G

Consider now an MPP with net-gate matrix M. This matrix M defines a graph $G = ((V(G), E(G))$ by taking the *intersection graph* of the rows, i.e., $V(G)$ is the set of nets (rows), and two nets are connected by an edge if they share a gate (i.e. there is a column with 1's in both rows). G is called the *net adjacency graph* [DKL87], or *incompatibility graph* [ALN88] since its edges (u, v) express that the nets u, v cannot be assigned the same track in a feasible layout.

For any gate permutation π of M, the associated augmented matrix M_π defines a collection of intervals (the augmented nets) of the linear order $G_{i_1}, \ldots, G_{i_n}$ defined by π on the gates. This collection of intervals defines an interval graph $H = (V(H), E(H))$ that *contains* G in the sense that $V(G) = V(H)$ and $E(G) \subseteq E(H)$, i.e. by *augmenting* the edge set of G. (This follows directly from the fact that two nets that share a gate in M share also a gate in M_π.) A feasible track assignment for M_π corresponds then obviously to a *coloring* of H, in which tracks correspond to color classes.

This shows that every feasible layout for M induces a coloring of an interval graph augmentation of G, the incompatibility graph of M. The converse is also true as the following lemma shows.

3.1 Lemma: *Let $G = (V(G), E(G))$ be the incompatibility graph of a net-gate matrix M. Let $H = (V(H), E(H))$ be an interval graph augmentation of G and let $P = (V(H), <)$ be any interval order associated with H. Then:*

a) *P induces a partial order on the gates $G_1, \ldots, G_n$ of M by putting*

   (3.3) $G_r < G_s$ *if there are nets $N_i \neq N_j$ incident to G_r and G_s,*
             respectively (i.e. $m_{ir} = m_{js} = 1$), and $N_i < N_j$ in P.

b) *Any linear extension $G_{j_1}, \ldots, G_{j_n}$ of the gate order of a) induces a permutation π such that the augmented net-gate matrix M_π is an interval representation of H and P.*

c) *Any coloring of H induces a track assignment of M_π by taking each color class as a track and ordering the nets of the color class according to P.*

The proof is straightforward and left to the reader. An example of these constructions is presented in Figure 3.2. An immediate consequence of these consideration is:

3.2 Theorem: *The minimum number of tracks of a feasible layout for a net-gate matrix M is equal to the smallest chromatic number of an interval graph augmentation H of the incompatibility graph G of M, i.e.*

(3.4) $t(M) = \min\{\chi(H) | H \text{ is an interval graph with } E(G) \subseteq E(H)\}$

We briefly discuss the computational complexity of the constructions of Lemma 3.1. An adjacency matrix of G can be constructed from M in $O(n \cdot m^2)$ time by looking at each column separately and inserting the corresponding edge entries in the adjacency matrix.

An interval representation of an interval graph H and an associated interval order P (in interval representation) can be constructed in $O(|V(H)| + |E(H)|)$ time by

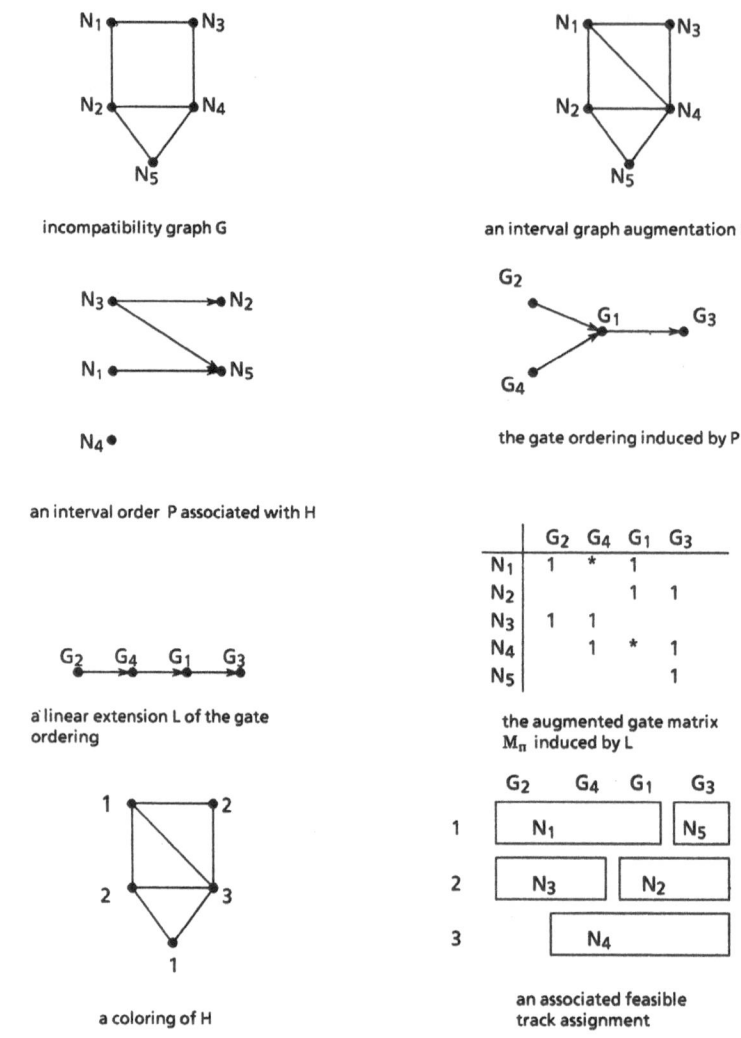

Figure 3.2. An illustration of Lemma 3.1 on the example of Figure 1.1

PQ-tree techniques [BL76, KM89]. The gate order induced by *P* can be obtained efficiently by scanning an interval representation of *P* from left to right and constructing an ordered partition of the gates as follows:

For the *i*-th right endpoint encountered in the scan, let the set $\mathbf{G}_i$ consist of all gates that are incident to nets ending (as intervals) at the current endpoint and do not belong to any previously constructed $\mathbf{G}_j$. The partition $\mathbf{G}_1, \ldots, \mathbf{G}_k$ thus constructed defines the gate ordering: all gates from G_i are incomparable among each other and precede all gates from $\mathbf{G}_{i+1}$ ($i = 1, \ldots, k-1$). This follows from the fact that all gates not yet considered at the current endpoint must belong to nets that come later in the interval representation and are thus successors in P of the currently ending nets. So linear extensions of the gate ordering are just permutations within the classes $\mathbf{G}_i$ of the partition. It follows that the partition and a linear extension (gate permutation) can be constructed in $O(n \cdot m)$ time.

An optimal coloring of an interval graph H can be obtained in $O(|V(H)|)$ time from an interval representation by scanning the interval representation from left to right [GLL82]. When the left endpoint of an interval is encountered in the scan, the corresponding vertex of H is assigned the smallest color from the set $\{1, 2, \ldots, n\}$ of colors that has not been assigned to intervals containing the current endpoint.

Obviously, the number of colors thus required is equal to the maximum number of intervals that intersect in a common point, i.e. the maximum size $\omega(H)$ of a clique of H. Since any coloring requires at least $\omega(H)$ colors, if follows that the coloring is optimal and $\chi(H) = \omega(H)$.

These arguments show that the hard core of the MPP is the construction of the right gate permutation, or, in terms of the interval graph augmentation, to find the right augmentation of the incompatibility graph to an interval graph. The track assignment problem or interval graph coloring problem can then be solved optimally by a simple linear-time algorithm.

Note that the interval graph coloring algorithm described here can of course also be directly carried out on augmented net-gate matrices (remember that they represent interval graphs). In this context, it is known as the *left-edge algorithm* (starting from the left edge of the rectangle described by the matrix) and it already occurs in channel routing applications of interval graphs [HS71]. It shows in particular that the minimum number of tracks for an augmented net-gate matrix M_π is equal to the maximum column sum of M_π.

Because of the easy solvability of a track assignment/interval graph coloring problem, we can rephrase our original problems as follows:

(MPP):
Given a 0-1 matrix M, find a permutation π of the columns of M such that the maximum column sum of the augmented matrix M_π is as small as possible.

Interval graph augmentation (IGA):
Given a graph G, find an augmentation of G to an interval graph H whose clique size $\omega(H)$ is as small as possible.

The smallest clique size $\omega(H)$ of an interval graph augmentation of G is also called the *interval thickness* of G. Because of its equivalence with MPP, we will denote it also by $t(G)$, and call it the *track number* of G.

So far, we have seen that every MPP can be transformed to an instance of IGA. The converse is, of course, also true, since very graph can be represented as the incompatibility graph of some MPP (e.g. by introducing a column for every edge of G with two 1-entries for the vertices joined by this edge).

This gives:

3.3 Theorem: *MPP and IGA are polynomially equivalent.*

The interpretation of MPP as IGA makes available the large body of algorithmic techniques for interval graphs, see e.g. [Go80], [Mö85], [Mö89]. Most of them are based on the following characterization of [FG65] of interval graphs.

3.4 Theorem: *A graph G is an interval graph iff its maximal cliques can be linearly ordered such that, for every vertex v, the maximal cliques containing v occur consecutively.*

Any such arrangement is called a *consecutive clique arrangement*. Such arrangements can be constructed and maintained by PQ-trees [BL76] or their specialized versions for interval graphs, the MPQ-trees [KM89]. These data structures will be useful for approximation algorithms discussed in Section 5.

Loosely speaking, the general idea of the interval graph augmentation approach to the MPP can be expressed as making the cliques of G consecutive (in the sense of Theorem 3.4) by extending or joining them to cliques of an interval graph while keeping the size of the new cliques small.

Path width. The path width of a graph was considered by Robertson and Seymour [RS83] in the first part of their series of papers on graph minors.

They define a *path decomposition* of a graph G as a sequence $X_1, \ldots, X_r$ of subsets of $V(G)$ such that

(3.5)   for every edge e of G, some X_i contains both ends of e

and

(3.6)   for $1 \le i \le j \le l \le r$, $X_i \cap X_l \subseteq X_j$

hold. The *path width* of G (denoted by $pw(G)$) is the minimum value of $k \ge 0$ such that G has a path decomposition $X_1, \ldots, X_r$ with $|X_i| \le k + 1$ $(i = 1, \ldots, r)$.

This notion is almost identical to interval graph augmentation. In fact, if G has a path decomposition $X_1, \ldots, X_r$ with $|X_i| \le k + 1$, then the graph H defined by letting $X_1, \ldots, X_r$ be its maximal cliques is an interval graph because of (3.6) and Theorem 3.4, and fulfills $E(G) \subseteq E(H)$ because of (3.5), and $\omega(H) = \max_i |X_i| \le k + 1$. Conversely, if H is an interval graph augmentation of G, then any consecutive clique arrangement $C_1, \ldots, C_r$ of H defines a path partition of G because of Theorem 3.4, and $|C_i| \le k + 1$ with $k = \omega(G) - 1$. This gives:

3.5 Proposition: *Determining the path width of a graph G is polynomially equivalent to IGA. In particular, $pw(G) = t(G) - 1$.*

This equivalence permits a direct translation of deep results from the Robertson-Seymour theory to gate matrix layout. The most important of these is related to the notion of the minor of a graph.

H is a *minor* of G if H can be obtained from G by deleting some vertices and/or edges, and/or contracting some edges. It is easy to see that

(3.7) $t(H) \leq t(G)$ if H is a minor of G.

This implies directly that for any fixed k, the class of graphs G with $t(G) \leq k$ is *closed under taking minors* (i.e., if G belongs to this class and H is a minor of G then H belongs also to this class). This is the starting point for the application of the following results of Robertson and Seymour.

3.6 Theorem [RS87]: *Let F be any set of graphs closed under minors. Then there are finitely many graphs $H_1, \ldots, H_r$ such that*

$$G \in F \Leftrightarrow G \text{ does not contain } H_i \text{ as a minor}, i = 1, \ldots, r.$$

This is a direct consequence of the proof of the Wagner Conjecture (no class of graphs has infinite antichains under the minor ordering) in [RS87] and the closedness property under taking minors.

3.7 Theorem [RS86]: *For any fixed graph H, it can be tested in polynomial time whether a graph G contains a minor isomorphic to H.*

The combination of these theorems yields the existence of a polynomial time algorithm for testing membership for any minor-closed family of graphs, thus in particular for the class of graphs with bounded path width.

However, since no proof of Wagner's conjecture can be entirely constructive [FRS87], Theorem 3.6 is a pure *existence result* for the finite family of forbidden minors. Moreover, though the algorithms for minor recognition have low degree polynomials as worst case bounds, their constants of proportionality are enormous, rendering them impractical for practical problems (see e.g. [Jo87a]). The general bound is $O(n^3)$ for a graph with n vertices, and even $O(n^2)$ if the family F excludes a planar graph. This is the case for the path width (see [RS83] and Proposition 3.12 below), and thus:

3.8 Theorem: *Within the class of graphs with bounded pathwidth k, k fixed, the interval graph augmentation problem can be solved in $O(n^2)$ time for a graph with n vertices.*

This result is interesting in view of the NP-hardness of the general problem when k is part of the input (see Section 4). A polynomial dynamic optimization algorithm for fixed k of order $O(n^{2k^2+4k+8})$ has also been obtained in the context of graph searching [EST87] (see also below). It is, however, open how to design a practical $O(n^2)$ algorithm.

The application of the Robertson-Seymour theory to gate matrix layout is discussed (among other problems) in a series of papers [FL87] [FL88a] [FL88b]. For a survey about computational implications of the Robertson-Seymour theory, we refer to [Jo87a].

Node searching. This problem formulation refers to a searching game on graphs introduced in [KP86] as a variant of the more investigated edge searching [Pa76].

In node searching, the edges of a graph represent a system of pipes or tunnels that are considered contaminated by a gas. The object of node searching is to clear all edges by a search. A *search* is a sequence of moves where a player places a *searcher* (also called *guard*) on a node of the graph that carries no searcher or deletes the searcher from a guarded node.

An edge is *cleared* if both its endpoints simultaneously carry a searcher. A cleared edge may be *recontaminated* if, at a later stage of the search, there is a path from an uncleared edge to the cleared edge without any searchers on it. So in order to avoid recontamination of cleared edges, the guarded nodes must after each move form a separating set that separates the still *unsearched part* of the graph (the not yet visited vertices) from the already *searched part* (all vertices that carried a searcher in the past).

A search is called *optimal* if the maximum number of searchers on the graph at any point is as small as possible. This number is called the *node-search number* of G, and denoted by $ns(G)$.

It was shown in [KP86] that there always is an optimal search without recontamination of cleared edges. This was used in [KP85] to show the following unexpected relationships to interval graph augmentation:

3.9 Theorem: *For any graph G, $ns(G) = t(G)$.*

The proof of this theorem is based on the following ideas. If H is an interval graph augmentation of G, then any consecutive arrangement $C_1, \ldots, C_k$ of the maximum cliques of H defines a search by letting the searchers move in this order through $C_1, \ldots, C_k$. If C_i is guarded, then searchers from $C_i - C_{i+1}$ and possibly new searchers may be moved to $C_{i+1} - C_i$ until C_{i+1} is guarded. It is easy to see that this defines a search without recontamination with $\omega(H)$ searchers.

In the converse direction, any recontamination-free search of G assigns to every node v of G the interval $[i,j]$ whose endpoints are the first and last step in the search at which v is occupied by a searcher. (Note that this definition makes sense since the search is recontamination-free.) Since every edge is searched, the intervals assigned to its endpoints intersect. So the interval representation induces an interval graph H with $E(G) \subseteq E(H)$. The maximum number k of searchers in this search is obviously just the maximum number of pairwise intersecting intervals, i.e. $\omega(H)$.

Node searching is also closely related to other linear graph layout problems and to pebbling games on graphs [KP86]. In particular, the interpretation as progressive pebbling game gives the following useful result [KP86] for investigating or generating searches.

Consider an acyclic orientation of the edges of G and a dynamic assignment of searchers to vertices that observes the rules

(3.8)   A vertex may accept a searcher only when all its immediate
            predecessors carry a searcher.

(3.9)   Every vertex is assigned a searcher exactly once.

Then:

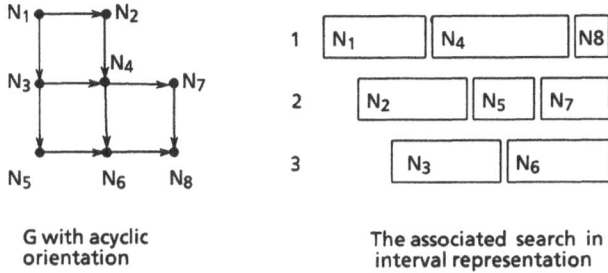

G with acyclic
orientation

The associated search in
interval representation

Figure 3.3. An illustration of a search

3.10 Proposition: *Every assignment observing (3.8) and (3.9) defines a recontamination-free search of G, and every recontamination-free search of G can be obtained in that way.*

The first part follows easily by induction and the observation that the "foremost" searchers (those that have an unvisited immediate successor) form a separating set in G that separates the searched part from the unsearched part. The other direction is obtained by considering the orientation of G defined by $u < v$ if u is visited by a searcher before v. An example is given in Figure 3.3.

We call such a search a *directed search*.

Another useful application is the combination of the search interpretation with a structural decomposition of graphs, the split decomposition [Cu82].

A *split* in an undirected graph G is a partition $V(G) = V_1 \cup V_2$ of the vertex set of G such that $|V_i| \geq 2$ $(i = 1, 2)$ and

(3.10)   The edges of of G going from V_1 to V_2 induce a complete bipartite graph.

3.11 Lemma: *Let $V(G) = V_1 \cup V_2$ be a split of G and let $A_i \subseteq V_i$ $(i = 1, 2)$ be the vertices of the associated complete bipartite graph. Then every recontamination-free search of G has a step at which all vertices from A_1 or all vertices from A_2 simultaneously carry a searcher. So in particular, $ns(G) \geq min\{|A_1|, |A_2|\}$.*

This can be seen as follows. If the statement is not true, then there is a first step of the search at which a searcher is deleted from the endpoint (u, say) of an already cleared edge $(u, v) \in A_1 \times A_2$. Since neither A_1 nor A_2 are completely visited at that step, there is an uncleared edge $(x, y) \in A_1 \times A_2$. But then (u, v) is recontaminated via the path $(x, y), (y, u)$.

Still another application is the following argument from [KP86], which, in the context of edge searching, is due to [Pa76]. It also shows that the search number is unbounded on the class of trees. This gives also the missing argument for the $O(n^2)$ algorithm for bounded tree width (the class of graphs with bounded tree width excludes some trees and thus a planar graph).

3.12 Lemma: *Let G contain a vertex v of degree 3 whose deletion separates G into three connected components G_1, G_2, G_3, each of which has node search number $ns(G_i) = t$. Then $ns(G) = t + 1$.*

It is easy to see that $t + 1$ searchers suffice. (Put a searcher on v and search G_1, G_2, G_3 with the remaining t searchers). To show that they are also necessary, assume that t searchers suffice for G.

Since t searchers are already required for each of the subgraphs G_1, G_2, G_3, there is a moment at which one of them (G_1, say) has already been searched, all searchers are on the second one (G_2, say), and the last of them is still unvisited. But then recontamination takes place between G_3 and G_1, a contradiction.

As a consequence, one obtains [KP86]:

3.13 Proposition: *The search number of a complete ternary tree T is equal to its height plus one.*

It was already mentioned that node searching is a variant of the more investigated edge searching. In edge searching, an edge is cleared by letting a searcher go through it (instead of by occupying both endpoints as in node searching). It is therefore possible [KP86] to obtain (optimal) node searchers on G from (optimal) edge searches on a slight modification of G (replace each edge of G by three parallel edges). Exploitation of this transformation and known results for edge searching on trees [MHGJP88] and dynamic programming formulations [EST87] yield a linear time search algorithm for trees (see also Section 4) and a polynomial time algorithm of order $O(n^{2k^2+4k+8})$ for graphs with search number at most k, k fixed.

Alternating paths. This final equivalence is considered in many papers on PLA folding. It views tracks in a layout of M as directed paths in the complement $\bar{G}$ of the incompatibility graph G of M. The directed edges of these paths may be considered as being added to G. This gives the following formulation.

Let G be the incompatibility graph of a MPP, and consider the edges of G as colored red. A *path partition* (or *multiple folding*) of G is a set of directed green edges F such that the following two conditions are satisfied.

(3.11)   The subgraph defined by the green edges is a collection of directed paths (i.e., indegree and outdegree of every vertex is at most 1). This constraint is called the *degree constraint*.

(3.12)   There exists no cycle of alternating directed green path segments and red edges (*alternating cycle*). This constraint is called the *cycle constraint*.

The *size* of a path partition is the number of green paths.

3.14 Theorem: *Determining a minimum size path partition is polynomially equivalent to interval graph augmentation.*

This can be seen as follows. From a path partition with t paths, one can construct an interval representation of an interval augmentation H of G with $\omega(H) = t$ by a left to right scan through the t paths. Initially, the intervals corresponding to the

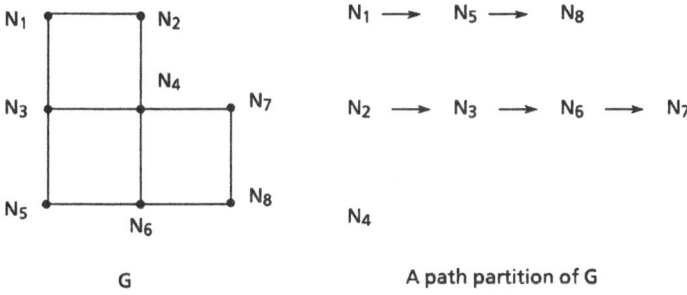

G   A path partition of G

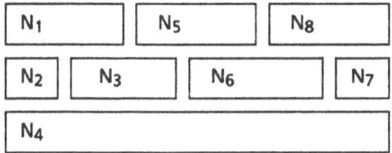

The associated interval representation
of Theorem 3.15

Figure 3.4. An illustration of a path partition

minimal vertices in the t paths are "opened". When an interval is closed, the interval of the next vertex in the corresponding path is opened etc. Given a collection of t opened intervals, the next interval to close corresponds to a vertex u such that there are no red edges (u, w) to a vertex w that occurs after a currently open interval v on a green path containing v. Note that there is such a vertex u because of (3.12).

Since exactly t intervals are open at any moment, $\omega(H) = t$. To see that H augments G, let $(u, v) \in E(G)$ and, w.l.o.g., let u be opened before v (as intervals). Then the choice defined above ensures that u is only closed after v is opened. Hence the corresponding intervals overlap. An example of this construction is given in Figure 3.4.

The converse direction is obvious since every optimal coloring of an optimal interval graph augmentation H of G defines a path partition of size $\chi(H) = \omega(H)$. This interpretation is particularly useful for the special cases of the MPP dealing with PLA folding.

There are several other equivalent or related graph theoretic notions. We just mention vertex separation, min cut linear arrangement, bandwidth and several modifications of these problems. While vertex separation is equivalent to node search [KP86], the other notions define bounds on $t(G)$ [KP86] [Bo88].

Restrictions of the MPP. The restrictions discussed in Section 2 have natural formulations within several of the different representations of this section.

For instance, the WMPP can be modeled by fixing two cliques of G as belonging to the first and last maximal clique of the interval augmentation to construct, or by requiring the searchers to start and finish their search on specified cliques of G.

For PLA folding problems, at most two vertices may share a common track. This means in the path partition formulation that the directed green paths reduce to directed green edges. This gives the most common formulation of PLA-folding, which is due to [HNS82].

3.15 Proposition: *Let G be the incompatibility graph of a PLA folding problem. Then the minimum number of tracks for a PLA folding is equal to $|V(G)| - s$, where s is the maximum number of green directed arcs that can be added to G such that*

(3.13)   *The green arcs form a matching in the complement of G (degree constraint),*

(3.14)   *There is no alternating cycle of directed green edges and undirected red edges of G (cycle constraint).*

Such a set F of green arcs is called a *folding set* or simply *folding*. So Proposition 3.15 states that the PLA-folding problem is equivalent to finding a folding set of maximum size.

Another characterization can be obtained from the observation that, in an optimal layout of the associated augmented matrix M_π, the rows can be permuted in such a way that the rows with two nets appear on top and that the rightmost 1's of the first nets in these rows define a "decreasing staircase".

This staircase pattern corresponds to the special subgraph $Z_{m,m}$ of $\bar{G}$ defined below.

A $Z_{m,m}$ (also called a *triangular clique* in [HK87]) is a bipartite graph $G = (U, V, E)$ with $U = \{a_1, \ldots, a_m\}$, $V = \{b_1, \ldots, b_m\}$, and $E = \{(a_i, b_j) | 1 \le i \le j \le m\}$.

This gives [HK87]:

3.16 Proposition: *Finding a maximum folding set in G is equivalent to finding a maximum size $Z_{m,m}$ as (partial, not induced) subgraph of $\bar{G}$.*

Both conditions can easily be sharpened for block folding and constrained folding problems. Call a folding set F resulting in a block folding a *block folding set*. Then one obtains the following characterization of block folding, see e.g. [RL88].

3.17 Proposition: *Let G be the incompatibility graph of a PLA folding problem. Then:*
(1) *A set F of directed green arcs added to G is a block folding set iff F satisfies the degree constraint (3.13) and*

(3.15)   *There is no red edge (u, v) from the head of green edge to the tail of another green edge.*

(2) *Finding a maximum block folding set in G is equivalent to finding a maximum size $K_{m,m}$ (the complete bipartite graph on $2m$ vertices) as (partial, not induced) subgraph of $\bar{G}$.*

Remember that in constrained PLA folding, the nets are preassigned to the two sides of the layout. Thus only incompatibility relations between these two sides are of importance, i.e., the incompatibility graph G can be assumed to be bipartite. Therefore, constrained PLA-folding is sometimes also called *bipartite folding* [EL84], [HK87].

The above conditions then specialize further for block folding.

3.18 Proposition: *Finding a maximum block folding in a bipartite graph G is equivalent to finding a maximum size $K_{m,m}$ as induced subgraph of the (bipartite) complement of G.*

4. Complexity Results

It was already mentioned several times that the general MPP is NP-hard. In fact, this has been obtained independently in many of the equivalent formulations, e.g. for interval graph augmentation in [KF79], for node search in [KP86], for directed node search in [Le82], and for path width in [ACP87].

The NP-hardness of the PLA folding problems (including the general MPP, but excluding block folding) is shown in [LVVS82] by a series of reductions from *matrix upper triangulation* (given an $n \times n$ 0-1 matrix A, is there a permutation of the rows and another of the columns such that resulting matrix is upper triangular?).

We will here sketch a different series of reductions that starts from constrained block folding and contains block folding and a sharper version for interval graph augmentation (i.e. the general MPP), which turns out to be NP-hard already on chordal graphs. Our starting point is (see e.g. [EL84]):

4.1 Theorem: *Constrained block folding is NP-hard.*

This follows in fact directly from Proposition 3.19 that gives the equivalence to "balanced complete bipartite subgraph" which is stated to be NP-complete in [GJ79] (the proof has appeared in [Jo87b]).

Preassignment of nets to sides can easily be enforced by adding two gates G_0, G_{n+1} that are connected to the nets from the left and right side, respectively. Then any (block) folding of the augmented problem can only fold nets incident to G_0 with nets incident to G_{n+1}, and the sides are (up to reversal of the layout) fixed by the position of G_0 and G_{n+1}. In view of Theorem 4.1, this gives:

4.2 Theorem: *Block folding is NP-hard.*

This result has been sharpened in [MW89] by a different reduction from GRAPH BISECTION. Exploiting techniques from [BCLS87], [WW89] they obtain:

4.3 Theorem: *Block folding is NP-hard even for graphs with degree at most k for any fixed $k \geq 3$.*

The reduction from block folding to IGA on chordal graphs is based on the following equivalent formulation of block folding.

Given: A graph G and an integer k.

Question: Is there a partition of $V(G)$ into three sets V_1, V_2, V_3 such that
              (i) every path from V_1 to V_3 goes through a vertex of V_2,
              (ii) $\min\{|V_1|, |V_3|\} \geq k$?

The answer to such an instance is obviously yes iff there exists a block folding set of G with k green edges (viz. from vertices to V_1 to vertices of V_3). Based on this formulation, the following result is obtained in [Gu89].

4.4 Theorem: *Gate matrix layout and pathwidth are already NP-hard on the class of chordal graphs.*

Recall that a graph is *chordal* (or *triangulated*) if every elementary cycle $v_1, v_2, \ldots,$ v_k, v_1 of length $k \geq 4$ possesses a chord, i.e. an edge (v_i, v_j) with $1 \leq i < j + 1 \leq k + 1$.

Chordal graphs form a natural generalization of interval graphs, see e.g. [Go80] for more information about chordal graphs. The chordal graphs needed in the proof are quite special. They consist of a set of maximal cliques that overlap in a central clique.

In more detail, let G, k be an instance of the above formulation of block folding such that G is w.l.o.g. connected. From G we construct such a special chordal graph H as follows. H contains a maximal clique C_0 (called the *central* clique) with vertex set $V(G)$. For each edge $(u, v) \in E(G)$, a maximal clique C_{uv} is added that consists of the vertices $u, v \in C_0$ and $|V(G)|$ additional vertices that are incident only to vertices in C_{uv}. Clearly, this graph is of the desired type.

From H one can construct a net-gate matrix M with incompatibility graph H by introducing a gate for each of the maximal cliques of H. Consider an augmented matrix M_π of M and let C^+ and C^- be the cliques (gates) before and after the central clique (gate) C_0 in M_π. Since every net corresponding to an original vertex of G that is incident to some gate from C^+ or C^- is also incident to C_0, the order of the gates in C^+ or C^- does not influence the number of tracks.

Let V^+ and V^- denote the nets of C_0 (vertices of G) incident to a gate from C^+ and C^-, respectively, and let $V_1 := V^+ - V^-$, $V_2 := V^+ \cap V^-$ and $V_3 := V^- - V^+$. It can then be shown that V_1, V_2, V_3 form a partition of $V(G)$ with the sbove disconnecting property, and that $\min\{|V_1|, |V_3|\} \geq k$ is equivalent to $t(M) \leq 2 \cdot |V(G)| - k$.

Conversely, every partition V_1, V_2, V_3 of G with the above properties can be transformed into an augmented matrix M_π by first taking all gates (cliques) C_{uv} with $u, v \in V_1 \cup V_2$ (in any order), followed by C_0 and all cliques C_{uv} with $u, v \in V_2 \cup V_3$. This proves the theorem. An example of this construction is given in Figure 4.1.

Note that if all maximal cliques that intersect the central clique are mutually disjoint, then the problem can be solved in $O(|V(G)|^2)$ time by a dynamic programming algorithm [Gu89]. This confirms that the borderline between easy and hard problems for subclasses of chordal graphs G depends essentially on the overlapping behavior of the maximal cliques of G (see also [ACP87]).

$V_1 = \{1\}$

$V_2 = \{2, 4\}$

$V_3 = \{3\}$

An associated partition
V_1, V_2, V_3 with $k = 1$

	C_0	C_{12}	C_{14}	C_{23}	C_{34}
1	1	1	1		
2	1	1		1	
3	1			1	1
4	1		1		1
5		1			
6		1			
7		1			
8		1			
9			1		
10			1		
11			1		
12			1		
13				1	
14				1	
15				1	
16				1	
17					1
18					1
19					1
20					1

	C_{12}	C_{14}	C_0	C_{23}	C_{34}
1	1	1	1		
2	1	*	1	1	
3			1	1	1
4	1	1	*		1
5	1				
6	1				
7	1				
8	1				
9		1			
10		1			
11		1			
12		1			
13				1	
14				1	
15				1	
16				1	
17					1
18					1
19					1
20					1

The net gate-matrix of H

The augmented matrix corresponding to V_1, V_2, V_3

Figure 4.1. An illustration of the proof of Theorem 4.4

A reduction from gate matrix layout to constrained PLA-folding can be obtained by turning an arbitrary graph G into a bipartite graph $H = (U, V, E)$ as follows:

$U := V(G)$, $V := \{v' | v \in V(G)\}$ (a copy of $V(G)$), and $(u, v') \in E$ (with $u \in U$ and $v' \in V$) iff $(u, v) \in E(G)$ or $u = v$.

Then any constrained folding set F for H corresponds uniquely to a collection of tracks for G by combining the edges of F to paths (tracks) (x_1, x_2'), (x_2, x_3'), (x_3, x_4'), ..., (k_k', x_1). Then $t(G) = |V(G)| - |F|$, i.e. maximizing $|F|$ in H corresponds to minimizing $t(G)$ in G, Hence:

4.5. Theorem: *Constrained PLA folding is NP-hard.*

Finally, the reduction from constrained PLA folding to PLA folding is achieved in the same way as from constrained block folding to block folding.

4.6. Theorem: *PLA folding is NP-hard.*

There are several other NP-hardness results related to gate matrix layout and PLA folding. We mentioned already the directed search problem [Le82]. Another such problem is

Orderability:

Given: A graph G of red edges, a set F of green edges of $\bar{G}$.

Question: Is there an orientation of the edges of F such that F is a folding?

Orderability is shown to be NP-complete in [HNS82]. It is solvable in polynomial time for constrained PLA folding [Ra88].

Other variants of the PLA-folding problem not discussed here are shown to be NP-hard in [ALN88].

We consider now some special classes of graphs on which the gate matrix layout problem can be solved in polynomial time. Most of the arguments leading to polynomial algorithms come from node searching (in particular Lemma 3.11 and Lemma 3.12) and demonstrate again the usefullness of this interpretation.

We will start with the class of trees. As mentioned before, the polynomial algorithm for edge searching on trees [MHGJP88] can be transformed by the principles of [KP86] to a polynomial algorithm for node searching on trees. This requires $O(n)$ time for determining $ns(G)$, and $O(n\log n)$ time for finding the associated search. We sketch here a different, equally fast algorithm with a much simpler correctness proof.

The algorithm peels the tree, i.e. it starts with the leaves and works its way towards the "center" of the tree. So at a typical step of the algorithm, certain subtrees T_1, $\ldots$, T_r of the tree T have already been investigated. Each of these trees T_i has a vertex v_i connecting it to the still unsearched part of T.

The peeling is done in *phases* $t = 1, 2, \ldots, ns(T)$. At the beginning of phase t, the vertices v_i are the leaves of the remaining tree, and every T_i requires t searchers, while every $T_i - v_i$ can be searched with $t - 1$ searchers. We then peel the remaining tree starting from the v_i until we reach new vertices u_j in which the search number must go up to $t + 1$ (or T has been searched completely). The current phase is completed when all v_i have been processed and are connected to some u_j in the already searched part of the tree.

All vertices investigated in phase t (expect the u_j which are the starting vertices for the next phase) receive the *label* label$(v) = t$, $t = 1, \ldots, ns(T)$. These labels have the following important property:

Let, for a vertex with label t, T_v be the connected component in the subgraph of all vertices with label at most t that contains v. Then:

(4.1) $ns(T_v) = \text{label}(v).$

An example of the peeling and the different phases is given in Figure 4.2.

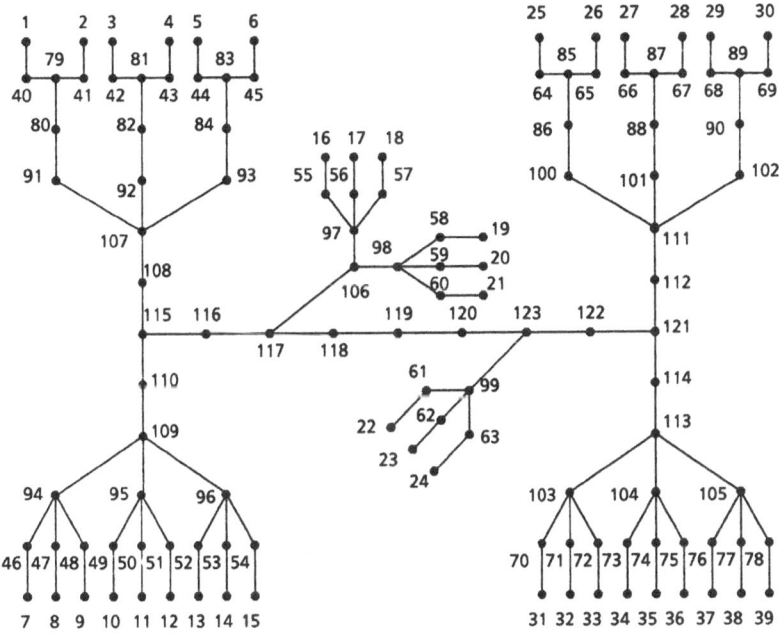

a) Processing order of the vertices

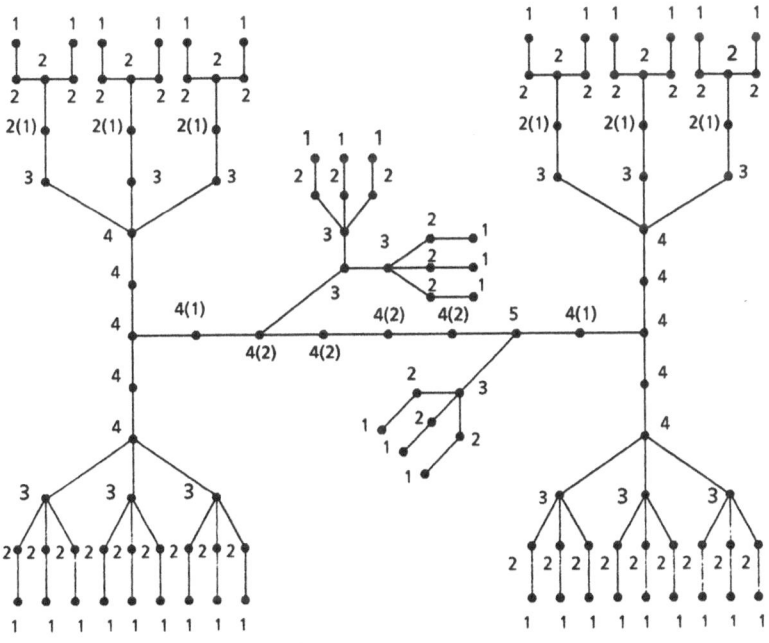

b) Pase labels with respect to the processing order of a)

Figure 4.2. An example tree for the algorithm

The essential part of the algorithm consists in identifying the new vertices u_j in which the search number must go up. There are several special cases for small t (e.g. the transition from vertex 80 to 91 in Figure 4.2), but the general procedure is as follows.

We first explore from every v_i the path (in the remaining tree) connecting it to the still unexplored part of T until we reach a vertex v (where $v = v_i$ is possible) with 2 unexplored neighbors or another neighbor u with label$(u) =$ label(v_i). In Figure 4.2, such paths are e.g. 109, 110 or 111, 112. After this *path exploration*, they are three cases.

We reach a vertex v with two unexplored neighbors. Due to the algorithm, these neighbors are conected to other subtrees with search number t, and so the search number must go up to $t + 1$ because of Lemma 3.12. Then v will be considered in a later phase. In Figure 4.2, $v = 99$ is such a vertex.

Three or more paths meet in a vertex u. Then the search number goes up in u to $t + 1$ because of Lemma 3.12. In Figure 4.2, this is e.g. the case for $u = 107$ (from 91, 92, 93) and $u = 111$ (from 100, 101, 102).

Exactly two path next meet in a vertex u. Then the search can be carried over with t searchers from one path to the other with u in the "interior" of the corresponding layout (i.e. there is no search with t searchers that ends in u). It may, however, still be possible to search part of the tree starting in u with at most $t - 1$ searchers while a searcher guards the vertex u. To this end, we consider the unexplored neighbor of u as a leaf and start a search from it to the unexplored part until we reach the first vertex where t searchers are required. This vertex is then a leaf for the next phase. We call this situation a *fork*.

In Figure 4.2, such forks are given by $u = 106$, $u = 115$, and $u = 121$. In $u = 106$, the search stops immediately because 117 has two unexplored neighbors. In $u = 115$, we start a search along the path 116, 117, etc., where 116 is treated as a leaf. The new phase labels of this search are given in brackets. Note that the fork in 106 influences this search since a guard must remain on 106, thus implying that only two searchers are available along the path 117, 118, etc. This is why the search stops in 120. The formal stopping argument is that the subtree induced by $\{116, 117, \ldots, 120, 123\}$ requires 4 searchers, which together with the fork at 115 and Lemma 3.12 increases the search number to 5.

The fork of 121 then meets the fork of 115 in 123 and brings the search number up to 5 because of Lemma 3.12 (every fork contributes two subgraphs of search number t).

These are the main ingredients of the algorithm. There are several additional remarks to be made.

The argument for an increase of the search number is always Lemma 3.12. This is in fact due to the "converse" of Lemma 3.12 [Pa76] stating that, for any tree T, $ns(T) \geq t + 1$ iff there is a vertex v at which there are three or more subtrees with search number t or more.

The path exploration in forks depends of course on the processing order of the vertices. For instance, exploring the fork in 121 first would explore the path until vertex 117.

The phrase labels and the information about forks (i.e. which paths are joined) can be directly used to obtain an optimal search. This is demonstrated in Figure 4.3 for the example of Figure 4.2.

Altogether, this gives:

4.7 Theorem: *The track number $t(T)$ of a tree T and an optimal layout can be obtained in $O(|V(T)|)$ time.*

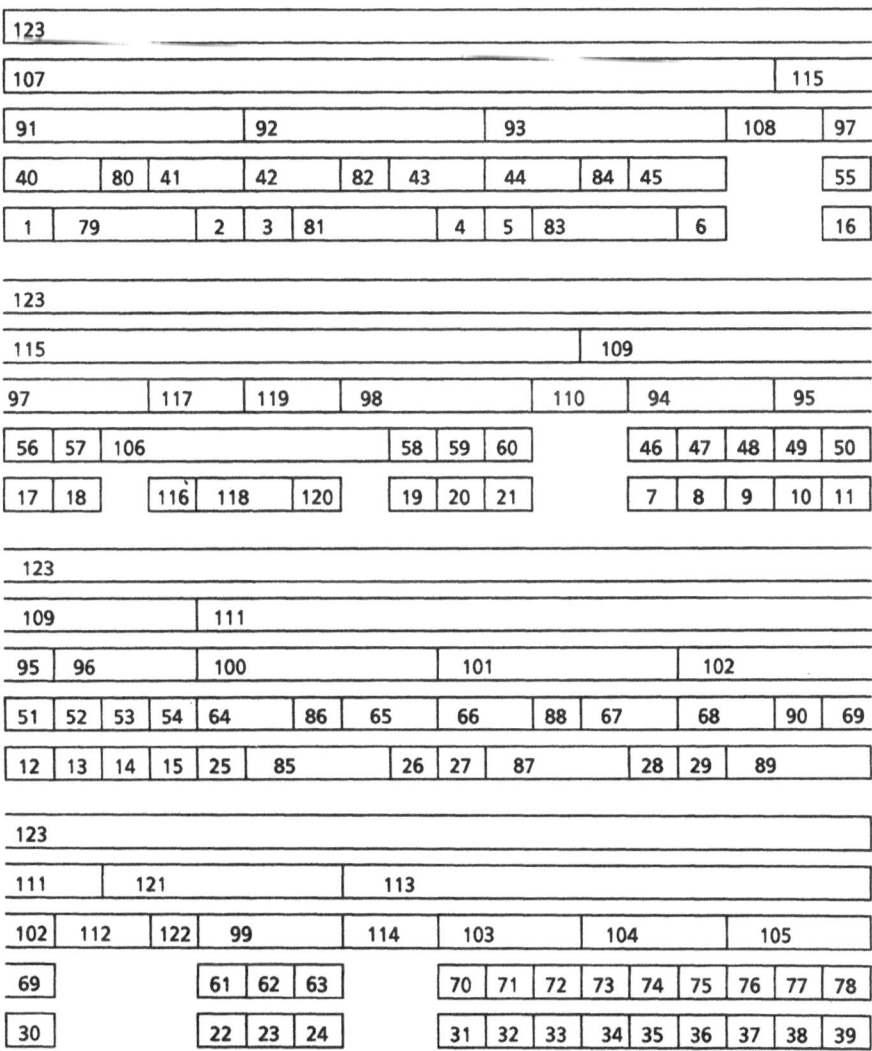

Figure 4.3. The layout corresponding to the search of Figure 4.2

The techniques underlying the tree search algorithm seem to be valid also for other "tree-structured" graphs such as (2-connected) outerpanar graphs, k-trees, and possibly partial k-trees. This is currently being investigated together with J. Gustedt and R. Müller from the TU Berlin.

Lemma 3.11 can be used to obtain a polynomial algorithm for cographs. Recall that a *cograph* (or *complement reducible graph*, see [CLS81] for details) can be defined recursively by

(4.2)   The one-vertex graph is a cograph

(4.3)   If $G_1 = (V_1, E_1)$ and $G_2 = (V_2, E_2)$ are cographs on disjoint sets V_1, V_2, then

$$G_1 + G_2 := (V_1 \cup V_2, E_1 \cup E_2) \qquad \text{and}$$

$$G_1 * G_2 := (V_1 \cup V_2, E_1 \cup E_2 \cup (V_1 \times V_2))$$

are also cographs.

Then Lemma 3.11 gives:

(4.4) $ns(G_1 + G_2) = \max\{ns(G_1), ns(G_2)\}$

(4.5) $ns(G_1 * G_2) = \min\{|V_1| + ns(G_2), |V_2| + ns(G_1)\}$

An associated interval representation of an optimal search can be defined similarly as in the tree algorithm by defining offsets in each operation. So if a sequence of operations "$*$" and "$+$" producing the whole graph is given (a parse tree defining such a sequence for G can be obtained in $O(|V(G)| + |E(G)|)$ time [CPS85]) one obtains:

4.8 Theorem: *The track number of a cograph G (given in "decomposed" form or by its parse tree) and an optimal layout can be obtained in $O(|V(G)|)$ time.*

The operation "$*$" can be seen as a special split in which (with the terminology of Lemma 3.11) $A = V_1$ and $B = V_2$. It is still open whether there is a polynomial time search algorithm for the larger class of graphs that can be recursively decomposed by splits. They are known as *distance-hereditary graphs* [BM86] or *completely separable* graphs [HM87], and contain both the class of trees and the class of cographs.

Finally, we mention the class of chordal graphs discussed in connection with Theorem 4.4. By using a dynamic programming approach similar to that for PARTITION in [GJ79], [Gu89] obtains:

4.9 Theorem: *If the maximal cliques $C_0, C_1, \ldots, C_m$ of G fulfill*

(4.6) $C_i \cap C_j \neq \emptyset \Rightarrow C_i \cap C_j = C_i \cap C_0,$

then $t(G)$ and an optimal layout can be obtained in $O(|V(G)|^2)$ time.

Condition (4.6) means that the cliques $C_1, \ldots, C_n$ have a special overlap structure with the "central" clique C_0. Similar arguments yield also a polynomial algorithm for *split graphs* [Go80], i.e. those chordal graphs whose complement is also chordal.

For PLA-folding problems, much less is known about polynomially solvable cases. An algorithm for constrained PLA-folding on trees has been obtained in [HK87]. Results of [Bo87] on the balanced complete bipartite subgraph problem (in particular the transformation applied there) show, when combined with Propositions 3.16 and 3.18, the polynomial solvability of constrained block folding and block folding on partial k-trees. It is easy to also obtain polynomial time dynamic programming algorithms for these cases.

5. Algorithms

Due to the VLSI-background, many algorithms have been proposed and studied in the literature. The majority of them can be classified as (sometimes a combination of) heuristics, branch-and-bound algorithms, or dynamic programming algorithms.

Representatives for the different technologies discussed here are [YKK75], [As82] for Weinberger arrays, [LVVS82], [Li83], [WHW85], [Leo86], [NFKY86], [DKL87] for gate matrix layout, and [LL84], [HDB86], [HK87], [KH87], [ALN88] for PLA folding (see [GL88] for additional references). The problems studied in these papers are usually not larger than 50×60 (in terms of the net-gate matrix), with the exception of 100×80 in [ALN88].

Only little is known about the performance of heuristics for these problems. By using standard arguments from [GJ79], the existence of an approximation algorithm for gate matrix layout with a constant absolute performance guarantee is ruled out in [DKL87]:

5.1 Theorem: *Unless $P = NP$, there is no approximation algorithm A for gate matrix layout with*

(5.1) $$A(I) \leq OPT(I) + K, \qquad K \in \mathbb{N} \text{ fixed}$$

for all instances I.

It is well known [GJ79] that this also rules out the existence of a fully polynomial approximation scheme for gate matrix layout.

The existence of approximation algorithms with a *constant relative* performance guarantee is open. There are, however, some indications that they might not exist.

For PLA-folding problems, such an indication is given in [RL88]. Call two problems *equivalent with respect to approximation* if both or none are approximable with constant relative performance guarantee in polynomial time. Then [RL88]:

5.2 Theorem: *PLA-folding, block folding, and constrained block folding are equivalent with respect to approximation, when viewed as maximization problems (maximize the size of a folding).*

The proof is based on transforming feasible solutions of one problem to feasible solutions of the other while preserving a constant relative performance guarantee. For instance, let algorithm A produce, for an instance I of PLA-floding, a solution with $A(I)$ folded pairs such that $A(I) \le K \cdot OPT(I)$, where $OPT(I)$ denotes the size of an optimal folding. Then we can transform this solution to a solution of block folding, with size $A'(I)$ by taking, in a decreasing staircase arrangement of the layout (cf. the remarks preceeding Proposition 3.17), the first nets of rows $1, 2, \ldots, \lfloor A(I)/2 \rfloor$ and fold them with the last nets of rows $\lfloor A(I)/2 \rfloor + 1, \ldots, A(I)$. Clearly, this gives a block-folding with $A'(I) = A(I)/2$ folded nets, and so the maximum size $OPT'(I)$ of a block folding fulfills $OPT'(I) \le OPT(I) \le K \cdot A(I) = 2K \cdot A'(I)$.

This equivalence result is combined in [RL88] with the following, unexpected result that provides the indication for non-approximatibility.

5.3 Theorem: *If there is an approximation algorithm A for block folding (in the maximization version) with relative performance guarantee K, $K \in \mathbb{N}$ fixed, then there is also one with relative performance guarantee $1 + \varepsilon$ for every fixed $\varepsilon > 0$.*

Note, however, that these results carry only through for the mximization version of the PLA-folding problems (maximize the size of a folding set). Both proofs fail for the—perhaps more natural—minimization of the number of tracks required in a layout.

Current work at the TU Berlin applies Lagrangean relaxation and subgradient methods combined with branch-and-bound techniques to the constrained block folding problem [Mü89]. Starting point is the formulation as a matching problem on a bipartite graph with side constraints (Proposition 3.17). Relaxation of the side constraints (3.15) and an appropriate reformulation then leads to a special well-solved min-cost flow problem whose use in branch-and-bound schemes seems to be promising.

For gate matrix layout, an indication for the non-existence of approximation algorithms with constant relative performance guarantee is obtained in [DKL87] by considering algorithms that are *on-line* with respect to the nets. This means that the nets are processed in an incremental fashion according to the following rules:

(1) A partial layout for the nets processed so far has already been constructed and may not be changed when the next net is processed.

(2) The next net N_i to be processed is chosen such that its addition to the partial layout causes the least increase in the number of tracks.

By designing a class of examples, [DKL87] show that such on-line algorithms cannot guarantee a constant relative performance error.

This definition of on-line seems, however, to be very restrictive, since the algorithms even fail to construct an optimal layout for interval graphs. (In fact, the class of examples of [DKL87] consists entirely of interval graphs.) The reason for this behavior is the rigidity of the already constructed partial layout.

We suggest here a less rigid approach that is still on-line and equally fast, but allows more flexibility in modifying the partial layout. The basic idea is to maintain not a partial layout, but the corresponding interval graph (see Section 3), and to represent it by its MPQ-tree.

The MPQ-tree (see [KM89] for details) is a data structure that represents an interval graph H and all associated interval orders (cf. Lemma 3.1) in $O(|V(H)|)$ space. It permits also fast updating when a vertex is added to H. Such an update will always recognize when $H + v$ is again an interval graph and modify the MPQ-tree accordingly. (So the examples from [DKL87] are solved optimally.) If $H + v$ is not an interval graph, then there are several possibilities to augment $H + v$ to an interval graph H^* by considering different interval orders associated with H. So keeping the maximum clique size small then means to permute or invert the nodes of the MPQ-tree representing H in such a way that adding v increases $\omega(H)$ as little as possible. This "local" optimization can be done in polynomial time.

Altogether, this leads to a class of on-line algorithms based on incremental interval graph generation by means of MPQ trees. Current work at the TU Berlin in this direction seems to be quite promising.

Acknowledgement

I would like to thank H. L. Bodländer, A. Proskurowski, and M. M. Syslo for various discussions and many helpful remarks.

References

[ALN88] C. Arbib, M Lucertini, and N. Nicoloso (1988). Optimal design of programmed logic arrays, preprint, Università degli Studi di Roma "La Sapienza".

[ACP87] S. Arnborg, D. G. Corneil, and A. Proskurowski (1987). Complexity of finding embeddings in a k-tree, SIAM J. Alg. Disc. Meth. 8, 277–284.

[As82] Asano, T. (1982). An optimum gate placement algorithm for MOS one-dimensional arrays, Journal of Digital Systems, 1–27.

[Bo87] H. L. Bodländer (1987). Dynamic programming on graphs with bounded tree width. Report RUU-CS-87-22, University of Utrecht.

[BM86] H. J. Bandelt and H. M. Mulder (1986). Distance-hereditary graphs, J. Comb. Th. B 41, 182–208.

[Bo88] H. L. Bodländer (1988). Some classes of graphs with bounded treewidth, Bulletin EATCS 36, 116–125.

[BL76] S. Booth and S. Lueker (1976). Testing for the consecutive ones property, interval graphs, and planarity using PQ-tree algorithms, J. Comput. Syst. Sci. 13, 335–379.

[BMHS84] R. K. Brayton, C. McMullen, G. D. Hachtel, and A. Sangiovanni-Vincentelli (1984). Logic Minimization Algorithms for VLSI Synthesis. Kluwer Academic Publishers, Boston, MA.

[BCLS87] T. N. Bui, S. Chaudhuri, F. T. Leighton and M. Sipser (1987). Graph bisection algorithms with good average case behavior, Combinatorica 7, 171–191.

[CLS81] D. G. Corneil, H. Lerchs, and L. Stewart Burlingham (1981). Complement reducible graphs, Discrete Appl. Math. 3, 163–174.

[CPS85] D. G. Corneil, Y. Pearl, and L. Stewart (1985). A linear recognition algorithm for cographs, SIAM J. Computing 14, 926–934.

[Cu82] W. H. Cunningham (1982). Decomposition of directed graphs, SIAM J. Alg. Discr. Methods 3, 214–228.

[DKL87] N. Deo, M. S. Krishnamoorty and M. A. Langston (1987). Exact and approximate solutions for the gate matrix layout problem, IEEE Trans. on Computer-Aided Design 6, 79–84.

[EL84] J. R. Egan and C. L. Liu (1984). Bipartite folding and partitioning of a PLA, IEEE Trans. CAD 3, 191–199.

[EST87] J. Ellis, I. Sudborough and J. Turner (1987). Graph separation and search number, Report DCS-66-IR, University of Victoria

[Ev79]   S. Even (1979). Graph Algorithms, Computer Science Press, Potomac, MD.
[FL87]   M. R. Fellows and M. A. Langston (1987). Nonconstructive advances in polynomial-time complexity. Infor. Proc. Letters 26, 157–162.
[FL88a]   M. R. Fellows and M. A. Langston (1988). Nonconstructive tools for proving polynomial-time decidability, J. ACM 35, 727–739.
[FL88b]   M. R. Fellows and M. A. Langston (1988). Layout permutation problems and well-partially-ordered sets. Proc. 5th MIT Conf. on Advanced Research in VLSI, 315–327.
[FM75]   H. Fleisher and L. I. Maissel (1975). An introduction to array logic, IBM J. Res. and Developm. 19, 98–109.
[FRS87]   H. Friedman, N. Robertson and P. D. Seymour (1987). The methamathematics of the graph minor theorem, in "Applications of Logic to Combinatorics", AMS Contemporary Mathematics Series, Amer. Math. Soc. Providence, R. I., to appear.
[FG65]   D. R. Fulkerson and O. A. Gross (1965). Incidence matrices and interval graphs, Pacific J. Math. 15, 835–855.
[GL88]   D. G. Gajski and Y-L. S. Lin (1988). Module generation and silicon compilation, in B. Preas and M. Lorenzetti (eds.) Physical Design Automation of VLSI Systems, Benjamin/Cummings, Menlo Park, CA, 283–345.
[GJ79]   M. R. Garey and D. S. Johnson (1979). Computers and Intractibility: A Guide to the Theory of NP-Completeness, Freemann, San Francisco.
[Go80]   M. C. Golumbic (1980). Algorithmic Graph Theory and Perfect Graphs, Academic Press, New York.
[Go85]   M. C. Golumbic (1985). Interval graphs and related topics, Discrete Math. 55, 113–121.
[GLL82]   U. I. Gupta, D. T. Lee and I. Y.-T. Leung (1982). Efficient algorithms for interval graphs and circular arc graphs, Networks 12, 459–467.
[Gu89]   J. Gustedt (1989). Path width for chordal graphs is NP-complete, preprint, TU Berlin.
[HNS82]   G. D. Hachtel, A. R. Newton, and A. L. Sangiovanni-Vincentelli (1982). An algorithm for optimal PLA folding, IEEE Trans. on CAD of Integrated Circuits and Systems, CAD-1 (2), 63–77.
[HM87]   P. L. Hammer and F. Maffray (1987). Completely separable graphs, preprint, Rutgers University.
[HS71]   A. Hashimoto, and J. Stevens (1971). Wire routing by optimizing channel assignment within large apertures, Proc. of 8th Design Automation Conf., 155–169.
[HK87]   T. C. Hu and Y. S. Kuo (1987). Graph folding and programmable logic array, Networks 17, 19–37.
[HDB86]   S. Y. Hwang, R. W. Dutton, and T. Blank (1986). A best-first search algorithm for optimal PLA folding, IEE Trans. on CAD, CAD-5 (3), 433–442.
[Jo87a]   D. S. Johnson (1987). The NP-completeness column: An ongoing guide, J. Algorithms 8, 285–203.
[Jo87b]   D. S. Johnson (1987). The NP-completeness column: An ongoing guide, J. Algorithms 8, 438–448.
[KF79]   T. Kashiwabara and T. Fujisawa (1979). NP-completeness of the problem of finding a minimum-clique-number interval graph containing a given graph as a subgraph, Proc. 1979 Intern. Symposium on Circuits and Systems, 657–660.
[KP85]   L. M. Kirousis and C. H. Papadimitriou (1985). Interval graphs and searching, Discrete Math. 55, 181–184.
[KP86]   L. M. Kirousis and C. H. Papadimitriou (1986). Searching and pebbling, Th. Comp. Science 47, 205–218.
[KM89]   N. Korte and R. H. Möhring (1989). An incremental linear-time algorithm to recognize interval graphs, SIAM J. Computing 18, 68–81.
[KH87]   Y. S. Kuo and T. C. Hu (1987). An effective algorithm for optimal PLA column folding INTEGRATION, the VLSI journal 5, 217–230.
[Le82]   T. Lengauer (1981). Black-white pebbles and graph separation, Acta Inf. 16, 465–475.
[Leo86]   H. W. Leong (1986). A new algorithm for gate matrix layout. Digest, Intl. Conf. on Computer-Aided Design, 316–319.
[LL84]   J. L. Lewandowski and C. L. Liu (1984). A branch and bound algorithm for optimal PLA folding. Proc. of the 21st Design Automation Conf., 425–433.
[Li83]   J. T. Li (1983). Algorithms for gate matrix layout, Proc. 1983 IEEE Int. Symp. Circuits and Systems, 1013–1016.
[LLa80]   A. Lopez and H. Law (1980). A dense gate matrix layout method for MOS VLSI, IEEE Trans. on Electronic Devices, ED-27 (8), 1671–1675.

[LVVS82]   M. Luby, U. Vazirani, V. Vazirani, and A. L. Sangiovanni-Vincentelli (1982). Some
             theoretical results on the optimal PLA folding problem, IEEE Internat. Confer. Circuits
             and Systems, 165–170.
[MS89]   F. Makedon and I. H. Sudborough (1989). On minimizing width in linear layouts, Discrete
             Appl. Math. *23*, 201–298.
[MC80]   C. Mead and L. Conway (1980). Introduction of VLSI Systems, Addison Wesley, Reading
             MA.
[MHGJP88]   N. Megiddo, S. L. Hakimi, M. R. Garey, D. S. Johnson, and C. H. Papadimitriou (1988).
             The complexity of searching a graph, J. ACM *35*, 18–44.
[Mö85]   R. H. Möhring (1985). Algorithmic aspects of comparability graphs and interval graphs,
             in I. Rival (ed.) Graphs and Order, Reidel, Dordrecht, 41–101.
[Mö89]   R. H. Möhring (1989). Computationally tractable classes of ordered sets, in I. Rival (ed.)
             Algorithms and Order, Kluwer Acad. Publ., Dordrecht, 105–194.
[Mü89]   R. Müller (1989), in preparation.
[MW89]   R. Müller and D. Wagner (1989). α-vertex separation is NP-complete for 3-regular graphs,
             preprint, TU Berlin.
[NFKY86]   K. Nakatani, T. Fujii, T. Kikuno, and N. Yoshita (1986). A heuristic algorithm for gate
             matrix layout, Digest Intl. Conf. on Computer-Aided Design, 324–327.
[OMKK79]   T. Ohtsuki, H. Mori, E. S. Kuh, T. Kashiwabara, and T. Fujisawa (1979). One-dimensional
             logic gate assignment and interval graphs, IEEE Trans. on Circuits and Systems, 675–684.
[Pa76]   T. D. Parsons (1976). Pursuit-evasion in a graph, in Y. Alavi and D. Lick (eds.), Theory
             and Applications of Graphs, Springer Verlag, Berlin, 426–441.
[Ra88]   S. S. Ravi (1988). A note on the orderability problem for PLA folding, to appear in Discrete
             Appl. Math.
[RL88]   S. S. Ravi and E. L. Loyd (1988). The complexity of near-optimal progrmmable logic array
             folding, SIAM J. Computing *17*, 696–710.
[RS83]   N. Robertson and P. D. Seymour (1983). Graph minors I. Excluding a forest, J. Comb.
             Th. B. *35*, 39–61.
[RS86]   N. Robertson and P. D. Seymour (1986). Graph minors XIII. The disjoint paths problem,
             preprint.
[RS87]   N. Robertson and P. D. Seymour (1987). Graph Minors XVI. Wagner's conjecture,
             preprint.
[SM83]   K. H. Schmidt and K. D. Mueller-Glaser (1983). NMOS dense gate matrix VLSI Design,
             IEEE J. of Solid-State Circuits, SC-18, 157–159.
[WW89]   D. Wagner and F. Wagner (1989). Graph separation is NP-hard even for graphs with
             bounded degree, preprint TU Berlin.
[Wei67]   A. Weinberger (1967). Large scale integration of MOS complex logic: a layout method,
             IEEE Journal of Solid-State Circuits, SC-2 (4), 182–190.
[Win82]   O. Wing (1982). Automated gate matrix layout, Intl. Symposium on Circuits and Systems,
             681–685.
[Win83]   O. Wing (1983). Interval-graph-based circuit layout, Proc. IEEE 1983, Int. Conf. on CAD,
             84–85.
[WHW85]   O. Wing, S. Huang, and R. Wang (1985) Gate matrix layout, IEEE Trans. on CAD, CAD-4
             (3), 220–231.
[YKK75]   H. Yoshizawa, H. Kawanishi, and K. Kami (1975). A heuristic procedure for ordering
             MOS arrays, Proc. 12th Design Automation Conference, 384–389.

Rolf H. Möhring
Technical University of Berlin
Fachbereich Mathematik
Strasse des 17. Juni 136
D-1000 Berlin

Computing, Supp. 7, 53–68 (1990)

Planar Graph Problems

Takao Nishizeki, Sendai

Abstract — Zusammenfassung

Planar Graph Problems. Classical and recent results are surveyed in the development of efficient algorithms for the following eleven famous problems on planar graphs: planarity testing, embedding, drawing, separators, vertex-coloring, independent vertex set, listing subgraphs, Hamiltonian cycle, network flows, and Steiner trees and forests. Also typical methods and techniques useful for computational problems on planar graphs are discussed. Furthermore open questions on planar graphs are mentioned.

AMS Subject Classification: 05.

Key words: Planar graphs, planarity testing, embedding and drawing, separators, vertex-coloring, edge-coloring, independent vertex sets, listing subgraphs, Hamiltonian cycles, network flows, Steiner trees.

Probleme anf planaren Graphen. In dieser Arbeit wird über klassische und jüngste Ergebnisse bei der Entwicklung effizienter Algorithmen für die folgenden elf wohlbekannten Probleme an planaren Graphen berichtet: Planaritätstests, Einbettung und Zeichnen, Separation, Knotenfärbung, Kantenfärbung, Unabhängige Knotenmengen, Auflisten von Untergraphen, Hamiltonsche Kreise, Netzwerkflüsse, Steiner Bäume und Wälder. Ferner werden typische Methoden und Techniken zur Behandlung planarer Graphen diskutiert. Einige offene Fragen bezüglich planarer Graphen werden erwähnt.

1. Introduction

Recent research efforts in computational graph theory have concentrated on designing efficient algorithms for solving combinatorial problems on graphs, and many efficient algorithms have been obtained for various problems, such as planarity testing, maximum matchings, and network flows. On the other hand many problems of practical importance have been shown NP-complete and appear to be intractable.

Planar graphs are those that can be drawn in the plane in such a way that vertices are represented by points, edges by lines connecting their endpoints, and no two such lines intersect except at common endpoints. Since planar graphs often appear in practical areas such as traffic networks and electrical circuits, it would be useful to design efficient algorithms for planar graphs. However most of NP-complete problems for general graphs remain NP-complete even for planar graphs, but some become tractable in a sense that there are efficient exact or approximate algorithms at least for a large class of planar graphs.

This paper surveys classical and recent results in the development of efficient algorithms for planar graph problems. In this paper a graph $G = (V, E)$ means an *undirected simple* graph with vertex set V and edge set E unless otherwise specified. We denote by n the number of vertices in G and by m the number of edges. We deal only with sequential algorithms for the following eleven famous problems: planarity testing, embedding, drawing, planar separators, vertex-coloring, edge-coloring, independent vertex set, listing subgraphs, Hamiltonian cycle, network flows, and Steiner trees and forests. Also typical methods and techniques useful for planar graph problems such as dualization, divide-and-conquer using planar separators, dynamic programming using planar embeddings, etc. are discussed. Furthermore, significant open questions on planar graphs are mentioned. We refer to [NC88] for more details on the planar graph problems.

2. Planarity Testing and Embedding

There are many practical situations in which one wishes to determine whether a given graph is planar, and if so, to find a planar embedding (drawing) of the graph. For example, in the layout of printed or VLSI circuits, one is interested in knowing whether a graph G representing a circuit is planar and if so, also in finding a planar embedding of G.

An input graph G in the planarity testing problem is represented by a set of n lists, called adjacency lists. The list $Adj(v)$ for vertex $v \in V$ contains all the neighbours of v. For each $v \in V$ an actual drawing of a planar graph G determines, within a cyclic permutation, the order of v's neighbours embedded around v. *Embedding* a planar graph G means constructing adjacency lists of G such that, in each $Adj(v)$, all the neighbours of v appear in clockwise order with respect to an actual drawing. Such a set of adjacency lists is called an *embedding* of G.

Two planarity testing algorithms which run in linear time are well-known: one by Hopcroft and Tarjan [HT74], and the other by Booth and Lueker [BL76]. The former called the "path addition algorithm" starts by finding a simple cycle and adding to it one simple path at a time. Each such new path connects two old vertices via new edges and vertices. Whole pieces are sometimes flipped over. The algorithm is the first one that tests the planarity of a given graph in linear time.

The latter called the "vertex addition algorithm" is conceptually simpler than the former. It was first presented by Lempel, Even and Cederbaum [LEC67], and improved later to a linear algorithm by Booth and Lueker [BL76] employing an "*st*-numbering" algorithm and a data structure called a "*PQ*-tree". The algorithm adds one vertex at each step. Previously embedded edges incident with this vertex are connected to it, and new edges incident with it are embedded and their ends are left unconnected. Sometimes whole pieces have to be reversed (flipped) around or permuted so that some ends occupy consecutive positions. If the representation of the embedded subgraph is updated with each alteration of the embedding, then the final representation will be an actual embedding of a given whole graph.

The "*st*-numbering" plays a crucial role in the testing algorithm. A numbering of the vertices of G by $1, 2, \ldots, n$ is called an *st*-numbering if the two vertices 1 and n are necessarily adjacent and each other vertex j is adjacent to two vertices i and k such that $i < j < k$. Every 2-connected graph G has an *st*-numbering, and an algorithm given by Even and Tarjan [ET76] finds an *st*-numbering in linear time.

A data structure called a *PQ*-tree is used in the vertex addition algorithm. A *PQ*-tree represents the permutations of a set S in which various subsets of S occur consecutively. Booth and Luekker gave a linear algorithm for manipulating *PQ*-trees [BL76].

Another linear-time planarity testing algorithm appeared in [DR82].

The aforementioned planarity testing algorithms can be modified to construct an embedding of a planar graph. Such a linear algorithm using *PQ*-trees appeared in [CNA85].

3. Drawing

The problem of drawing a planar graph often arises in applications, including the Design Automation of VLSI circuits. Wagner [Wag36] and Fáry [Fár48] independently showed that every planar graph can be drawn in the plane in such a way that the edges are straight line segments and the vertices are points.

A *convex drawing* of a planar graph is a straight-line drawing in which all the face boundaries are convex polygons. Clearly the complete bipartite graph $K_{2,n-2}$, $n \geq 6$, has no convex drawing. Thus not every planar graph has a convex drawing. Tutte [Tut60] proved that every 3-connected planar graph has a convex drawing, and established a necessary and sufficient condition for a planar graph to have a convex drawing. Furthermore he gave a "barycentric mapping" method for finding a convex drawing, which requires solving a system of $O(n)$ linear equations [Tut63]. The system of equations can be solved in $O(n^3)$ time and $O(n^2)$ space using the ordinary Gaussian elimination method, or in $O(n^{1.5})$ time and $O(n \log n)$ space using the sparse Gaussian elimination method which relies on the planar separator algorithms [LRT79]. Thus the barycentric mapping leads to an $O(n^{1.5})$ time convex drawing algorithm.

Chiba, Yamanouchi and Nishizeki [CYN84] gave two linear algorithms for the convex drawing problem: drawing and testing algorithms. The former, based on a short proof of Tutte's result given by Thomassen [Tho80], draws a given planar graph G convex if possible: it extends a given convex polygonal drawing of an outer facial cycle of G into a convex drawing of G. The latter algorithm tests the possibility: it determines whether a given planar graph has a convex drawing or not. Chiba et al. [CYN84] showed that the convexity testing of a graph G can be reduced to the planarity testing of a certain graph constructed from G.

Every planar graph can be augmented to a maximal planar graph by adding new edges; the resulting graph necessarily has a convex drawing. Thus all the convex drawing algorithms above immediately yield straight-line drawing algorithms.

Some papers study the problem of producing aesthetically desirable drawings of planar graphs or trees [SR83, CON85]. Obviously there are no absolute criteria that accurately capture our intuitive notion of a nice drawing of planar graphs. The linear algorithm of Chiba, Onoguchi and Nishizeki [CON85] obtains a pleasing drawing that satisfies the following property as far as possible: the complement of 3-connected components, together with inner faces and the complement of the outer face, are convex polygons.

All the drawing algorithms of planar graphs above have a drawback: vertices tend to bunch together and they require high precision real arithmetic relative to the size n of a graph. In fact it had been an open question whether or not every planar graph has a straight-line drawing on a grid of side length bounded by n^k for some fixed k. Recently de Fraysseix, Pach and Pollack [DPP88] solved this open problem affirmatively, and gave an $O(n \log n)$ algorithm which draws any given planar graph on the $2n-4$ by $n-2$ grid. Chrobak [CP88] improved the time complexity to $O(n)$.

Eades and Tamassia have extensively surveyed graph drawing algorithms [ET87].

4. Planar Separator Algorithm

The "divide-and-conquer" is one of the efficient approaches for solving computational problems on graphs. In this method, the original graph is divided into two or more smaller graphs. The problems for subgraphs are solved by applying the same method recursively, and then the solutions for the subgraphs are combined to give the solution to the original problem. The planar separator theorem of Lipton and Tarjan provides a basis for this approach [LT79]. The theorem asserts that any planar graph of n vertices can be divided into components of roughly equal size by removing only $O(\sqrt{n})$ vertices. They also gave a linear algorithm for finding such a separator. Miller [Mil86] generalized the planar separator theorem to that for a cycle separator. The latter separator can be used to simplify algorithms for some applications [JV82, Ric86].

Lipton and Tarjan obtained the following form of their separator theorem: every planar graph of n vertices contains a set C of $O\left(\sqrt{\dfrac{n}{\varepsilon}}\right)$ vertices whose removal leaves no connected component with more than εn vertices, where ε is any constant such that $0 < \varepsilon < 1$. Furthermore they showed that the set C can be found in $O(n \log n)$ time. Using this theorem, one can obtain approximation algorithms with time complexity $O(n \log n)$ and worst-case ratio $1 - O(1/\sqrt{\log \log n})$ for the maximum induced subgraph problem with respect to the following properties (among others): (1) independent, (2) bipartite, (3) forest, and (4) outerplanar [LT80, CNS81b].

The planar separator theorem has many applications. The layout of graphs, such as trees, X-trees and k-dimensional meshes, for VLSI are discussed in [Lei80]. Generalization of the "nested dissection" method for carrying out sparse Gaussian elimination on a system of linear equations is discussed in [LRT79]. Applications to the problems of nonserial dynamic programming, pebbling, lower bounds on Boolean circuit size and embedding of data structures can be found in [LT80].

Using the planar separator theorem, Frederickson obtained algorithms which solve the single-source shortest path problem for planar graphs in $O(n\sqrt{\log n})$ time and the all pair shortest path problem in $O(n^2)$ time [Fre87].

5. Vertex-Coloring

A *(vertex-)coloring* of a graph is an assignment of colors to the vertices so that adjacent vertices get distinct colors. A *k-coloring* of a graph uses at most k colors. The smallest integer k such that a graph G has a k-coloring is called the *chromatic number* of G and is denoted by $\chi(G)$.

The *vertex-coloring problem*, i.e., coloring a graph G with $\chi(G)$ colors, has practical applications in production scheduling, construction of time tables, etc. Since the problem is NP-hard [GJ79], it is unlikely that it admits a polynomial algorithm. One might expect that there would be an efficient approximate algorithm which uses a number of colors, not necessarily $\chi(G)$ but close to $\chi(G)$. However a polynomial algorithm that guarantees to color a graph with at most $a\chi(G) + b$ colors, $a < 2$, will imply a polynomial algorithm to color every graph G with $\chi(G)$ colors [GJ76]. In other words, getting close within a factor of two to the optimum is as hard as achieving it.

The situation for planar graphs is much more favorable. The famous four-color theorem proved by Appel and Haken says that every planar graph is 4-colorable [AH77]. We now sketch the outline of the proof. A graph is *k-chromatic* if it is not $(k - 1)$-colorable but k-colorable. A *configuration* is an induced subgraph of a planar graph. A configuration is *reducible* if no minimal 5-chromatic planar graph can contain it. A set of configurations is *unavoidable* if every planar graph contains at least one of them. In order to prove that every planar graph is 4-colorable, one has to find an *unavoidable set of reducible configurations*. Making use of the so-called discharging method and fast electronic computers, Appel and Haken eventually found an unavoidable set of over 1900 reducible configurations.

The proof of the four-color theorem leads to an algorithm of 4-coloring a planar graph. The algorithm runs in $O(n)$ recursive steps; at each step the algorithm detects in a graph one of over 1900 reducible configurations belonging to the unavoidable set, and recurses to a smaller graph. Since all the configurations contain at most 13 vertices, one recursive step can be done in time proportional to n, but the coefficient is no less than 1900. Thus the 4-coloring algorithm runs in $O(n^2)$ time, although it does not seem practical. The problem of finding a linear-time 4-coloring algorithm remains open.

In contrast, one can easily prove the five-color theorem that every planar graph has a 5-coloring, and there are linear algorithms which color every planar graph with at most five colors [MST80, CNS81a, Fre84, Will85]. Although most of the standard texts on graph theory use the Kempe-chain argument in proving the theorem, the proof on which the linear algorithms are based uses an argument of "identification of vertices" [Wils85]. The proof is by induction on n and goes as follows. The Euler's formula implies that every planar graph has a vertex of degree at most five. Consider first the case when there is a vertex v of degree at most four. The deletion of v leaves a planar graph $G - v$ having $n - 1$ vertices, which is 5-colorable by the inductive hypothesis. Then v can be colored with any color not used by the (at most four) neighbours, completing the proof of this case. In the remaining case there is a vertex v of degree five. Since the subgraph of G induced by the five neighbours of v is not K_5, v has two nonadjacent neighbours x and y. Delete vertex v from G, identify vertices x and y, and let G' be the resulting graph. Since G is planar, so is G'. Furthermore no loop is produced in G' since x and y are nonadjacent in G. Since G' has $n - 2$ vertices, the hypothesis implies that G' has a 5-coloring, which naturally induces a 5-coloring of $G - v$ in which x and y are colored with the same color. Assigning to v any color other than the (at most four) colors of the neighbours, we get a 5-coloring of G, completing the proof.

The proof have immediately yields a recursive algorithm which colors every planar graph G with at most five colors. Clearly the time required by vertex-identifications dominates the running time of the algorithm. One can easily identify two vertices in time proportional to the sum of their degrees. However the same vertex may appear in identifications $O(n)$ times, so a direct implementation of the algorithm would require $O(n^2)$ time. There are essentially two types of linear algorithms. The first one given by Chiba, Nishizeki and Saito [CNS81a] runs in several stages, in each of which a set of vertex-identifications are performed in linear time but at least some fixed percentage of the vertices are eliminated. The second one given by Matula, Shiloach and Tarjan [MST80] and later simplified by Frederickson [Fre84] is to recurse after each identification, choosing identification that requires constant time to perform. Both approaches involve a clever exploitation of properties of planar graphs.

We add one more remark on the vertex-coloring problem. The problem remains NP-complete even for planar graphs [GJS76]. However every planar graph is 4-colorable, and it is easy to check whether a graph is 2-colorable, i.e., bipartite or not. Thus the problem of deciding whether a given planar graph is 3-chromatic or 4-chromatic is indeed NP-complete.

6. Edge-Coloring

In this section we survey the edge-coloring problem for planar graphs. The problem is to color the edges of a given graph G using as few colors as possible, so that no two adjacent edges receive the same color. The minimum number of colors is called the *chromatic index* of G and denoted by $\chi'(G)$. Holyer showed that the edge-coloring problem is NP-hard [Hol81], and therefore it seems unlikely that a polynomial algorithm exists for the problem.

Let Δ denote the maximum degree of a graph G, then trivially $\Delta \leq \chi'(G)$. On the other hand, by the Vizing's classical result, $\chi'(G) \leq \Delta + 1$ for every simple graph G [FW77, Viz64]. Special cases which can be colored with Δ colors are bipartite graphs, cubic bridgeless planar graphs (whose edge-coloring in three colors is equivalent to the four-color problem), and planar graphs with $\Delta \geq 8$ [CH82, FW77, GK82, Viz65].

The fastest known algorithm for edge-coloring a simple graph with $\Delta + 1$ colors runs in $O(\Delta m \log n)$ or $O(m \sqrt{n} \log n)$ time [GNK84]. One of the algorithms in [GNK84] edge-colors with Δ colors a planar simple graph with $\Delta \geq 8$ in $O(n^2)$ time. For planar simple graphs with $\Delta \geq 9$ the time complexity can be improved to $O(n \log n)$ [CNi89].

It is not known whether the edge-coloring problem remains NP-complete for planar graphs.

Concerning multigraphs, Goldberg and Seymour have a conjecture that the bound

$$\chi'(G) \leq \max\{r(G), \Delta + 1\}$$

would hold for any multigraph G [Gol73, Sey79b]. Here $r(G)$ is a trivial lower bound on $\chi'(G)$:

$$r(G) = \max_{H \subseteq G} \left\lceil \frac{m(H)}{\lfloor n(H)/2 \rfloor} \right\rceil,$$

where H runs over all subgraphs of G having at least three vertices, $m(H)$ is the number of edges in H, and $n(H)$ the number of vertices in H. This is one of the most important open problems in the area of edge-coloring. The conjecture was verified for the case of outerplanar graphs [Mar86]. The best upper bound known for multigraphs [NK85] is:

$$\chi'(G) \leq \max\{r(G), \lfloor 1.1\Delta + 0.8 \rfloor\}.$$

7. Independent Set

A set of vertices in a graph is *independent* if no two vertices in the set are adjacent. The *maximum independent set problem,* in which one would like to find a maximum independent set in a given graph, is NP-hard, and remains so even for the class of planar graphs. There are however efficient approximation algorithms for planar graphs, which find large independent set.

An approximation algorithm is often evaluated by the *worst-case ratio*: the smallest ratio of the size of an approximation solution to the size of an optimal solution, taken over all problem instances. If a polynomial time algorithm existed with any constant worst-case ratio > 0 for the maximum independent set problem on graphs, then one could design a polynomial time algorithm with any constant worst-case ratio < 1 [HU79]. This fact does not imply that there exists no polynomial time approximation algorithm with a constant worst-case ratio > 0 for the problem on a special class of graphs, such as planar graphs. In fact, Lipton and Tarjan's $O(n \log n)$ time approximation algorithm [LT80] mentioned in Section 4 has a worst-ratio $1 - O(1/\sqrt{\log \log n})$, asymptotically tending to 1 as $n \to \infty$. Such a ratio is called an "asymptotic worst-case ratio". On the other hand, some approximation algorithms have an "absolute worst-case ratio", which does not dependent on the size n of a graph. For example, the 4-coloring algorithm, derived from the proof of the four-color theorem, immediately yields an approximation algorithm for the problem with the worst-case ratio $\frac{1}{4}$: Simply output the largest class of vertices colored with the same color. On the hand the 5-coloring algorithm achieves the absolute worst-case ratio $\frac{1}{5}$. Moreover the algorithm of Albertson or Chiba et al. guarantees the worst-case ratio $\frac{2}{9}$ for the problem [Alb74, CNS83]. It is still open to prove, without use of the four-color theorem, the fact that every planar graph contains an independent set of size $\geq \frac{1}{4}n$[Alb76].

Chiba, Nishizeki and Saito [CNS82] gave an $O(n \log n)$ time approximation algorithm with absolute worst-case ratio $\frac{1}{2}$. For a given planar graph of any number n of vertices, the algorithm finds, in $O(n \log n)$ time, an independent vertex set that is necessarily larger than half a maximum independent set. The idea of the algorithm is to reduce a given planar graph to a planar graph of minimum degree $\delta = 5$ by modifying the graph around vertices of minimum degree. A planar graph of $\delta = 5$ cannot have a large independent set: the size is necessarily less than $\frac{2}{3}n$. It is easy to find in such a graph an independent vertex set that is necessarily larger than half a maximum independent set. For example one may use the 5-coloring algorithm. Recently Chrobak and Naor improved the time complexity of the approximation algorithm to $O(n)$ [CNa88]. The algorithm of Lipton and Tarjan [LT80] can also guarantee the absolute worst-case ratio $\frac{1}{2}$, but the number n of vertices must be huge, say $2^{2^{400}}$.

Baker [Bak83] gave an elegant approximation algorithm which works for various computational problems on planar graphs, including the maximum independent set problem. The algorithm attains the worst-case ratio $\dfrac{k}{k+1}$ and runs in $O(8^k kn)$ time for any positive integer k. Thus her algorithm realizes the worst-case ratio of both types; absolute and asymptotic. For example, letting $k = 1$ one can get a linear time algorithm having the absolute worst-case ratio $\frac{1}{2}$, while letting $k = \log \log n$ one can get an $O(n(\log n)^3 \log \log n)$ time algorithm with the asymptotic worst-case ratio of $(\log \log n)/(1 + \log \log n)$ tending to 1. Her algorithm uses the dynamic programming approach based on planar embedding.

8. Listing Subgraphs

Listing certain kind of subgraphs of a graph such as cliques, triangles, cycles et al. arises in many applications. Itai and Rodeh were the first to give a linear-time algorithm for listing all triangles in a planar graph [IR78]. They used depth-first search.

Chiba and Nishizeki [CN85b] found a simple strategy for edge-searching a graph, which is useful for various subgraph listing problems. The algorithm chooses a vertex v in a graph and scans the edges of the subgraph induced by the v's neighbours to find the pattern subgraphs containing v. The main feature of the strategy is to repeat the searching above for each vertex v in decreasing order of degree and to delete v after v is processed so that no duplication occurs. The procedure above requires $O(a(G)m)$ time for a graph G, where $a(G)$ is the arboricity of G, that is, the minimum number of edge-disjoint spanning forests into which G can be decomposed. Using the strategy, they presented algorithms which list in $O(a(G)m)$ time all the triangles C_3 or all the quadrangles C_4 in G. Since every planar graph satisfies $a(G) \leq 3$, the algorithms run in linear time for such graphs. Based on this approach, they also gave an $O(la(G)^{l-2}m)$ time algorithm for listing all the cliques K_l in a graph G. The algorithm lists all K_4 contained in a planar graph in linear time. Since a planar graph contains no $K_l, l \geq 5$, the problem for finding all the cliques in a planar graph can be solved in linear time. Papadimitriou and Yannakakis [PY81] reported another linear algorithm for the problem, based on breadth-first search. Their algorithm however does not work correctly but can be corrected easily. Matula and Beck [MB83] obtained another linear algorithm for detecting a triangles in a planar graph, based on what they call "smallest-last ordering".

Richards [Ric86] gave $O(n \log n)$ algorithms for detecting both a C_5 or a C_6 in a planar graph. The algorithm uses a divide-and-conquer approach which relies on the Lipton-Tarjan separator algorithm [LT79]. It is open whether there exists a linear algorithm for detecting C_5 or C_6 and whether there exists an $O(n \log n)$ algorithm for detecting a $C_l, l \geq 7$. Another important open problem is whether there is a linear algorithm to detect a triangle in a graph.

Sysło gave a cycle vector space algorithm for listing all cycles of a planar graph [Sys81].

9. Hamiltonian Cycle

A *Hamiltonian cycle* of a graph G is a cycle which contains all the vertices of G. The Hamiltonian cycle problem asks whether a given graph contains a Hamiltonian cycle. It is NP-complete even for 3-connected cubic planar graphs [GJT76, Kar72], 2-connected cubic bipartite planar graphs [ANS80], or maximal planar graphs [Chv85]. However the problem becomes polynomial-time solvable for 4-connected planar graphs: Tutte proved that such a graph contains a Hamiltonian cycle [Tut56, Tut77]. Based on Tutte's proof, Gouyou-Beauchamps obtained an $O(n^3)$ algorithm which finds a Hamiltonian cycle in such a graph [Gou82]. Asano, Kikuchi and Saito presented a linear algorithm for the problem on 4-connected maximal planar graphs [AKS84]. Chiba and Nishizeki [CN89] constructed a linear algorithm for 4-connected planar graphs, based on Thomassen's short proof of Tutte's theorem [Tho83, CN85a].

A *Hamiltonian walk* in a connected graph is a shortest closed walk that passes through every vertex at least once, and the length is the total number of edges traversed by the walk. A Hamiltonian cycle is obviously a Hamiltonian walk. A trivial lower bound and a trivial upper bound are known on the length, $h(G)$, of a Hamiltonian walk of a connected graph G: $n \leq h(G) \leq 2(n - 1)$. A nontrivial upper bound on the length of a Hamiltonian walk for maximal planar graphs was obtained [ANW80]:

$$h(G) \begin{cases} \leq \frac{3}{2}(n - 3) & \text{if } n \geq 11; \\ = n & \text{otherwise.} \end{cases}$$

Since the proof in [ANW80] is constructive, it immediately yields an $O(n^2)$ algorithm for finding a closed spanning walk of length $\leq \frac{3}{2}(n - 3)$ in maximal planar graphs [NAW83]. The algorithm uses a divide-and-conquer approach involving a partition of a graph at a separation triple, which forms a triangle in a maximal planar graph. One can improve the time complexity to $O(n)$ by using two linear-time algorithms: the algorithm for finding a Hamiltonian cycle in 4-connected planar graphs, and the traingle listing algorithm [NC88]. The upper bound on $h(G)$ is conjectured to be improved to $h(G) \leq \frac{4}{3}(n - 2)$ if $n \geq 11$.

10. Network Flows

The network flow problem and its variants have been extensively studied. The original and most classical problem is that of finding a maximum flow of a single commodity in an arbitrary graph. The key theorem in flow theory is the Max Flow-Min Cut theorem of Ford and Fulkerson [FF56], which holds for single commodity and two commodity flows [Hu69]. There are efficient algorithms for finding a maximum single commodity flow; an $O(mn \log(n^2/m))$ time algorithm is the best known one for sparse graphs [ST83]. Two commodity flows in undirected graphs can be found by solving two single-commodity flow problems, hence in $O(mn \log(n^2/m))$ time [Ita78, Sak73, Sey79a].

The so-called *uppermost path algorithm* can find a maximum single commodity flow in a planar graph with source s and sink t both on the outer boundary B [FF56, IS79]. The algorithm starts with zero flow and pushes flow as much as possible through the "uppermost path" on B connecting s and t. Thereby, at least one edge becomes saturated. Such an edge is deleted, and the process is repeated using the uppermost path of the resulting graph. One can observe that the algorithm merely executes the shortest path computation on the dual of G [Has81]. Thus the maximum single-commodity flow can be found in $O(T(n))$ time, where $T(n)$ denotes the time required for finding the single-source shortest paths in a planar undirected graph with nonnegative edge weights having n vertices. If the usual Dijkstra's algorithm [AHU74, Joh77] is used, then $T(n) = O(n \log n)$. If Frederickson's algorithm which relies on the planar separator theorem [Fre87] is used, then $T(n) = O(n\sqrt{\log n})$. The uppermost path algorithm does not work when s and t are not on the same face boundary, but an $O(n \log n)$ algorithm for such a planar undirected graph and an $O(n^{1.5} \log n)$ algorithm for such a planar directed graph are known [Rei83, HJ85, JV82, Fre87].

The situation is different with regard to flows of more than two commodities. In general the multicommodity *integral* flow problem is NP-complete. No simple polynominal-time algorithm is known even for the multicommodity (*real-valued*) flow problem on graphs. Recently Tardos obtained a strongly polynomial-time algorithm to solve combinatorial linear programs including the multicommodity flow problem [Tar86]. However, it employs a polynomial-time linear programming algorithm, and hence neither has a polynomial-time bound of lower order nor is easy to implement. Therefore simple efficient algorithms are useful in practice even if they are valid only for planar graphs.

It has been established that the Max Flow-Min Cut theorem of multicommodity type holds for the following five classes of planar undirected graphs [OS81, Oka83, Sey81, Sch88b]:

C_1: all sources and sinks are located on a specified face boundary [OS81];

C_{12}: all sources and sinks are located on two specified face boundaries with each source-sink pair on the same boundary [Oka83];

C_{01}: some source-sink pairs are located on a specified face boundary, and all the other pairs share a common sink located on the boundary (their sources may be located anywhere) [Oka83];

C_a: All the sources can be joined with the corresponding sinks without violating planarity [Sey81]; and

C_{12r}: All the sources $s_1, s_2, \ldots, s_k$ appears on the boundary of the outer face in clockwise order, and all the sinks $t_1, t_2, \ldots, t_k$ appear on some other face boundary in counterclockwise order [Sch88b].

Efficient algorithms for the first four classes were obtained [MNS85, MNS86, SNS88, Has84]. All the algorithms reduce the flow problem on a planar undirected graph to the shortest path or cycle problem on an undirected or directed graph obtained from the dual of the given undirected graph. Multicommodity flows for C_1, C_{12}, C_{01} can be found by solving $O(n)$ times the single-source shortest path problem for a planar graph. Hence one can find flows in $O(kn + nT(n))$ time, where k is the number of source-sink pairs. On the other hand, multicommodity flows for C_a can be found by solving $O(n)$ times a weighted matching problem on a certain graph. Using the planar separator algorithm, one can solve the matching problem in $O(n^{1.5} \log n)$ time. Thus the flows for C_a can be found in $O(n^{2.5} \log n)$ time [MNS86]. Recently Barahona showed that flows for C_a can be found in $O(n^{1.5} \log n)$ time by solving once the Chinese postman problem in the dual planar graph [Bar87]. Using the same idea, he showed that the max cut problem can be solved in $O(n^{1.5} \log n)$ time for planar graphs.

The Max Flow-Min Cut theorem was shown to hold for certain kinds of planar directed graphs [NI88].

The edge-disjoint path problem is to find edge-disjoint paths connecting specified pairs of vertices in a graph. The problem can be formulated as a multicommodity integral flow problem. Recently many results have been obtained for the edge-disjoint path problem on planar graphs or plane grids [Fra85, KM86, MNS85, MP86, SNS88].

11. Steiner Tree and Forests

The *Steiner minimum tree problem* on a weighted graph $G = (V, E)$ with a set N of special vertices called *terminals* is to find a tree of minimum weight which interconnects the terminals of N (possibly using some vertices in $V - N$). The Steiner minimum tree problem is known to be strongly NP-hard for planar graphs [GJ79], and polynomially-solvable for planar graphs if the terminals lie on a fixed number of faces [EMV87]. See [Win87] for an extensive survey of works on this problem.

The *Steiner forest problem* on an unweighted graph with sets (called *nets*) of terminals is to find a forest, that is, vertex-disjoint trees, each of which interconnects all the terminals of a net. The problem does not require to minimize the number of used edges, and hence is a generalization of the *vertex-disjoint path problem*. Since the vertex-disjoint path problem is NP-hard even for planar graphs [Lyn75] or plane grids [KL82], so is the Steiner forest problem for planar graphs. Robertson and Seymour showed that the problem is solvable in polynomial time if all the terminals lie on only two faces of a planar graph [RS86]. Suzuki, Akama and Nishizeki improved the time complexity to $O(n \log n)$ [SAN88]. They also give an $O(n \log n)$ algorithm for finding a maximum number of internally vertex-disjoint paths connecting two specified vertices in a planar graph [SAN88]. The internally disjoint path algorithm employs a divide-and-conquer approach without using the plannar separator algorithm, and plays a crucial role in their Steiner forest algorithm. Schrijver showed that the Steiner forest problem is solvable in polynomial time if all the terminals lie on a fixed number of faces in a planar graph [Sch88a].

Acknowledgement

This work was partly supported by Grant in Aid for Scientific Research of the Ministry of Education, Science, and Culture, under grant number: General Research (C) 01550275 (1989).

References

[AHU74]   A. V. Aho, J. E. Hopcroft, and J. D. Ullman, The Design and Analysis of Computer Algorithms, Addison-Wesley, Reading, Mass., 1974.
[ANS80]   T. Akiyama, T. Nishizeki and N. Saito, NP-completeness of the Hamiltonian cycle problem for bipartite graphs, J. Information Processing, *3*, 2, pp. 73–76, 1980.
[Alb76]   M. O. Albertson, A lower bound for the independence number of a planar graph, J. Combinatorial Theory, Series B, *20*, pp. 84–93, 1976.
[Alb74]   M. O. Albertson, Finding an independent set in a planar graph, in "Graphs and Combinatorics," ed., R. A. Bari and F. Harary, Springer-Verlag, Berlin-Heidelberg-New York: pp. 173–179, 1974.
[AH77]   K. Appel and W. Haken, Every planar map is four colourable, Part I: discharging, Illinois J. Math. *21*, pp. 429–490, 1977.
[AKS84]   A. Asano, S. Kikuchi and N. Saito, A linear algorithm for finding Hamiltonian cycles in 4-connected maximal planar graphs, Discrete Appl. Math., 7, pp. 1–15, 1984.
[ANW80]   T. Asano, T. Nishizeki and T. Watanabe, An upper bound on the length of a Hamiltonian walk of a maximal planar graph, J. Graph Theory, *4*, 3, pp. 315–336, 1980.
[Bak83]   B. S. Baker, Approximation algorithms for NP-complete problems on planar graphs, 24th Ann. Symp. on Found. of Compt. Sci., pp. 265–273, 1983.

[Bar87] F. Barahona, Planar multicommodity flows, maximum cut and the Chinese postman prob-
         lem, Rept. 87454-OR, Institut für Operations Research, Universität Bonn, 1987.

[BL76]   K. S. Booth and G. S. Lueker, Testing the consecutive ones property, interval graphs, and
         graph planarity using PQ-tree algorithms, J. Comput. Syst. Sci., 13, pp. 335–379, 1976.

[CN85a] N. Chiba and T. Nishizeki, A theorem on paths in planar graphs, J. Graph Theory, 10,
         pp. 449–450, 1985.

[CN85b] N. Chiba and T. Nishizeki, Arboricity and subgraph listing algorithms, SIAM J. Comput.,
         14, 1, pp. 210–223, 1985.

[CN89]   N. Chiba and T. Nishizeki, The Hamiltonian cycle problem is linear-time solvable for
         4-connected planar graphs, J. Algorithms, 10, pp. 187–211, 1989.

[CNA85] N. Chiba, T. Nishizeki, S. Abe and T. Ozawa, A linear algorithm for embedding planar graphs
         using PQ-trees, J. Comput. Syst. Sci., 30, 1, pp. 54–76, 1985.

[CNS81a] N. Chiba, T. Nishizeki and N. Saito, A linear 5-coloring algorithm of planar graphs,
         J. Algorithms, 2, pp. 317–327, 1981.

[CNS82] N. Chiba, T. Nishizeki and N. Saito, An approximation algorithm for the maximum indepen-
         dent set problem on planar graphs, SIAM J. Comput., 11, 4, pp. 663–675, 1982.

[CNS83] N. Chiba, T. Nishizeki and N. Saito, An efficient algorithm for finding an independent set
         in planar graphs, Networks, 13, pp. 247–252, 1983.

[CNS81b] N. Chiba, T. Nishizeki and N. Saito, Applications of the Lipton and Tarjan's planar separator
         theorem, J. Information Processing, 4, 4, pp. 203–207, 1981.

[CON85] N. Chiba, K. Onoguchi and T. Nishizeki, Drawing plane graphs nicely, Acta Informatica,22,
         pp. 187–201, 1985.

[CYN84] N. Chiba, T. Yamanouchi and T. Nishizeki, Linear algorithms for convex drawings of planar
         graphs, in "Progress in Graph Theory", eds., J. A. Bondy and U. S. R. Murty, Academic
         Press, Toronto, pp. 153–173, 1984.

[CNa88] M. Chrobak and J. Naor, Sequential and parallel algorithms for computing a large indepen-
         dent set in planar graphs, manuscript, 1988.

[CNi89] M. Chrobak and T. Nishizeki, Improved edge-coloring algorithms for planar graphs,
         J. Algorithms, to appear.

[CP88]   M. Chrobak and T. Payne, A linear-time algorithm for drawing graphs on a grid, submitted
         for publication, 1988.

[Chv85] V. Chvátal, Hamiltonian cycles, in "The Travelling Salesman Problem", eds. E. L. Lawler,
         J. K. Lenstra, A. H. G. Rinnooy Kan, and D. B. Shmoys, John Wiley & Sons, pp. 403–429,
         1985.

[CH82]   R. Cole and J. Hopcroft, On edge coloring bipartite graphs, SIAM J. Comput., 11, 3,
         pp. 540–546, 1982.

[DPP88] H. de Fraysseix, J. Pach and R. Pollack, Small sets supporting Fáry embeddings of planar
         graphs, Proc. 20th ACM Symp. on Theory of Computing, pp. 426–433, 1988.

[DR82]   H. de Fraysseix and P. Rosenstiehl, A depth-first search characterization of planarity, Annals
         of Discrete Mathematics, 13, pp. 75–80, 1982.

[ET87]   P. Eades and R. Tamassia, Algorithms for drawing graphs: An annotated bibliography,
         manuscript, 1987.

[EMV87] R. E. Erickson, C. L. Monma and A. F. Veinott, Jr., Send-and-split method for minimum-
         concave-cost network flows, Math. of Operations Research, 12, 4, pp. 634–664, 1987.

[ET76]   S. Even and R. E. Tarjan, Computing an st-numbering, Theor. Comput. Sci., 2, pp. 339–344,
         1976.

[Fár48] I. Fáry, On straight lines representations of planar graphs, Acta Sci., Math. Szeged, 11,
         pp. 229–233, 1948.

[FF56]   L. R. Ford and D. R. Fulkerson, Maximal flow through a network, Canad. J. Math., 8,
         pp. 399–404, 1956.

[Fra85] A. Frank, Edge-disjoint paths in planar graphs, J. Combinat. Theory, Series B, 39, 2,
         pp. 164–178, 1985.

[Fre87] G. N. Frederickson, Fast algorithms for shortest paths in planar graphs, with applications,
         SIAM J. Comput., 16, 6, pp. 1004–1022, 1987.

[Fre84] G. N. Frederickson, On linear-time algorithms for five-coloring planar graphs, Information
         Processing Letters, 19, pp. 219–224, 1984.

[FW77]   S. Fiorini and R. J. Wilson, Edge-Colourings of Graphs, Pitman, London, 1977.

[GK82]   H. N. Gabow and O. Kariv, Algorithms for edge coloring bipartite graphs and multigraphs,
         SIAM J. Comput., 11, 1, pp. 117–129, 1982.

[GNK84] H. N. Gabow, T. Nishizeki, O. Kariv, D. Leven, and O. Terada, Algorithms for edge-coloring
         graphs, submitted to a journal, 1984.

[GJ79]   M. R. Garey and D. S. Johnson, Computers and Intractability, W. H. Freeman and Company, San Francisco, 1979.

[GJ74]   M. R. Garey and D. S. Johnson, The complexity of near-optimal graph coloring J. Assoc. Comput. Mach., 23, pp. 43–49, 1974.

[GJS76]   M. R. Garey, D. S. Johnson and L. Stockmeyer, Some simplified NP-complete graph problems, Theor. Comput. Sci., pp. 237–267, 1976.

[GJT76]   M. R. Garey, D. S. Johnson and R. E. Tarjan, The planar Hamiltonian circuit problem is NP-complete, SIAM J. Comput., 5, pp. 704–714, 1976.

[Gol73]   M. K. Goldberg, On multigraphs with almost maximal chromatic class (in Russian), Diskret Analiz, 23, pp. 3–7, 1973.

[Gou82]   D. Gouyou-Beauchamps, The Hamiltonian circuit problem is polynomial for 4-connected planar graphs, SIAM J. Comput., 11, pp. 529–539, 1982.

[Has81]   R. Hassin, Maximum flow in (s, t) planar networks, Inf. Proc. Lett., 13, 3, p. 107, 1981.

[Has84]   R. Hassin, On multicommodity flows in planar graphs, Networks, 14, pp. 225–235, 1984.

[HJ85]   R. Hassin and D. B. Johnson, An $O(n \log^2 n)$ algorithm for maximum flow in undirected planar networks, SIAM J. Comput., 14, 3, pp. 612–624, 1985.

[Hol81]   I. J. Holyer, The NP-completeness of edge colourings, SIAM J. Comput., 10, pp. 718–720, 1981.

[HT74]   J. E. Hopcroft and R. E. Tarjan, Efficient planarity testing, J. Assoc. Comput. Mach., 21, pp. 549–568, 1974.

[HU79]   J. E. Hopcroft and J. D. Ullman, Introduction to Automata Theory, Languages, and Computation, Addison-Wesley, Reading, Mass., 1979.

[Hu69]   T. C. Hu, Integer Programming and Network Flows, Addison-Wesley, Reading, Mass., 1969.

[Ita78]   A. Itai, Two-commodity flow, J. Assoc. Comput. Mach., 25, 4, pp. 596–611, 1978.

[IR78]   A. Itai and M. Rodeh, Finding a minimum circuit in a graph, SIAM J. Comput., 7, 4, pp. 413–423, 1978.

[IS79]   A. Itai and Y. Shiloach, Maximum flows in planar networks, SIAM J. Comput., 8, 2, pp. 135–150, 1979.

[Joh77]   D. B. Johnson, Efficient algorithms for shortest paths in sparse networks, J. Assoc. Comput. Mach., 24, pp. 1–13, 1977.

[JV82]   D. B. Johnson and S. M. Venkatesan, Using divide and conquer to find flows in directed planar networks in $O(n^{3/2} \log n)$ time, Proc. 20th Ann. Allerton Conf. on Communication, Control, and Computing, Univ. of Illinois, pp. 898–905, 1982.

[Kar72]   R. M. Karp, Reducibility among combinatorial problems, in "Complexity of Computer Computations", eds. R. E. Miller and J. W. Thacher, Plenum Press, New York, pp. 85–104, 1972.

[KM86]   M. Kaufmann and K. Mehlhorn, Routing through a generalized switchbox, J. Algorithms, 7, pp. 510–531, 1986.

[KL82]   M. R. Kramer and J. van Leeuwen, Wire-routing is NP-complete, Report No. RUU-CS-82-4, Department of Computer Science, University of Utrecht, Utrecht, the Netherlands, 1982.

[Lei80]   C. E. Leiserson, Area-efficient graph layout (for VLSI), Carnegie-Mellon University, CMU-CS-80-138, 1978.

[LEC67]   A. Lempel, S. Even and I. Cederbaum, An algorithm for planarity testing of graphs, in "Theory of Graphs," Int. Symp. Rome, July 1966, ed. P. Rosenstiehl, Gordon and Breach, New York, pp. 215–232, 1967.

[LRT79]   R. J. Lipton D. J. Rose and R. E. Tarjan, Generalized nested dissection, SIAM J. Numer. Anal., 16, 2, pp. 346–358, 1979.

[LT79]   R. J. Lipton and R. E. Tarjan, A separator theorem for planar graphs, SIAM J. Appl. Math., 35, pp. 177–189, 1979.

[LT80]   R. J. Lipton and R. E. Tarjan, Applications of a planar separator theorem, SIAM J. Comput., 9, 3, pp. 615–627, 1980.

[Lyn75]   J. F. Lynch, The equivalence of theorem proving and the interconnection problem, ACM SIGDA Newsletter 5 : 3, p. 31–65, 1975.

[Mar86]   O. Marcotte, On the chromatic index of multigraphs and a conjecture of Seymour (I), J. Combinat. Theory, Series B, 41, 3, pp. 306–331, 1986.

[MNS85]   K. Matsumoto, T. Nishizeki and N. Saito, An efficient algorithm for finding multicommodity flows in planar networks, SIAM J. Comput., 14, pp. 289–301, 1985.

[MNS86]   K. Matsumoto, T. Nishizeki and N. Saito, Planar multicommodity flows, maximum matchings and negative cycles, SIAM J. Comput., 15, 2, pp. 495–510, 1986.

[MB83]   D. W. Matula and L. L. Beck, Smallest-last ordering and clustering and graph coloring algorithms, J. Assoc. Comput. Mach., 30, pp. 417–427, 1983.

[MST80]   D. W. Matula, Y. Shiloach and R. E. Tarjan, Two linear-time algorithms for five-coloring a planar graph, Manuscript, 1980.

[MP86]   K. Mehlhorn and F. P. Preparata, Routing through a rectangle, J. Assoc. Comput. Mach., *33, 1*, pp. 60–85, 1986.

[Mil86]   G. Miller, Finding small simple cycle separators for 2-connected planar graphs, J. Comput. Syst. Sci., *32*, pp. 265–279, 1986.

[NI88]   H. Nagamochi and T. Ibaraki, Max-Flow Min-Cut theorem for the multi-commodity flows in certain planar directed networks (in Japanese), Trans. Inst. Elect. Inf. Comm. Eng., Japan, J71-A, *1*, pp. 71–82, 1988.

[NAW83]   T. Nishizeki, T. Asano and T. Watanabe, An approximation algorithm for the Hamiltonian walk problem on a maximum planar graph, Discrete Applied Math., *5*, pp. 211–222, 1983.

[NC88]   T. Nishizeki and N. Chiba, Planar Graphs: Theory and Algorithms, North-Holland, Amsterdam, 1988.

[NK85]   T. Nishizeki and K. Kashiwagi, An upper bound on the chromatic index of multigraphs, in "Graph Theory with Applications to Algorithms and Computer Science", eds. Y. Alavi et al., John Wiley & Sons, New York, pp. 595–604, 1985.

[Oka83]   H. Okamura, Multicommodity flows in graphs, Discrete Appl. Math., *6*, pp. 55–62, 1983.

[OS81]   H. Okamura and P. D. Seymour, Multicommodity flows in planar graphs, J. Combinat. Theory, Series B, *31*, pp. 75–81, 1981.

[PY81]   C. H. Papadimitriou and M. Yannakakis, The clique problem for planar graphs, Information Processing Letters, *13, 4, 5*, pp. 131–133, 1981.

[Rei83]   J. H. Reif, Minimum s-t cut of a planar undirected network in $O(n \log^2(n))$ time, SIAM J. Compt., *12*, pp. 71–81, 1983.

[Ric86]   D. Richards, Finding short cycles in planar graphs using separators, J. Algorithms, *7*, pp. 382–394, 1986.

[RS86]   N. Robertson and P. D. Seymour, Graph minors. VI. Disjoint paths across a disc, Journal of Combinatorial Theory, Series B, *41*, pp. 115–138, 1986.

[Sak73]   M. Sakarovitch, Two commodity network flows and linear programming, Math. Prog., 4, pp. 1–20, 1973.

[Sch88a] A. Schrijver, Disjoint homotopic trees in a planar graph, Manuscript, 1988.

[Sch88b] A. Schrijver, The Klein bottle and multicommodity flows, Manuscript, 1988.

[Sey79a] P. D. Seymour, A short proof of the two-commodity flow theorem, J. Comb. Theory, Series B, *26*, pp. 370–371, 1979.

[Sey79b] P. D. Seymour, On multi-colorings of cubic graphs, and conjectures of Fulkerson and Tutte, Proc. London Math. Soc., *3, 38*, pp. 423–460, 1979.

[Sey81]   P. D. Seymour, On odd cuts and planar multicommodity flows, Proc. London Math. Soc., (3), *42*, pp. 178–192, 1981.

[ST83]   D. D. Sleator and R. E. Tarjan, A data structure for dynamic trees, J. Comput. Syst. Sci., *26*, pp. 362–390, 1983.

[SR83]   K. J. Supowit and E. M. Reingold, The complexity of drawing trees nicely, Acta Informatica, *18*, pp. 377–392, 1983.

[SAN88]   H. Suzuki, T. Akama and T. Nishizeki, Finding Steiner forests in planar graphs, submitted to a journal, 1988.

[SNS88]   H. Suzuki, T. Nishizeki, and N. Saito, Algorithms for multicommodity flows in planar graphs, Algorithmica, *4*, pp. 471–501, 1989.

[Sys81]   M. M. Sysło, An efficient cycle vector space algorithm for listing all cycles of a planar graph, SIAM J. Comput., *10, 4*, pp. 797–808, 1981.

[Tar86]   É. Tardos, A strongly polynomial algorithm to solve combinatorial linear programs, Oper. Res., *34*, pp. 250–256, 1986.

[Tho80]   C. Thomassen, Planarity and duality of finite and infinite graphs, J. Combinat. Theory, Series B, *29*, pp. 244–271, 1980.

[Tho83]   C. Thomassen, A theorem on paths in planar graphs, J. Graph Theory, 7, pp. 169–176, 1983.

[Tut56]   W. T. Tutte, A theorem on planar graphs, Trans. Amer. Math. Soc., *82*, pp. 99–116, 1956.

[Tut77]   W. T. Tutte, Bridges and Hamiltonian circuits in planar graphs, Aequationes Mathematica, *15*, pp. 1–33, 1977.

[Tut60]   W. T. Tutte, Convex representations of graphs, Proc. London Math. Soc. (3), *10*, pp. 304–320, 1960.

[Tut63]   W. T. Tutte, How to draw a graph, Proc. London Math. Soc., *13*, pp. 743–768, 1963.

[Viz65]   V. G. Vizing, Critical graphs with a given chromatic class (in Russian), Discret Analiz, *5*, pp. 9–17, 1965.

[Viz64]   V. G. Vizing, On an estimate of the chromatic class of a p-graph (in Russian), Discret Analiz,
           3, pp. 23–30, 1964.
[Wag36]   K. Wagner, Bemerkungen zum Vierfarbenproblem, Jber. Deutsch. Math.-Verein., 46,
           pp. 26–32, 1936.
[Will85]   M. H. Williams, A linear algorithm for coloring planar graphs with five colours, The
           Computer J., 28, 1, pp. 78–81, 1985.
[Wils85]   R. J. Wilson, Introduction to Graph Theory, 3rd ed., Longman, London, 1985.
[Win87]   P. Winter, Steiner problem in networks: A survey, Networks, 17, pp. 129–167, 1987.

Takao Nishizeki
Department of Electrical
Communications
Faculty of Engineering
Tohoku University
Sendai 980
Japan

Computing Suppl. 7, 69–91 (1990)

Basic Parallel Algorithms in Graph Theory

Ernst W. Mayr, Frankfurt a. M.

Abstract — Zusammenfassung

Basic Parallel Algorithms in Graph Theory. We discuss some of the more common machine models for parallel computation and their variants, as well as some relevant basic results from parallel complexity theory. We then describe a few of the very basic and fundamental "tricks" and techniques to obtain efficient parallel algorithms. Finally, we survey work on parallel algorithms for a number of graph theoretic problems.

AMS Subject Classifications: 68E10, 68Q10.

Key words: parallel computation, parallel machine model, fundamental programming techniques, parallel graph theoretic algorithms

Fundamentale Parallelalgorithmen in der Graphentheorie. Wir diskutieren einige der gebräuchlicheren Maschinenmodelle und ihre Varianten für Parallelrechnung, sowie einige wichtige und grundlegende Resultate aus der parallelen Komplexitätstheorie. Anschließend beschreiben wir eine Auswahl von elementaren und wichtigen "Tricks" und Methoden für effiziente parallele Algorithmen. Zum Schluß geben wir einen Überblick über parallele Algorithmen für eine Reihe graphentheoretischer Probleme.

1. Introduction

Advances in VLSI technology have made it possible to build (and buy) computers with a large number of processors and blocks of memory. Using parallel computation, one hopes to circumvent or avoid many of the problems caused by the so-called von-Neumann bottleneck of one serial CPU. First experiences with parallel machines and algorithms have also shown that in order to achieve efficient parallel computation and make optimal use of the available hardware, many careful decisions have to be made when designing the parallel architecture as well as the parallel algorithms supposed to run on it.

While the apparent potential of parallel computation is certainly large and promising, there are also obvious problems to utilize this potential. One reason for this frustration may lie in the fact that there is no "standard" parallel architecture for which to design efficient algorithms. This is in contrast to the situation in the sequential world where there is a (more or less) unique model (called *Random Access Machine* by theoreticians) for which most algorithms are designed, at least as a first stage.

Another difficulty stems from the fact that many fundamental sequential programming techniques or algorithms, most of them considered by now probably as straightforward, are rather difficult if not impossible to parallelize.

In this paper, we first discuss some of the more common models for parallel computation and their variants, in particular the Parallel Random Access machine model which though theoretical and somewhat idealistic, is a good model to express parallel algorithms in. It separates the issue of finding parallelism in problems or developing highly parallel algorithms from more implementation dependent problems like inter-processor communication and network congestion. We also present some of the complexity theoretic background relevant to parallel computation. It provides some means to characterize those problems that are efficiently parallelizable on the one hand, and problems that in all likelihood have no efficient parallel solutions on the other.

Then we present a number of very basic and fundamental programming techniques and little routines that are tools for the development of many efficient parallel algorithms and applications. Where possible, we state such simple parallel procedures in a pseudoformal parallel programming language. Finally, we survey some classes of graph theoretic problems and parallel algorithms for them. We conclude by discussing some of the limits of our current knowledge on efficient parallel computation.

2. Machine Models, Basic Complexity Results

2.1. Models of Parallel Computation

As we have already mentioned there is a large number of parallel machine models, varying considerably in power and programmability. In Table 1 we give a short list of such models. The list is not intended to be exhaustive, and it also gives pointers for more detailed descriptions of the models.

We shall base most of our discussions onto a theoretical machine model for parallel computation called the *Parallel Random Access Machine*, or *PRAM* (see, e.g., [31] [43]). In this model, there is an unbounded number of identical processors which are basically *Random Access Machines* (or *RAM's*), as defined in [4], and an unbounded number of global, shared memory cells. Each processor can execute its own program (though, in most cases, all processors will have the same program), and the processors work synchronously, controlled by a global clock. Each processor can access any memory cell in one step.

Depending on whether simultaneous access to the same memory cell by more than one processor is permitted or not, several variants of the PRAM model have been defined. We do not consider conflicts between read and write operations since we always assume that the read operations are performed in the first half of a memory access cycle and the write operations in the second. The *concurrent read exclusive write* variant of the basic model (CREW-PRAM) allows that more than one processor read the same memory cell in one step, but it disallows *concurrent writes* to

Table 1. Models of Parallel Computation

1. data flow	[8] [28] [45]
2. actors	[52]
3. vector machines	[88]
4. local area networks	[77]
5. fixed (multistage) interconnection networks	[99] [104]
6. VLSI	[71] [76]
7. Parallel Random Access Machine .	[31] [43]
8. Boolean circuits	[87] [94] [108]
9. unbounded fan-in circuits	[17] [101]
10. alternating Turning machines	[16] [93] [94]

the same memory cell. The *exclusive read exclusive write* variant (EREW-PRAM), on the other hand, forbids concurrent access completely.

While there are no (logical problems with simultaneous read access to the same memory cell by more than one processor, some precautions have to be taken for simultaneous write access. Depending on the method used to resolve such conflicts, we further distinguish the following variants of the *concurrent read concurrent write* PRAM (CREW-PRAM):

1. in the COMMON CREW-PRAM, all processors writing concurrently to the same memory cell have to write the same value;
2. in the ARBITRARY CRCW-PRAM, if several processors write concurrently to a memory cell, some arbitrary processor succeeds;
3. in the PRIORITY CRCW-PRAM, if several processors write concurrently to a memory cell, the processor with the highest index succeeds.

It should be clear that the sequence of machine models given by

EREW-CREW-COMMON CRCW-ARBITRARY CRCW-PRIORITY CRCW

forms a hierarchy of machine models of increasing power in the sense that any model in the list can be (trivially) emulated by any other model further down in the list, without incurring any time loss.

It is also not too hard to see that a PRIORITY CRCW-PRAM using n processors can be simulated by an n processor EREW-PRAM in such a way that the simulation of every step of the CRCW-PRAM requires $O(\log n)$ steps of the EREW-PRAM. The simulation is based on the following idea: Instead of directly accessing their desired memory cells, the n processors instead write a description of their request as well as their own index to some appropriate array of length n. This array can then be lexicographically sorted by an EREW-PRAM algorithm in $O(\log n)$ time [21] by memory address and processor index. A simple computation then determines the outcome of the memory access by every processor and writes it into the corresponding array element from which it can be read by the processor. For a more detailed description, we refer the reader to [29], [68], and [107].

The shared memory feature of the PRAM model is somewhat idealistic. A more realistic machine model consists of a *network* of (identical) processors with *memory modules* attached to them. The processors are connected via point-to-point communication channels. Each processor can directly access only cells in its own memory module, and it has to send messages to other processors in order to access data in their modules. To respect technological constraints, the number of channels per processor is usually bounded or a very slowly growing function of the number of processors. Examples for such networks of processors are the Hypercube [98] or Connection Machine [53], the Cube-Connected-Cycles network [89], or the Ultracomputer (RP3) [86] [97]. The latter is an example for a *multistage interconnection network*, where an array of processors is connected to an array of memory modules by a switching network consisting of several stages of small switches.

2.2. Basic Complexity Theoretic Concepts

There are at least two goals one wishes to achieve with parallel computation: *speedup* and *efficiency*. Speedup is the ratio between the sequential running time $T_s(n)$ and the parallel running time $T_p(n)$, measured for problem instances of size n. Efficiency is the ratio between the *work* performed by the sequential algorithm (which is, of course, equal to its running time) and the work performed by the parallel algorithm, which is given by its number of processors times its running time. We are interested in problems for which we can find parallel algorithms with large speedup using a reasonable number of processors.

One way of formalizing this approach is given by the complexity class $\mathcal{NC}$ [87]. It is the class of all those problems that a PRAM with a polynomial number of processors can solve in polylogarithmic time. More formally, for every problem in $\mathcal{NC}$, there are constants c and k and a PRAM algorithm that requires $O(n^c)$ processors and $O(\log^k n)$ time on instances of size n. Note that since the most powerful PRAM model we have listed (the PRIORITY CRCW-PRAM) can be simulated by the least powerful model (the EREW-PRAM) with an $O(\log n)$ slowdown, the definition of $\mathcal{NC}$ is independent of the specific PRAM model. It even turns out that some fixed interconnection networks (including all those mentioned above) can simulate $\mathcal{NC}$ algorithm with only a polylogarithmic slowdown, thus making the definition of $\mathcal{NC}$ even more robust.

It has also become customary to call problems in $\mathcal{NC}$ "efficiently parallelizable", in the same manner as problems in $\mathcal{P}$ (polynomial time) are called "feasible." We should note, however, that parallel algorithms requiring a number of processors which is a high degree polynomial, or running in time $O(\log^k n)$, for some large constant k, are certainly impractical, even though formally they are in $\mathcal{NC}$.

This leads us to also put some emphasis on the efficiency of parallel algorithms, as defined above. We call a parallel algorithm *optimal* if it runs in polylogarithmic time and with efficiency $\Omega(1)$, and we call it *efficient* if it runs in polylogarithmic time with efficiency $\Omega(\log^{-k} n)$, for some constant k.

2.3. $\mathscr{P}$-completeness

While many problems are efficiently parallelizable and have optimal or efficient parallel solutions, other problems and algorithmic techniques seem harder or impossible to parallelize. The complexity theoretic concept of a problem being $\mathscr{P}$-complete may be useful in characterizing such cases. $\mathscr{P}$ is the well-known class of problems solvable in polynomial time on a sequential machine, like a Turing machine or Random Access Machine. To study hardest problems in $\mathscr{P}$, we look at those problems in $\mathscr{P}$ to which all other problems in $\mathscr{P}$ can be reduced in an efficient manor. Formally, we call some problem $C \in \mathscr{P}$ $\mathscr{P}$-complete if for every other problem $A \in \mathscr{P}$, there is a function f computable by a Turing machine in logarithmic space such that

$$x \in A \text{ iff } f(x) \in C.$$

It is quite easy to see that $\mathscr{NC}$ is a subset of $\mathscr{P}$. The reason is that a (sequential) Turing machine can simulate an $\mathscr{NC}$ computation by first simulating the first step of the polynomially many processors, then the second step, and so on. The simulation overhead per step is at most a polynomial in the number of simulated processors, as long as all operands used by the parallel algorithm remain reasonably small. This is certainly the case for $\mathscr{NC}$ algorithms which can run only for a polylogarithmic number of steps.

A somewhat more difficult construction can be used to show that, by a different type of simulation, $\mathscr{NC}$ algorithms can be simulated by space efficient Turning machines. More precisely, if the running time of the $\mathscr{NC}$ algorithm is $O(\log^k n)$, then $O(\log^{2k} n)$ space suffices. For details of this construction, we refer the interested reader to [31] and [14].

The space efficient simulation of parallel algorithms implies in particular that $\mathscr{NC}$ is a subset of POLYLOGSPACE, where the latter is the class of all problems solvable by a Turing machine whose workspace is bounded by a polylogarithmic function of the input size. If we now assume that some $\mathscr{P}$-complete problem A is in $\mathscr{NC}$ then $\mathscr{P}$ is equal to $\mathscr{NC}$ and, in addition, $\mathscr{P} \subseteq$ POLYLOGSPACE. This would mean that every problem in $\mathscr{P}$ could be solved using *very little* space (but not necessarily, of course, simultaneously polynomial time). Though there is no known proof ruling out this situation, it is widely agreed to be highly unlikely. We therefore take the fact that a given problem has been shown $\mathscr{P}$-complete as strong evidence that it is not in $\mathscr{NC}$, not efficiently parallelizable. A fortiori, it won't admit efficient or optimal parallel algorithms as we have defined them.

Finally, we discuss the *generic* $\mathscr{P}$-complete problem. It is the so-called *circuit value problem* (CVP). Define a circuit to be a directed acyclic graph whose nodes have indegree at most two. The indegree zero nodes are the inputs to the circuit, they are labelled with values in $\{0,1\}$, representing the input to the circuit. Indegree one nodes are labelled with NOT, they represent NOT-gates, and indegree two nodes are each labelled AND or OR, representing the corresponding Boolean gates. The nodes of the graph with outdegree zero are called outputs of the circuit. The circuit value problem requires to determine, given a circuit together with a designated output node, whether the value of this designated output as "computed by the circuit" usng the obvious rules is 1. It has been shown [69] that CVP is $\mathscr{P}$-complete, as are some important special cases [42].

3. Some Fundamental Techniques

In this section, we present some of the more fundamental programming techniques and procedures. They are used in many applications dealing with combinatorial or graph theoretic problems. Table 2 contains a (not necessarily complete) list of some of these techniques.

In the following, we shall describe some of these fundamental techniques in more detail.

3.1. Doubling

The job at hand is to compute the sum (or some other associative function) of n numbers, a_0 through a_{n-1}. Suppose initially that n processors are available. Then we could first compute the sum of all even-odd pairs and store it at the even positions, then add up pairs of these sums, and so on. The scheme of this type of computation is depicted in Figure 1. Note that all operations on one level of the

Table 2. Fundamental Parallel Algorithms

---

1. doubling
2. pointer jumping *or* path doubling
3. parallel prefix
4. list ranking
5. Euler contour path
6. numbering of trees

$a_1 + a_2 +$ $\ldots$ $+ a_{15} + a_{16}$

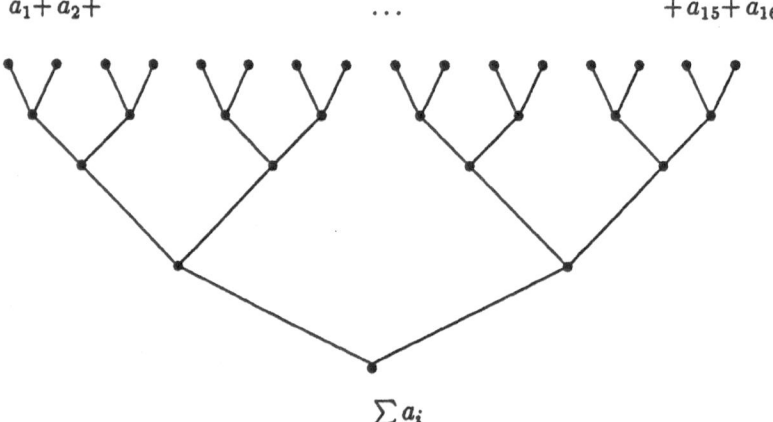

$$\sum a_i$$

Figure 1. Summation by Doubling

procedure *census_function*(n, s, res, ∘);
int *n*; gmemptr *s*, *res*; binop ∘;
co *n* is the number of elements in the input array starting at position *s* in global memory; *res* is the index of the global memory cell receiving the result; ∘ is an associative binary operator **oc**
begin
    local type_of_S: *save*, int: *mask*, gmemptr *myindex*;
    if PROC_NUM < *n* **then**
       mask := 1; *myindex* := *s* + PROC_NUM; *save* := $M_{myindex}$,
       while *mask* < *n* **do**
          if (PROC_NUM AND (2 ∗ *mask* − 1) = 0) **and** PROC_NUM + *mask* < *n* **then**
             $M_{myindex} := M_{myindex} \circ M_{myindex+mask}$
          fi;
          mask := 2 ∗ *mask*
       od;
       if PROC_NUM = 0 **then** $M_{res} := M_s$ **fi**;
       if *myindex* ≠ *res* **then** $M_{myindex} := save$ **fi**
    fi;
    return
end *census_function*.

Figure 2. PRAM Algorithm for Census Functions

tree can be performed in parallel since they access disjoint sets of variables. What we double in every iteration is the number of inputs whose sum we have already collected in a single variable.

Another name used with regard to this doubling technique is census functions. In Figure 2, we give a detailed PRAM program to compute a census function. The program works as follows. Every processor uses a local variable *save* which is needed in order to make the procedure free of side-effects. All processors participating in the computation have indices less than *n*. The variable *mask* serves to select those processors that perform nontrivial computations in any given step. The algorithm starts with *mask* set to 1. The processor with index PROC_NUM takes care of the variable with index *s* + PROC_NUM. We just call this index *myindex*, and the processor initially saves away whatever there is in cell *myindex*. The variable *mask* is used to do the doubling. The expression (PROC_NUM AND (2 ∗ *mask* − 1) = 0) is just a way of saying: What do the last few bits of PROC_NUM look like? In the first step, only those processors are active whose PROC_NUM is even, *i.e.* the last bit of their PROC_NUM is zero. Every such processor combines its value with that of the next processor. We also assume that all processors for which the **if**-condition does not hold perform an appropriate number of no-op steps such as to stay synchronized with the active processors. Then we double *mask*. Thus, *mask* will be two in the next iteration of the loop, and we will collect the pairs into sums over 4 elements each. Then we'll double *mask* again, to collect pieces of 4 into pieces of 8, and so on. In the end, we just have to do some cleanup. The first processor takes care of storing the whole sum into the result position, and then all other processors restore the global memory cell that they initially saved away.

The algorithm as just presented is not optimal, however, since it uses *n* processors. To obtain an optimal solution, we first group the items to be summed into contiguous groups of length log *n*. We then assign one of *n*/log *n* processors to each of these groups. Each processor first sums up sequentially the elements in its own

group. On the resulting sums, we perform a census-function computation with
$n/\log n$ processors as described above. The total time requirement is still $O(\log n)$,
thus providing an optimal solution.

3.2. Pointer Jumping and Path Doubling

The next technique, pointer jumping or path doubling, is useful for finding paths
from vertices to their respective roots in in-forests. An in-forest is a collection of
in-trees which in turn are trees with all edges oriented towards the root. We assume
that by some preceding computation the in-forest is stored in the global memory of
the PRAM in such a way that for every node in the forest there is a pointer to its
immediate ancestor in its tree. The pointer for the root just points to the root itself.
We also assume that there is a unique processor associated with every node in the
forest. The problem consists of finding, for every node in the in-forest, the root of
the tree to which it belongs.

To simplify the following description, we shall identify each processor with the node
it is associated with. In the first step, every node finds its grandparent by reading
the location pointed to by its own parent pointer. It then replaces the parent pointer
by a pointer to the grandparent. In the second step, every processor again reads the
location given by its pointer, and substitutes the value found there. Thus, after two

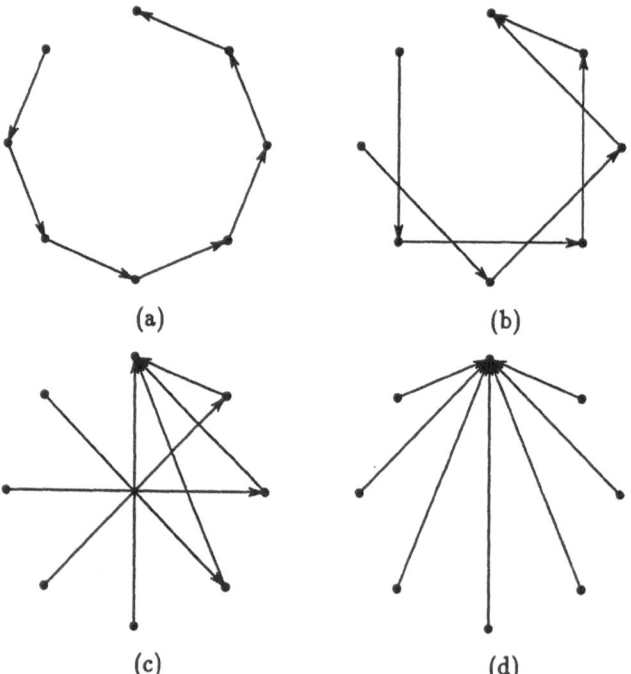

Figure 3. Example for Path Doubling

steps, every node knows its ancestor four generations away, after three steps its ancestor eight generations away, and so on. If there are n nodes in the forest, then $\log n$ steps suffice and every node will know the root of the tree to which it belongs. Figure 3 gives a simple example consisting of a single path with eight nodes.

We note that in the path doubling algorithm no write conflicts will occur since only the processor associated with a node will update the pointer belonging to that node. However, in general, there will be read conflicts since several pointers can point to the same node, as becomes immediately clear if we consider in-trees which are not just paths as in the example.

We should also like to emphasize that for the path doubling algorithm the pointers need not be stored in contiguous positions, and, of course, not in order as they are shown in Figure 3 for clarity only.

3.3. Parallel Prefix

Parallel prefix computation is a very essential technique, representing a generalization of the doubling technique considered earlier [70]. Again, we are given an array of quantities a_0, a_1 ..., a_{n-1}, and we are required to compute the partial sums $\sum_{j=0}^{i} a_j$, for $j = 0, \ldots, n - 1$. And again, the connective could be any binary associative operator. Figure 4 shows an EREW program for the parallel prefix problem.

The inputs for the procedure are the length of the array and the starting location for the array of results. Also, $\circ$ stands for the binary operator. We need a few local variables, *save* and *save2*, to save away values. The variable *span* serves the same purpose as *mask* in the previous program, *i.e.* it denotes the distance spanned in a given step.

```
procedure parallel_prefix(n, start, result, ∘);
int n; gmemptr start, result; binop ∘;
co n is the length of the input list, start indicates the place in memory where the input list begins and
result where the list of results a₀ ∘ ··· ∘ aᵢ is to be put; ∘ is an associative binary operator oc
begin
    local type_of_S: save, save2; int: span, gmemptr myindex;
    if PROC_NUM < n then
        span := 1;
        myindex := start + PROC_NUM;
        save := M_myindex co save global cell M_myindex since it may get changed during the computation oc;
        while span ≤ PROC_NUM do
            M_myindex := M_myindex-span ∘ M_myindex;
            span := 2 * span
        od;
        save2 := M_myindex;
        M_myindex := save co restore the original input values oc;
        myindex := result + PROC_NUM;
        M_myindex := save2
    fi
end parallel_prefix.
```

Figure 4. Parallel Prefix Algorithm

Again, the algorithm as given is not optimal. A similar technique as above, grouping the input elements into segments of length log n, can also be used here to achieve an optimal implementation. We leave the rather straightforward details to the reader.

3.4. List Ranking

The list ranking problem is: given an array of n pointers which form a simply linked list, determine, for each element in the array, its distance from the end of the list.

Typically, the list ranking problem occurs as a subproblem in other algorithms. We want to point out that generally, of course, the list is not stored in a monotone fashion in contiguous memory cells. Obviously, the problem becomes trivial in this case. We may, however, assume that all list pointers are stored in a contiguous array. Should this not be the case initially, we can use the parallel prefix routine to "compactify" the representation of the list in memory. We leave the details of this operation to the reader.

As a possible application, think of the following (we shall see something similar when we discuss the contour path technique below): some parallel computation produces a simply linked list of numbers, and we wish to compute the sum of these numbers for all initial segments of the list. This looks like a parallel prefix problem. To apply our algorithm presented above, however, we first have to arrange the numbers in list order in a contiguous array in memory. It should be clear that this is a straightforward task once we have solved the list ranking problem since the rank of an element in the list can be used to easily determine its position in the array.

We should like to mention that for sequential computation, list ranking is a rather trivial problem. We just go through the list, push every element onto a stack, and after arriving at the end the list, we pop the elements from the stack and just count.

For parallel computation, there is also no problem if we are given n processors. We can simply apply the path doubling routine, keeping a count of the distance each pointer covers.

Theorem 1 *The list ranking problem can be solved in $O(\log n)$ time on an EREW-PRAM using $n/\log n$ processors.*

To actually prove this theorem would exceed the limits of this presentation. For two different solutions, both optimal, we refer the reader to [7] [23] [25]. Here, we give a brief sketch of one of the algorithms.

Sketch of Optimal List Ranking Algorithm:

1. break array into stacks of height $\lceil \log n \rceil$;
2. select top remaining element in each stack;
3. determine chains formed by selected elements;
4. splice out distinguished elements in singleton chains;

5. have processor at tail of non-singleton chain splice out all elements in chain, one per step;

6. if there are elements left, go to step 2; otherwise stop.

First, we break the array into roughly $n/\log n$ segments. The list could be completely disintegrated at this point since the list pointers are completely independent of the segments. There are $n/\log n$ stacks and the basic idea is to assign one processor to each stack. Each processor selects the top element of its stack. If the (generally two) neighbors of a selected element are not also selected, the processor removes the element from the list, leaving appropriate information with the neighbors. The bad case occurs if some of the selected top elements form a subchain of the list. Then the simple splicing technique does not work. Instead, each of the chains is handled by the processor at the tail of the chain. It will work on the chain sequentially, one element per step. The other processors will go on and select the next element in their stack. The problem with this method is that the first processor which gets a long chain is busy for a long time dealing with this chain. Since, in the worst case, the length of such a chain could be $\Omega(n/\log n)$ we have to make sure that no long chains are created. The technique used for this purpose is called *deterministic coin tossing*, and is described in detail in [22] (see also [40]). It is of independent interest.

3.5. Euler Contour Path

The Euler contour path technique is useful for many computations on trees. We assume here that the (rooted) tree is given in form of a list of children for every node in the tree. The contour path technique replaces each arc of the tree by a pair of pointers. Intuitively, the first pointer will correspond to an arc pointing in the same direction as the tree edge, the second pointer to an arc in the opposite direction. The second pointer of each edge is made to point to the first pointer of the next edge in each adjacency list, except for the last edge in the list whose second pointer points to the second pointer of its parent edge. Also, the first pointer of each edge points to the first pointer of the first edge of the child's adjacency list, unless the child is a leaf in which case the first pointer of the edge is hooked to its second pointer. Pictorially, the path generated by the pointers follows the contour of the tree when drawn in the plane in the canonical way. It is called the *Euler contour path*.

Since all computations necessary to construct the pointer structure for the contour path are local it is very straightforward to obtain an optimal EREW-PRAM algorithm for this problem. It runs in $O(\log n)$ time on $n/\log n$ processors.

3.6. Numbering of Trees

There are several important numbering schemes for trees which are useful for many computational problems. Examples are the *pre-order*, *in-order*, and *post-order* numberings. While pre- and post-order numberings can be defined for any rooted tree, the in-order numbering only applies to binary trees. As an example, we describe an optimal EREW-PRAM routine to obtain a post-order numbering. The other problems are quite similar and left as an exercise.

In the post-order numbering of an n-node tree, the vertices of the tree receive unique labels from 0 to $n - 1$. The labels in the first subtree of a node are all smaller than the labels of the second subtree, these are all smaller than the labels in the third subtree, and so on, and finally the labels in the last subtree of the node are smaller than the label of the node itself.

One optimal algorithm for post-order numbering trees works as follows. It first constructs the Euler contour path for the tree. Using list ranking, it "flattens" out the path into a linear array. It then associates 1 with edges on the contour path pointing towards the root (second pointers of the corresponding tree edge), and 0 with the other edges. Then it performs a prefix computation on this array of 0's and 1's. The post-order number of a vertex x of the tree is then the prefix sum up to the last edge on the contour path entering x.

The pre-order numbering can be computed similarly, interchanging 0's and 1's. Given these two numberings, it is possible, for instance, to compute the number of descendents for every node in the tree.

Methods for other types of tree computations, like the height of the vertices, are discussed further below.

3.7. Other Techniques

There are quite a few more fundamental parallel programming techniques which we cannot present here in detail. One is the computation of *lowest common ancestors* in (rooted) trees, a routine used for many other applications. There is an optimal ($n/\log n$ processors and $O(\log n)$ time) EREW-PRAM algorithm that performs the following task: Given an n-vertex tree (in form of adjacency lists for every interior vertex), it performs some $O(\log n)$ time precomputation using $n/\log n$ processors such that after this precomputation every query of the form "what is the lowest common ancestor of vertices x and y?" can be answered by one processor in constant time. For a detailed description of this alogirthm see [96].

Another basic problem, of extreme significance in sequential as well as parallel computation, is sorting: Given n keys from some ordered universe, arrange them in an array in ascending order of their value. We shall be interested here only in comparison based algorithms, for which the only operations allowed on keys are pairwise comparisons. The sequential complexity of sorting is well-known to be $\Theta(n \log n)$ [64]. Sorting can also be performed optimally on an EREW-PRAM, due to an n processor $O(\log n)$ time algorithm given in [21].

Tree contraction is another very powerful programming technique, applicable to a large number of combinatorial problems on trees. We shall describe it in more detail in the next section.

4. Some Graph Theoretic Applications

So far, we have seen basic programming techniques which are all optimal. In this section, we shall first study some more optimal techniques which are a bit more involved. We shall then present a selection of fundamental graph theoretic problems for which currently no optimal solutions are known.

4.1. Tree Contraction

The probably simplest example for tree contraction is the evaluation of a parse tree for some arithmetic expression. The leaves of such a tree correspond to variables (with values) or constants, the interior nodes to arithmetic operators like $+$, $-$, $\times$, $/$. The structure of the tree is determined by the precedence of the operators and the parenthesis structure in the arithmetic expression. Also, each vertex in the tree can be associated with the subexpression given by its subtree, and with the value of this subexpression.

As another application, we could reduce the problem of computing the size of the subtrees of a given tree to a tree evaluation problem. We associate the value 1 with each leaf and the operation "sum up the values of your children and add 1 to this sum" to each internal vertex. It should be clear that the values obtained for the vertices are just the size of their respective subtrees.

A similar example is computing the *height* of the vertices in a tree, *i.e.* for each vertex the longest distance to a leaf. It corresponds to a tree evaluation problem with 0 associated with the leaves and the operation "add 1 to the maximum value of your children" with the interior vertices.

It should be obvious that such a tree can be evaluated in parallel proceeding level by level. We first evaluate all those internal vertices which have only leaves as children. We then remove the leaves and iterate. The number of iterations for this algorithm is given by the height of the tree. Unfortunately, there are trees with n vertices and height $\Omega(n)$.

The first efficient parallel algorithm for tree contraction was presented in [80]. Later, quite a host of optimal EREW-PRAM algorithms for the problem were given, e.g. in [1] [24] [38] [35], and [65]. One of the simplest methods is that presented in [65].

We assume that we are given a binary tree such that every non-leaf has exactly two children. For the purpose of simplicity, we also assume that the algebraic domain under consideration are the rationals with $+$, $-$, $\times$, and $/$. Intuitively, the algorithm proceeds as follows in stages: In every stage, it selects half of the leaves in such a way that the two children of a vertex are never selected at the same time. It then deletes the selected leaves, replaces the parent of each selected leaf by the subtree rooted at the other child, and updates the operation to be performed at the root of this subtree appropriately. To keep track of the operations that have to be performed at every node, it turns out for our case that it suffices to associate a quadruple $q_x = (a_x, b_x, c_x, d_x)$ with every node x in the tree, with the following intuitive understanding, where we also use q_x to denote the rational function

$$q_x: z \mapsto \frac{a_x z + b_x}{c_x z + d_x}.$$

1. Let a be the value associated with some leave x in the original tree. Then $q_x = (a, 0, 0, 1)$.
2. Every node x in a tree will have a value $v(x)$. If x is a leaf then

$$v(x) = q_x(1);$$

if x is an interior vertex with children y and z and operation $\circ$ attached to it, then its value is

$$v(x) = q_x(v(y) \circ v(z)).$$

Let x be a parent in the tree whose left child y becomes selected, and whose right child is z. Then the subtree rooted at z replaces the subtree rooted at x, and we rename z to z'. We want to have $v(z') = v(x)$. An easy computation shows that since y is a leaf with a known value, and since rational functions are closed under composition, there is a quadruple $q_{z'} = (a_{z'}, b_{z'}, c_{z'}, d_{z'})$ computable with a constant number of arithmetic operations from q_x, q_y, and q_z such that

$$v(x) = q_{z'}(v(r) \circ_{z'} v(s))$$

where r and s are the two children of z (or z') and $\circ_{z'}$ is its associated operation, if z is an interior vertex, and

$$v(x) = q_{z'}(1)$$

if z (or z') is a leaf.

Thus, whenever a leaf gets selected and removed by the algorithm, the quadruple of its sibling, which replaces its parent, can be updated in a constant number of arithmetic operations. It turns out that in order to avoid memory access conflicts, the subdivision of every stage in the following algorithm works (it guarantees that if x, y, and z are as above then no processor removing some selected leaf other than y will touch z).

Sketch of Optimal Tree Contraction Algorithm:

1. use Euler contour path technique to label the leaves from left to right by 0 through $n - 1$;
2. assign quadruples q_x to all leaves x and quadruples $(1, 0, 0, 1)$ to all interior nodes;
3. perform $\lceil \log n \rceil$ stages consisting of
   (a) remove all even numbered leaves that are a left child, replace their parent by their sibling and update the quadruple of the sibling;
   (b) perform the same operations on all even numbered leaves that are a right child;
   (c) divide the label of every remaining leaf by 2, without remainder.

If there are $n/\log n$ processors, the first two steps require time $O(\log n)$. The time for stage i is

$$\max\left\{O\left(\frac{n/2^i}{n/\log n}\right), O(1)\right\}$$

and the time for all $\lceil \log n \rceil$ stages is therefore $O(\log n)$.

When the algorithm terminates it has reduced the original tree to a singleton node whose associated quadruple gives the value belonging to the root of the original tree. The algorithm can also be modified to compute the value of all subtrees of the original tree, using the same number of processors and the same asymptotic running time.

4.2. Connected Components, Spanning Trees

A basic task for graph theoretic algorithms is often the computation of the connected components of a given graph. Sequentially, various graph traversal techniques like breadth first or depth first search can be used to obtain linear time algorithms. Since no $\mathcal{NC}$ algorithms are currently known for depth first search in general graphs, and since all current $\mathcal{NC}$ algorithms for breadth first search in general graphs employ transitive closure techniques requiring basically $M(n)$ processors (where $M(n)$ is the sequential time needed to multiply two $n \times n$ matrices), we have to use other techniques to obtain efficient parallel algorithms for the connected components problem, or the closely related problem of computing a spanning forest.

One such approach is based on the following idea. Given an arbitrary graph G, we first put each vertex of G into a singleton set. The algorithm then proceeds in stages. In each stage, a set of edges is selected whose two endpoints are in different sets. For each edge, the two sets belonging to its endpoints are merged into one set. To facilitate the edge selection and merging routines, for each set a pointer structure is maintained. The pointer structure forms an in-tree, with the root representing the whole set. Each node can find the set it is currently in by following the path in the in-tree up to the root. It is advantageous for the algorithm if it can keep the trees shallow by occasionally redirecting pointers like in well-known sequential UNION-FIND structures (see, e.g., [102]).

By a careful implementation of these ideas, an EREW-PRAM algorithm using $n^2/\log n$ processors and $O(\log^2 n)$ time can be obtained [54] [85] [19] [66]. If we use instead the more powerful ARBITRARY CRCW-PRAM model then a running time of $O(\log n)$ can be achieved using $m + n$ processors, where n is the number of vertices and m the number of edges in the graph. By our simulation result for variants of the PRAM model, we thus obtain another EREW-PRAM algorithm running in $O(\log^2 n)$ time, using $m + n$ processors. Using a more sophisticated approach [23] [11] the ARBITRARY CRCW-PRAM algorithm can be improved to run on $O((m + n)\alpha(m, n)/\log n)$ processors where $\alpha(m, n)$ is the inverse of Ackermann's function, well-known from the sequential UNION-FIND problem [102].

There is also an optimal **ARBITRARY CRCW-PRAM** algorithm for the connected components problem that runs in time $O(\log n)$. However, this algorithm is not deterministic, it is *randomizing* and uses internal coin-flipping [34].

All the algorithms for the connected components problem can be modified in a very straightforward manner to construct a spanning forest for the given graph, within the same processor and time bounds.

There is also a nice extension of a connected components/spanning tree algorithm to find the biconnected components of a graph [103]. The complexity of this algorithm is dominated by the part that finds connected components.

4.3. (Open) Ear Decomposition

We have already remarked above that some efficient sequential graph traversal techniques don't seem to be efficiently parallelizable, like depth first search. Hence, other methods to decompose a given graph into simpler parts had to be developed. One such method is the *ear decomposition* technique proposed in [73].

Definition 4.1 *An (open) ear decomposition of a graph $G = (V, E)$ is a sequence P_0, $P_1, \ldots, P_r$ of simple, edge-disjoint paths, with P_0 a cycle and only the endpoints of P_i, $i > 0$, on earlier paths. In an open ear decomposition, the endpoints of each P_i, $i \geq 1$, have to be distinct.*

It turns out that a graph has an ear decomposition iff it is 2-edge-connected, and it has an open ear decomposition iff it is biconnected.

The notion of an ear decomposition can also be defined for digraphs.

An (open) ear decomposition can be found by an efficient algorithm along the following lines [75] [78], where the input is an arbitrary (undirected) 2-edge-connected graph $G = (V, E)$:

Sketch of Ear Decomposition Algorithm

1. find spanning tree for G;
2. root the spanning tree, number it in preorder;
3. label each non-tree edge with the (preorder number of the) least common ancestor of its endpoints;
4. assign consecutive numbers to the non-tree edges in non-decreasing order of their labels;
5. number each tree edge with the minimal number of a non-tree edge whose fundamental cycle it is contained in.

The running time of this algorithm is dominated by the requirements of the first step for finding a spanning tree. The remaining steps can be implemented using a number of the optimal fundamental techniques described in the previous section and earlier in this section. The algorithm, as given above, finds an ear decomposition but not necessarily an open ear decomposition even when it exists. The algorithm can, however, be slightly modified by using a somewhat more elaborate numbering scheme for the non-tree edges to obtain open ear decompositions for biconnected graphs. For more details, see the references given above.

Ear decomposition and open ear decomposition have found a number of applications. As an example, it is quite easy to construct an st-numbering for a graph, given an open ear decomposition. Let G be a biconnected graph with n vertices, and let s and t be two vertices of G connected by an edge. In an st-numbering, the vertices of G have distinct labels, s being labelled 1, t being labelled n, such that every vertex other than s and t has both a neighbor with a larger and a neighbor with a smaller label. For the details of an efficient parallel st-numbering algorithm, see [75].

st-numberings have in turn been used in [63] as a subroutine in an efficient parallel algorithm for testing planarity and finding planar embeddings.

Ear decomposition techniques also play an important part in some $\mathcal{NC}$ algorithms for testing k-vertex-connectivity of (undirected) graphs, for $k = 3$ [79] [91] [32] and $k = 4$ [57].

4.4. More Graph Problems and Algorithms

In this subsection, we are going to mention briefly a number of other graph problems and subclasses of graphs for which $\mathcal{NC}$ or random $\mathcal{NC}$ ($\mathcal{RNC}$) algorithms have been developed.

Euler tours for general graphs: We have discussed the Euler contour path technique for trees. There are also parallel algorithms for Euler tours in general undirected or directed graphs (of course, not all graphs have Euler tours). Two efficient parallel algorithms are given in [9] and [10].

Maximal independent sets: An *independent* (or *stable*) set in a graph is a subset of the vertices such that no two of them are connected by an edge of the graph. Such a set is maximal if no other independent set properly contains it. The first $\mathcal{NC}$ algorithm for the maximal independent set problem was given in [61]. Other, more efficient algorithms appear in [74], [41], and [5]. The latter algorithm uses randomization.

Matching problems: There are various types of matching problems: determining whether a graph has a *perfect* matching, constructing a *maximum* matching (*i.e.*, a matching of maximal cardinality), and constructing a *maximal* matching (*i.e.*, a matching that is not properly contained in any other matching). There is an efficient $\mathcal{NC}$ algorithm for the maximal matching problem [56] (also see [55] for a fast and simple randomizing algorithm for the same problem). For the general maximum matching problem, only $\mathcal{RNC}$ algorithms are known [60] [82] [58] [33]. These algorithms strongly rely on methods for determining the rank of certain matrices related to the Tutte matrix, with polynomial entries. All currently known efficient methods for these rank tests use randomization. In certain cases, and for certain subproblems, however, the randomization can be avoided [44]. There are also subclasses of graphs for which deterministic $\mathcal{NC}$ algorithms for the maximum matching problem have been found, e.g. for regular bipartite graphs [72], for strongly chordal graphs [26], for dense graphs [27], for $K_{3,3}$-free graphs [106] (more precisely, for computing the number of perfect matchings in such graphs), and for co-comparability graphs (complements of partial orders) [50]. Matching algorithms are also used as subroutines for some flow problems, as in [60] and [2].

Depth first search: Assume a (connected) graph is given by a standard adjacency list representation. Then the canonical sequential depth first search algorithm finds a uniquely determined DFS tree for the graph. To construct the same tree in parallel seems to be hard since it is $\mathcal{P}$-complete to determine whether a given edge is contained in this tree, or even in its first branch [92] [6]. There are, however, $\mathcal{RNC}$ algorithms for constructing DFS trees in general undirected graphs [2] and directed graphs [3]. There are also $\mathcal{NC}$ algorithms for the DFS problem for planar graphs [100] [40].

Graph coloring problems: Of course, optimal graph coloring in $\mathcal{NP}$-complete in general. However, planar graphs can always be colored using at most four colors, and for many special cases or relaxed problems (which do not necessarily require that the coloring be optimal) efficient parallel algorithms have been found. For a selection, see, e.g., [12] [15] [20] [36] I39] [40] [46] [47] [59] [83].

Chordal graphs: The recognition, representation, and many combinatorial problems for chordal and strongly chordal graphs can be solved by $\mathcal{NC}$ algorithms, as e.g. in [18] [26] [30] [62] [84].

(Co-)comparability graphs: The maximum matching problem can be solved for the complements of partial orders by an $\mathcal{NC}$ algorithm derived from an $\mathcal{NC}$ algorithm for the 2-processor scheduling problem [49] [50] [51], as can some combinatorial problems for such graphs which are $\mathcal{NP}$-complete for general graphs [51] [67].

Interval graphs, series-parallel graphs, reducible flow graphs, outerplanar graphs: A number of $\mathcal{NC}$ and efficient parallel algorithms have been shown for graphs in these families. They include [13] [67] [81] [48][90] [37].

Finally, [95] and [105] contain some more $\mathcal{NC}$ and efficient parallel algorithms for various graph problems.

5. Conclusion

In the preceding sections, we have seen a number of very efficient or even optimal algorithms for the PRAM model of parallel computation. To be able to use the potential of parallelism more and on a wider range, currently some of the most important shortcomings seem to be:

● we need an efficient method replacing sequential graph traversal schemes like breadth first and depth first search; all current schemes are bound to require a large number of processors since they basically compute transitive closures; as we have seen, there are even some attempts to parallelize depth first search, though there is also evidence that this might be impossible.

- the only fast parallel algorithms for matching (and related problems like certain flow problems) which are currently known rely heavily on randomization; it is very desirable to find efficient deterministic parallel algorithms for these problems; it seems, however, that some new approach is necessary.
- the machine model for parallel computation has to become more standardized; real parallel architectures have to be developed and real parallel programming systems for them that are highly independent to free the programmer from idiosyncrasies of the underlying architecture and let him concentrate on extracting and specifying the parallelism in an algorithm instead.
- efficient and optimal algorithms need to be developed for more realistic parallel machine models, like certain fixed (multistage) interconnection networks or architectures like the binary hypercube. We are very confident that progress is happening here since a number of such machines is available in practice.

Finally, we'd like to say that this paper is intended as a survey of very basic issues in parallel computation for graph theoretic, combinatorial problem. By our own admission, it is incomplete. However, we hope that we could show what potential parallelism carries, and where some of the important current problems lie.

References

[1] K. Abrahamson and N. Dadoun and D. G. Kirkpatrick and T. Przytycka. A simple parallel tree contraction algorithm. Technical Report 87-30, Department of Computer Science, University of British Columbia, Vancouver; August 1987.

[2] A. Aggarwal and R. J. Anderson. A random $\mathcal{NC}$ algorithm for depth first search. Combinatorica, 8(1):1–12, 1988.

[3] A. Aggarwal and R. J. Anderson and M.-Y. Kao. Parallel depth-first search in general directed graphs. In Proceedings of the 21st Annual ACM Symposium on Theory of Computing (Seattle, Washington, May 15–17, 1989), pages 297–308, 1989.

[4] A. V. Aho and J. E. Hopcroft and J. D. Ullman. The design and analysis of computer algorithms. Addison-Wesley 1974.

[5] N. Alon and L. Babai and A. Itai. A fast and simple randomized parallel algorithm for the maximal independent set problem. J. Algorithms, 7(4):567–583, 1986.

[6] R. Anderson and E. W. Mayr. Parallelism and the maximal path problem. Inf. Process. Lett., 24(2):121–126, 1987.

[7] R. J. Anderson and G. L. Miller. Deterministic parallel list ranking. In Proceedings of the 3rd Aegean Workshop on Computing: VLSI Algorithms and Architectures, AWOC 88. Corfu, Greece, June/July 1988, pages 81–90, 1988.

[8] Arvind, et al. The Tagged Token Date Flow Architecture. Technical Memo 229, Laboratory of Computer Science, MIT, 1983.

[9] M. Atallah and U. Vishkin. Finding Euler tours in parallel. J. Comput. Syst. Sci., 29(3):330–337, 1984.

[10] B. Awerbuch and A. Israeli and Y. Shiloach. Finding Euler circuits in logarithmic parallel time. In Proceedings of the 16th Ann. ACM Symposium on Theory of Computing (Washington, D. C.), pages 249–257, 1984.

[11] B. Awerbuch and Y. Shiloach. New connectivity and MSF algorithms for shuffle-exchange network and PRAM. IEEE Trans. Comput., C-36(10):1158–1163, 1987.

[12] F. Bauernöppel and H. Jung. Fast parallel vertex colouring. In L. Budach, editor, Proceedings of the International Conference on Fundamentals of Computation Theory (Cottbus, GDR), pages 28–35. LNCS 199, Berlin, Heidelberg, New York, Tokyo: Springer-Verlag, 1985.

[13] A. A. Bertossi and M. A. Bonucelli. Some parallel algorithms on interval graphs. Discrete Appl. Math., 16:101–111, 1987.

[14] A. Borodin. On relating time and space to size and depth. SIAM J. Comput., 6(4):733–744, 1977.

[15] J. F. Boyar and H. J. Karloff. Coloring planar graphs in parallel. J. Algorithms, 8(4):470–479, 1987.

[16] A. Chandra, D. Kozen, and L. Stockmeyer. Alternation. J. ACM, 28(1):114–133, 1981.

[17] A. K. Chandra and L. J. Stockmeyer and U. Vishkin. A complexity theory for unbounded fan-in parallelism. In Proceedings of the 23rd Ann. IEEE Symposium on Foundations of Computer Science (Chicago, IL), pages 1–13, 1982.

[18] N. Chandrasekharan and S. S. Iyengar. $\mathcal{NC}$ algorithms for recognizing chordal graphs and k-trees. Technical Report 86-020, Department of Computer Science, Louisiana State University, 1986.

[19] F. Y. Chin and J. Lam and I. Chen. Efficient parallel algorithms for some graph problems. Commun. ACM, 25(9):659–665, 1982.

[20] M. Chrobak and M. Yung. Fast parallel and sequential algorithms for edge-coloring planar graphs (extended abstract). In J. H. Reif, editor, Proceedings of the 3rd Aegean Workshop on Computing: VLSI Algorithms and Architectures, AWOC 88. Corfu, Greece, June/July 1988, pages 11–23. LNCS 319, New York, Berlin, Heidelberg: Springer-Verlag, 1988.

[21] R. Cole. Parallel merge sort. In Proceedings of the 27th Ann. IEEE Symposium on Foundations of Computer Science (Toronto, Canada), pages 511–516, 1986.

[22] R. Cole and U. Vishkin. Deterministic coin tossing with applications to optimal parallel list ranking. Inf. Control, 70(1):32–53, 1986.

[23] R. Cole and U. Vishkin. Approximate and exact parallel scheduling with applications to list, tree, and graph problems. In Proceedings of the 27th Ann. IEEE Symposium on Foundations of Computer Science (Toronto, Canada), pages 478–491, 1986.

[24] R. Cole and U. Vishkin. Optimal parallel algorithms for expression tree evaluation and list ranking. In Proceedings of the 3rd Aegean Workshop on Computing: VLSI Algorithms and Architectures, AWOC 88. Corfu, Greece, June/July 1988, pages 91–100 1988.

[25] R. Cole and U. Vishkin. Faster optimal parallel prefix sums and list ranking. Inf. Comput., 81(3):334–352, 1989.

[26] E. Dahlhaus and M. Karpinski. The matching problem for strongly chordal graphs is in $\mathcal{NC}$. Technical Report 855, Institut für Informatik, Universität Bonn, 1986.

[27] E. Dahlhaus and M. Karpinski. Parallel construction of perfect matchings and Hamiltonian cycles on dense graphs. Technical Report. Institut für Informatik, Universität Bonn, 1987.

[28] J. Dennis. Data Flow Supercomputers. Computer, 18:42–56, 1980.

[29] D. M. Eckstein. Simultaneous memory access. Technical Report TR-79-6, Computer Science Department, Iowa State University, Ames, Iowa, 1979.

[30] A. Edenbrandt. Chordal graph recognition is in $\mathcal{NC}$. Inf. Process. Lett., 24:239–241, 1987.

[31] S. Fortune and J. Wyllie. Parallelism in random access machines. In Proceedings of the 10th Ann. ACM Symposium on Theory of Computing (San Diego, CA), pages 114–118, 1978.

[32] D. Fussel and V. Ramachandran and R. Thurimella. Finding triconnected components by local replacements. In G. Ausiello, M. Dezani-Ciancaglimi, S. Ronchi Della Rocca, editors, Proceedings of the 16th International Colloquium on Automata, Languages and Programming (Stresa, Italy, July 1989), pages 379–393. LNCS 372, Berlin Heidelberg New York: Springer-Verlag, 1989.

[33] Z. Galil. Sequential and parallel algorithms for finding maximum matchings in graphs. Ann. Rev. Comput. Sci., 1:197–224, 1986.

[34] H. Gazit. An optimal randomized parallel algorithm for finding connected components in a graph. In Processings of the 27th Ann. IEEE Symposium on Foundations of Computer Science (Toronto, Canada), pages 492–501, 1986.

[35] H. Gazit and G. L. Miller and S.-H. Teng. Optimal tree contraction in the EREW model. In Tewksbury, Stuart K. and Bradley W. Dickinson and Stuart C. Schwartz, editors, Concurrent Computations: Algorithms, Architecture, and Technology, pages 139–156, Plenum Press, 1988.

[36] A. M. Gibbons and A. Israeli and W. Rytter. Parallel $O(\log n)$ time edge-colouring of trees and Halin graphs. Inf. Process. Lett., 27(1):43–51, 1988.

[37] A. Gibbons and W. Rytter. A fast parallel algorithm for optimal edge-colouring of outerplanar graphs. Research Report RR80, Department of Computer Science, University of Warwick, 1986.

[38] A. Gibbons and W. Rytter. An optimal parallel algorithm for dynamic expression evaluation and its applications. In Proceedings of the Symposium on Foundations of Software Technology and Theoretical Computer Science, pages 453–469, 1986.

[39] A. V. Goldberg and S. A. Plotkin. Parallel (Delta + 1)-coloring of constant-degree graphs. Inf. Process. Lett., 25(4):241–245, 1987.

[40] A. Goldberg and S. Plotkin and G. Shannon. Parallel symmetry-breaking in sparse graphs. In Proceedings of the 19th Annual ACM Symposium on Theory of Computing (New York City, May 25–27, 1987), pages 315–324, 1987.

[41] M. Goldberg and T. Spencer. A new parallel algorithm for the maximal independent set problem. SIAM J. Comput., *18*(2):4–19–427, 1989.

[42] L. Goldschlager. The monotone and planar circuit value problems are log-space complete for $\mathscr{P}$. SIGACT News, *9*(2):25–29, 1977.

[43] L. Goldschlager. A unified approach to models of synchronous parallel machines. In Proceedings of the 10th Ann. ACM Symposium on Theory of Computing (San Diego, CA), pages 89–94, 1978.

[44] D. Y. Grigoriev and M. Karpinski. The matching problem for bipartite graphs with polynomially bounded permanents is in $\mathscr{NC}$. In Proceedings of the 28th Ann. IEEE Symposium on Foundations of Computer Science (Los Angeles, CA, October 12–14, 1987), pages 166–172, 1987.

[45] J. R. Guard and I. Watson and J. R. W. Glauert. A multi-layered data flow computer architecture. Technical Report, Univ. Manchester, 1978.

[46] T. Hagerup and M. Chrobak and K. Diks. Optimal parallel 5-colouring of planar graphs. SIAM J. Comput., *18*(2):288–300, 1989.

[47] X. He. Efficient parallel and sequential algorithms for 4-coloring perfect planar graphs. Technical Report 87-14, Department of Computer Science, State University of New York at Buffalo, 1987.

[48] X. He and Y. Yesha. Parallel recognition and decomposition of two terminal series parallel graphs. Inf. Comput., *75*(1):15–38, 1987.

[49] D. Helmbold and E. Mayr. Two processor scheduling is in $\mathscr{NC}$. SIAM J. Comput., *16*(4):747–759, 1987.

[50] D. Helmbold and E. Mayr. Applications of parallel scheduling to perfect graphs. In Tinhofer, G., Schmidt, G., editors, Proceedings of the International Workshop WG '86, Bernried, FRG, June 1986. Graph-Theoretic Concepts in Computer Science, pages 188–203. LNCS 246, Berlin Heidelberg: Springer-Verlag, 1987.

[51] D. Helmbold and E. Mayr. Applications of parallel scheduling algorithms to families of perfect graphs. This issue

[52] C. Hewitt and H. Baker. Laws for communicating parallel processes. AI Working Paper 134A, A. I. Laboratory, MIT, May 1977.

[53] W. D. Hillis. The Connection Machine. MIT Press, Cambridge MA, 1985.

[54] D. S. Hirschberg and A. K. Chandra and D. V. Sarwate. Computing connected components on parallel computers. Commun. ACM *22*:461–464, 1979.

[55] A. Israeli and A. Itai. A fast and simple randomized parallel algorithm for maximal matching. Inf. Process. Lett., *22*(2):77–80, 1986.

[56] A. Israeli and Y. Shiloach. An improved parallel algorithm for maximal matching. Inf. Process. Lett. *22*(2):57–60, 1986.

[57] A. Kanevsky and V. Ramachandran. Improved algorithms for graph four-connectivity. In Proceedings of the 28th Ann. IEEE Symposium on Foundations of Computer Science (Los Angeles, CA, October 12–14, 1987), pages 252–259, 1987.

[58] H. J. Karloff. A Las Vegas $\mathscr{RNC}$ algorithm for maximum matching. Combinatorica, *6*(4): 387–392, 1986.

[59] H. J. Karloff and D. B. Shmoys. Efficient parallel algorithms for edge coloring problems. J. Algorithms, *8*(1):39–52, 1987.

[60] R. M. Karp and E. Upfal and A. Wigderson. Constructing a perfect matching is in random $\mathscr{NC}$. Combinatorica, *6*(1):35–48, 1986.

[61] R. M. Karp and A. Wigderson. A fast parallel algorithm for the maximal independent set problem. J. ACM, *32*(4):762–773, 1985.

[62] P. N. Klein. Efficient parallel algorithms for chordal graphs. In Proceedings of the 29th Ann. IEEE Symposium on Foundations of Computer Science (White Plains, NY, October 24–26, 1988), pages 150–161, 1988.

[63] P. N. Klein and J. H. Reif. An efficient parallel algorithm for planarity. In Proceedings of the 27th Ann. IEEE Symposium on Foundations of Computer Science (Toronto, Canada), pages 465–477, 1986.

[64] D. E. Knuth. The art of computer programming, Vol. 3: Sorting and searching. Addison-Wesley, Reading, MA, 1973.

[65] S. R. Kosaraju and A. L. Delcher. Optimal parallel evaluation of tree-structured computations by ranking (extended abstract). In J. H. Reif, editor, Proceedings of the 3rd Aegean Workshop on Computing: VLSI Algorithms and Architectures, AWOC 88. Corfu, Greece, June/July 1988, pages 101–110. LNCS 319, New York, Berlin, Heidelberg: Springer-Verlag, 1988.

[66] V. Koubek and J. Krsnáková. Parallel algorithms for connected components in a graph. In L. Budach, editor, Proceedings of the International Conference on Fundamentals of Computation Theory (Cottbus, GDR), pages 208–217. LNCS 199, Berlin, Heidelberg, New York, Tokyo: Springer-Verlag, 1985.

[67] D. Kozen and U. V. Vazirani and V. V. Vazirani. $\mathcal{NC}$ algorithms for comparability graphs, interval graphs and testing for unique perfect matching. In Proceedings Fifth Conference on Foundations of Software Technology and Theoretical Computer Science, pages 496–503. LNCS 206, Berlin-Heidelberg-New York: Springer-Verlag,1985.

[68] L. Kucera. Parallel computation and conflicts in memory access. Inf. Process. Lett., 14(2):93–96, 1982.

[69] R. Ladner. The circuit value problem is log-space complete for $\mathcal{P}$. SIGACT News, 7(1):583–590, 1975.

[70] R. Ladner and M. Fischer. Parallel prefix computation. J. ACM, 27(4):831–838, 1980.

[71] C. E. Leiserson. Area-efficient graph layouts (for VLSI). In Proceedings of the 21st Ann. IEEE Symposium on Foundations of Computer Science, pages 270–281, 1980.

[72] G. Lev and N. Pippenger and L. G. Valiant. A fast parallel algorithm for routing in permutation networks. IEEE Trans. Comput., C-30(2):93–100, 1981.

[73] L. Lovász. Computing ears and branchings in parallel. In Proceedings of the 26th Ann. IEEE Symposium on Foundations of Computer Science (Portland, OR), pages 464–467, 1985.

[74] M. Luby. A simple parallel algorithm for the maximal independent set problem. SIAM J. Comput., 5(4):1036–1042, 1986.

[75] Y. Maon and B. Schieber and U. Vishkin. Parallel ear decomposition search (EDS) and st-numbering in graphs. Theor. Comput. Sci., 47(3):277–298, 1986.

[76] C. A. Mead and L. A. Conway. Introduction to VLSI systems. Reading, Mass.: Addison-Wesley, 1980.

[77] R. Metcalfe and D. Boggs. Ethernet: Distributed packet switching for local computer networks. Commun. ACM, 19:395–404. 1976.

[78] G. L.Miller and V. Ramachandran. Efficient parallel ear decomposition with applications. Manuscript, MSRI, Berkeley, 1986.

[79] G. L. Miller and V. Ramachandran. A new graph triconnectivity algorithm and its parallelization. In Proceedings of the 19th Annual ACM Symposium on Theory of Computing (New York City, May 25–27, 1987), pages 335–344, 1987.

[80] G. Miller and J. Reif. Parallel tree contraction and its application. In Proceedings of the 26th Ann. IEEE Symposium on Foundations of Computer Science (Portland, OR), pages 478–489. 1985.

[81] A. Moitra and R. Johnson. Parallel algorithms for maximum matching and other problems on interval graphs. Technical Report 88-927, Department of Computer Science, Cornell University, 1988.

[82] K. Mulmuley and U. V. Vazirani and V. V. Vazirani. Matching is as easy as matrix inversion. Combinatorica, 7(1):105–120, 1987.

[83] J. Naor. A fast parallel coloring of planar graphs with five colors. Inf. Process. Lett., 25(1):51–53, 1987.

[84] J. Naor and M. Naor and A. A. Schäffer. Fast parallel algorithms for chordal graphs. SIAM J. Comput., 18(2):327–349, 1989.

[85] D. Nath and S. N. Maheshwari. Parallel algorithms for the connected components and minimal spanning tree problems. Inf. Process. Lett., 14(1):7–11, 1982.

[86] G. Pfister. The architecture of the IBM research parallel processor prototype (RP3). Technical Report RC 11210 Computer Science, IBM Yorktown Heights, 1985.

[87] N. Pippenger. On simultaneous resource bounds. In Proceedings of the 20th Ann. IEEE Symposium on Foundations of Computer Science (San Juan, PR), pages 307–311, 1979.

[88] V. Pratt and L. Stockmeyer. A characterization of the power of vector machines. J. Comput. Syst. Sci., 12:198–221, 1976.

[89] F. Preparata and J. Vuillemin. The cube-connected-cycles: A versatile network for parallel computation. In Proceedings of the 20th Ann. IEEE Symposium on Foundations of Computer Science (San Juan, PR), pages 140–147, 1979.

[90] V. Ramachandran. Fast parallel algorithms for reducible flow graphs. In Tewksbury, Stuart K. and Bradley W. Dickinson and Stuart C. Schwartz, editors, Concurrent Computations: Algorithms, Architecture, and Technology, pages 117–138, Plenum Press, 1988.

[91] V. Ramachandran and U. Vishkin. Efficient parallel triconnectivity in logarithmic time (extended abstract). In J. H. Reif, editor, Proceedings of the 3rd Aegean Workshop on Computing: VLSI Algorithms and Architectures, AWOC 88. Corfu, Greece, June/July 1988, pages 33–42. LNCS 319, New York, Berlin, Heidelberg: Springer-Verlag, 1988.

[92] J. H. Reif. Depth-first search is inherently sequential. Inf. Process. Lett., *20*(5):229–234, 1985.

[93] W. L. Ruzzo. Tree-size bounded alternation. J. Comput. Syst. Sci., *21*(2):218–235, 1980.

[94] W. L. Ruzzo. On uniform circuit complexity. J. Comput. Syst. Sci., *22*(3):365–383, 1981.

[95] C. Savage and J. Ja'Ja. Fast efficient parallel algorithms for some graph problems. SIAM J. Comput., *10*(4):682–691, 1981.

[96] B. Schieber and U. Vishkin. On finding lowest common ancestors: Simplification and parallelization (extended abstract). In J. H. Reif, editor, Proceedings of the 3rd Aegean Workshop on Computing: VLSI Algorithms and Architectures, AWOC 88. Corfu, Greece, June/July 1988, pages 111–123. LNCS 319, New York, Berlin, Heidelberg: Springer Verlag, 1988.

[97] J. Schwartz. Ultracomputers. ACM Transactions on Programming Languages and Systems, *2*(4): 484–521, 1980.

[98] C. Seitz. The cosmic cube. CACM, *28*(1):22–33, 1985.

[99] H. J. Siegel. A model of SIMD machines and a comparison of various interconnection networks. IEEE Trans. Comput. *C-28*:907–917, 1979.

[100] J. R. Smith. Parallel algorithms for depth-first searches. I Planar graphs. SIAM J. Comput., *15*(3): 814–830, 1986.

[101] L. J. Stockmeyer and U. Vishkin. Simulation of parallel random access machines by circuits. SIAM J. Comput., *13*(2):409–422, 1984.

[102] R. E. Tarjan. Efficiency of a good but not linear set union algorithm. J. ACM, *22*:215–225, 1975.

[103] R. E. Tarjan and U. Vishkin. An efficient parallel biconnectivity algorithm. SIAM J. Comput., *14*(4):862–874, 1985.

[104] C. D. Thompson. Generalized connection networks for parallel processor interconnection. IEEE Trans. Comput., *C-27*:1119–1125, 1978.

[105] Y. H. Tsin and F. Y. Chin. Efficient parallel algorithms for a class of graph theoretic problems. SIAM J. Comput., *13*(3):580–599, 1984.

[106] V. V. Vazirani. $\mathcal{NC}$ algorithms for computing the number of perfect matchings in $K_{3,3}$-free graphs and related problems. Technical Report, Computer Science Department, Cornell University, 1987.

[107] U. Vishkin. Implementation of simultaneous memory address access in models that forbid it. J. Algorithms, *4*(1):45–50, 1983.

[108] I. Wegener. The complexity of Boolean functions. Stuttgart: B. G. Teubner; New York: Wiley, 1987.

Ernst W. Mayr
Fachbereich Informatik
Johann Wolfgang Goethe-Univ.
D-6000 Frankfurt a. M.
Federal Republic of Germany

Computing Suppl. 7, 93–107 (1990)

Applications of Parallel Scheduling Algorithms to Families of Perfect Graphs*

David Helmbold, Santa Cruz, Calif., and **Ernst W. Mayr**, Frankfurt a. M.

Abstract — Zusammerfassung

Applications of Parallel Scheduling Algorithms to Families of Perfect Graphs. We combine a parallel algorithm for the two processor scheduling problem, which runs in polylog time on a polynomial number of processors, with an algorithm to find transitive orientations of graphs where they exist. Both algorithms together solve the maximum clique problem and the minimum coloring problem for comparability graphs, and the maximum matching problem for co-comparability graphs. The transitive orientation algorithm can also be used to identify permutation graphs, another important subclass of perfect graphs.

AMS Subject Classifications: 68C15, 68E10, 68Q10.

Key words: Two processor scheduling, maximum clique, maximum matching, transitive orientation

Anwendungen parallelen Schedulingalgorithmen in Familien von perfekten Graphen. Wir kombinieren einen parallelen Algorithmus für das Zwei-Prozessor-Scheduling-Problem, der in polylogarithmischer Zeit und mit einer polynomialen Anzahl von Prozessoren läuft, mit einem Algorithmus für die transitive Orientierung von Graphen, falls eine solche existiert. Durch diese Kombination können wir das Clique-Problem und das Färbungsproblem für Vergleichbarkeitsgraphen und das Maximum-Matching-Problem für ihre Komplemente lösen. Der Algorithmus für die transitive Orientierung kann auch dazu benutzt werden, um Permutationsgraphen zu erkennen, eine weitere wichtige Unterklasse der perfekten Graphen.

1. Introduction

We present parallel algorithms for graph problems, in particular for several interesting subclasses of perfect graphs. Our main result is a deterministic $\mathcal{NC}$ algorithm for solving the two processor unit execution time scheduling problem, answering an important open problem posed in [27]. We also present an $\mathcal{NC}$ algorithm for transitively orienting comparability graphs. By combining these two results, we obtain an $\mathcal{NC}$ algorithm for the maximum cardinality matching problem on co-comparability graphs (the complements of comparability graphs) and nearly co-comparability graphs. Known fast parallel algorithms for general graphs rely heavily on randomization [16]. Our transitive orientation algorithm also gives us $\mathcal{NC}$ algorithms for several additional problems, such as identifying permutation graphs

---

* This work was supported in part by a grant from the AT&T Foundation, and NSF grant DCR-8351757.

and finding the maximum weighted clique and optimal colorings in comparability graphs. Comparability, co-comparability, and permutation graphs are all important subclasses of perfect graphs.

The most fundamental scheduling problems involve unit time execution tasks with precedence constraints restricting the order of execution [2]. When the number of processors varies, the scheduling problem is $\mathcal{NP}$-complete [26][20]. At present there are no published polynomial time algorithms for a fixed number of processors greater than two. The first polynomial time algorithm for the two processor case was published in [6]. Faster algorithms for the same problem were obtained by Coffman and Graham [3], and later, Gabow [7, 8] found an asymptotically optimal algorithm. Recently, Vazirani and Vazirani have published a randomized parallel solution [27]. Like Fujii et al. [6] they use the connection between matching and two processor scheduling, so their algorithm relies on an $\mathcal{RNC}$ matching subroutine such as [16] or [21].

In contrast, our scheduling algorithm [14] is deterministic and does not require the aid of a matching subroutine. Therefore we are able to exploit the relationship between matching and two processor scheduling in the other direction, obtaining a deterministic parallel maximum matching algorithm for co-comparability graphs.

The only ingredient required to convert our scheduling algorithm into a matching result is an $\mathcal{NC}$ transitive orientation subroutine. This routine takes an undirected graph and directs the edges so that the resulting digraph is transitively closed. The graphs with transitive orientations are called comparability graphs. The complements of comparability graphs are co-comparability graphs. Kozen, Vazirani and Vazirani, in independent work [18], coupled a different transitive orientation routine with our two processor scheduling algorithm to achieve an $\mathcal{NC}$ matching algorithm on co-comparability graphs. Our transitive orientation subroutine is also the key element in algorithms presented in this paper which test for permutation graphs and find maximum weighted cliques or optimal (minimal) colorings of comparability graphs.

The remainder of the paper is organized as follows. Section 2 discusses some fundamental concepts of parallel computation and states our main results. In Section 3, we present our $\mathcal{NC}$ algorithm for the general two processor unit execution time scheduling problem. Section 4 contains the parallel algorithm for recognizing transitively orientable graphs, and constructing such orientations whenever they exist. Then, in Section 5, we show how to combine the results of the two previous sections to obtain our $\mathcal{NC}$ algorithm for the maximum cardinality matching problem on co-comparability graphs. The final section mentions some conclusions and open problems. A preliminary version of our results appeared in [13].

2. Main Theorems and Applications

In this section we give some definitions, state our main results, and prove several important consequences.

As our model of parallel computation, we use the *Parallel Random Access Machine* or PRAM as defined in [5]. A PRAM consists of an unbounded number of identical processors running synchronously, stepped by a global clock. Each processor can be thought of, for the purpose of this paper, as an ordinary RAM [1], with local memory. A PRAM also contains an unbounded number of global memory cells which every processor can access in one timestep. We allow that several processors read the same memory cell simultaneously. However, several processors must not write simultaneously to the same memory cell, *i.e.* we use the so-called concurrent-read-exclusive-write model. Every processor has stored, in one of its registers, its unique processor index. All processors execute the same program. Since the instructions may depend on the processor index, the effect of an instruction will in general vary from processor to processor.

When measuring the complexity of parallel algorithms (for the PRAM model), we are mainly interested in the amount of time an algorithm uses, and the number of processors it employs. Time will be the number of parallel steps taken by the PRAM, and the number of processors will be the highest index of a processor active during the computation.

The class of parallel algorithms running in time which is bounded by a polynomial in the logarithm of the size of the input, and using a number of processors polynomial in the input size, has experienced considerable interest. One reason is that the algorithms in this class are considered very fast (the "speedup" over their sequential counterparts is exponential), and they use a "reasonable" amount of hardware, *i.e.* processors. Another reason is that this class is very robust under (reasonable) variations in the definitions of the underlying machine model. The class is commonly referred to as $\mathcal{NC}$, owing to its original definition for the boolean circuit model of parallel computation in [23].

A *perfect graph* is an undirected graph where the chromatic number and maximum clique size of every vertex induced subgraph coincide. A *precedence graph* is an acyclic, transitively closed digraph, or equivalently, a partial order. We use (a, b) to denote an undirected edge, and $\langle a, b \rangle$ to denote a directed edge or *arc* from vertex a to vertex b. Thus if arcs $\langle a, b \rangle$ and $\langle b, c \rangle$ are in a precedence graph, then so is the arc $\langle a, c \rangle$. A *comparability graph* is an undirected graph with the property that every edge can be assigned a direction such that the resulting graph is a precedence graph. The complement of a comparability graph is a *co-comparability* graph. Some graphs, such as a simple three-cycle, are both comparability and co-comparability graphs.

The undirected graph $G = (V, E)$ is a *permutation graph* if there exists a pair of permutations on the vertices such the edge $(v, v') \in E$ if and only if v precedes v' (or v' precedes v) in both permutations. Permutation graphs are equivalent to the comparability graphs of partial orders with dimension two. A graph is both a comparability graph and a co-comparability graph if and only if it is a permutation graph [24]. Permutation graphs, comparability graphs and co-comparability graphs are all non-trivial subclasses of perfect graphs [10].

An instance of the two processor scheduling problem is given by a precedence graph $\vec{G} = (V, \vec{E})$. Each vertex represents a task whose execution requires unit time on either of two identical processors. If there is an arc from task t to task t', then task t must be completed before task t' can be started. A schedule is a mapping from tasks to integer timesteps such that at most two tasks are mapped to any timestep and for all tasks t and t' if t must precede t' ($t \prec t'$) then t is mapped to an earlier timestep than t'. The length of a schedule is the number of timesteps used. An optimal schedule is one of shortest length.

The maximum matching problem on co-comparability graphs and the two processor scheduling problem are closely related. If G is a co-comparability graph and $\vec{\bar{G}}$ is a transitive orientation of G's complement, then the pairs of tasks mapped to the same timestep in an optimal two processor schedule of $\vec{\bar{G}}$ correspond to a maximum cardinality matching in G. Furthermore, there is a sequential algorithm for converting any maximum cardinality matching for G into an optimal two processor schedule for $\vec{\bar{G}}$ [6]. In [27] it was conjectured that this process is inherently sequential, but with our two processor scheduling algorithm it can be solved quickly in parallel.

Theorem 1 *Two processor scheduling is in $\mathcal{NC}$.*

Proof: We outline an $O(\log^2 n)$ time algorithm in Section 3. Further details can be found in [14]. □

Theorem 2 *There is an $\mathcal{NC}$ algorithm which detects whether an undirected graph is transitively orientable, and if so finds a transitive orientation.*

Proof: We present such an algorithm in Section 4. See also [18]. □

Corollary 2.1 *There is an $\mathcal{NC}$ algorithm which detects whether or not a graph is a permutation graph.*

Proof: Graph G is a permutation graph if and only if both G and $\bar{G}$ are comparability graphs [24]. Therefore, by running our transitive orientation algorithm on both G and $\bar{G}$, we can determine whether G is a permutation graph. □

Corollary 2.2 *There is an $\mathcal{NC}$ algorithm which finds a maximum node-weighted clique in comparability graphs.*

Proof: Given a comparability graph G, we find a transitive orientation, $\vec{G}$. Examine any k-path in $\vec{G}$. A k-path is a directed path containing exactly k vertices. Because $\vec{G}$ is transitively closed, the nodes on the k-path form a k-clique in G. Similarly, every k-clique in G is a k-path in $\vec{G}$. Thus the problem of finding a maximum node-weighted clique in G reduces to finding a maximum weight path in $\vec{G}$. Since $\vec{G}$ is a DAG, standard parallel techniques (*i.e.*, max-plus closure) can be used to find a heaviest path in $\vec{G}$. □

Corollary 2.3 *There is an $\mathcal{NC}$ algorithm which finds a minimal node-coloring of comparability graphs.*

Proof: Given a comparability graph G, we find a transitive orientation, $\vec{G}$. We say that a vertex v is on level i in $\vec{G}$ if the longest (directed) path from v to a sink contains exactly i vertices. Clearly any pair of nodes on the same level are not adjacent in G, so they can be assigned the same color. Every node on level $i > 1$ is a predecessor of at least one node on level $i - 1$. Therefore, if $\vec{G}$ has k levels then $\vec{G}$ has a path of length k and G has a k-clique. Since no coloring can use fewer colors than the size of the largest clique, using a distinct color for every level yields an optimal coloring. □

Theorem 3 *Finding maximum matchings on co-comparability graphs is in $\mathcal{NC}$.*

Proof: One such algorithm is given in section 5. □

This theorem is extended to nearly co-comparability graphs in section 5.

Corollary 3.1 *Maximum matchings on permutation graphs and partial orders of dimension 2 can be constructed in $\mathcal{NC}$.*

Proof: As stated above, these graphs are co-comparability graphs. ⊔

Corollary 3.2 *Maximum matchings on interval graphs can be found in $\mathcal{NC}$.*

Proof: Interval graphs are a (true) subclass of co-comparability graphs [9]. □

3. Two Processor Scheduling

In this section, we consider the scheduling problem for task systems with arbitrary precedence constraints, unit execution time per task, and two identical processors. Our scheduling algorithm for this problem is built around a routine that, for any precedence graph, computes the length of the graph's optimal schedule(s). This length routine is applied repeatedly in order to actually find an optimal schedule for the input graph.

Let $G = (V, \prec)$ be the precedence graph we are interested in. If $t \prec t'$ then t is a *predecessor* of t' and t' is a *successor* of t. For any pair of tasks, $t, t' \in V$, define $V_{t'}^t$ to be the set of tasks which are both successors of t and predecessors of t', and let $G_{t'}^t$ be the subgraph of G induced by $V_{t'}^t$. The *schedule distance* between tasks t and t', $SD(t, t')$, is defined to be the length of any optimal schedule for $G_{t'}^t$. If $t \nprec t'$ then $SD(t, t') = 0$.

Lemma 3.1 *Let $t, t' \in V$, and let S be a set of tasks such that for all $\hat{t} \in S$:*

  i. $t \prec \hat{t} \prec t'$;
  ii. $SD(t, \hat{t}) \geq k$; *and*
  iii. $SD(\hat{t}, t') \geq l$.

Then $SD(t, t') \geq k + l + \lceil |S|/2 \rceil$.

Proof: Count the number of timesteps required to schedule those tasks between t and t'. There must be at least k timesteps before the first task in S is scheduled. It takes at least $\lceil |S|/2 \rceil$ timesteps to complete the tasks in S. After the last task in S has been completed, at least l additional timesteps are required. Therefore $SD(t, t') \geq k + l + \lceil |S|/2 \rceil$. □

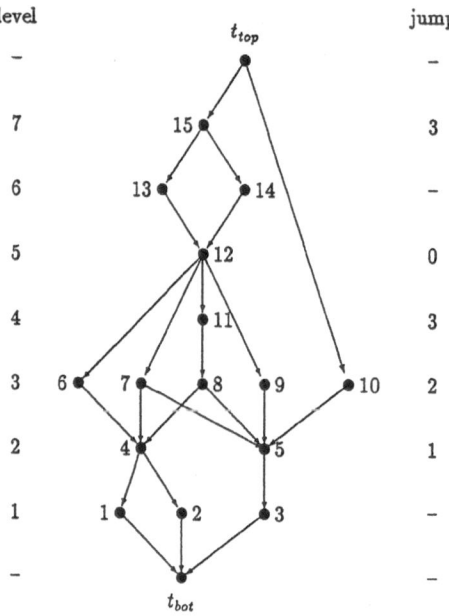

Figure 1. This is a precedence graph containing fifteen tasks (transitive arcs have been omitted). The special tasks t_{top} and t_{bot} are added when computing the length of optimal schedules for G. The levels of the original graph are on the left and the jump sequence is on the right.

The distance algorithm (see Figure 2) uses a doubling method similar to the standard transitive closure routine in order to compute the schedule distances between all pairs of tasks in a precedence graph $G = (V, \prec)$. It initially guesses that the scheduling distance between each pair of tasks is at least zero. By repeatedly applying Lemma 3.1 to each pair of tasks in parallel the algorithm refines its guesses. Below we prove that after $\log|V|$ iterations, the algorithm's guess for each pair of tasks has converged to the schedule distance. The distance algorithm has a straightforward implementation on an n^5 processor PRAM taking $O(\log^2 n)$ time.

Lemma 3.2 *The schedule distance algorithm always computes the schedule distance between every pair of tasks.*

Proof: Lemma 3.1 guarantees that the distances computed by the algorithm are never greater than the schedule distances.

In [3] it is shown how to construct sets of tasks $\chi_0, \chi_1, \ldots, \chi_k$ for any precedence graph such that:

- those tasks in any χ_i are predecessors of all tasks in χ_{i-1}; and
- the length of any optimal schedule for G is $\sum_i \lceil |\chi_i|/2 \rceil$ (See Figure 3).

Our algorithm does not compute the χ_i's directly, we simply use their existence to prove that the distances the algorithm does compute converge to the schedule distance.

$$d_0(*,*) := 0;$$
$$\textbf{for} \quad i := 1 \textbf{ to } \lceil \log n \rceil \textbf{ do}$$
$$\quad\quad \textbf{for} \quad \text{all } t, t' \text{ with } t \prec t' \textbf{ do in parallel}$$
$$\quad\quad\quad\quad \textbf{for} \quad \text{all } 0 \leq k, l < n - 1 \textbf{ do in parallel}$$
$$\quad\quad\quad\quad\quad\quad S_{t,t',k,l} := \{s : t \prec s \prec t', d_{i-1}(t,s) \geq k, d_{i-1}(s,t') \geq l\};$$
$$\quad\quad\quad\quad\quad\quad d_i(t,t') := \max_{S_{t,t',k,l} \neq \varnothing} \{d_{i-1}(t,t'), k + l + \lceil |S_{t,t',k,l}|/2 \rceil\};$$
$$SD(*,*) := d_{\lceil \log n \rceil}(*,*)$$

Figure 2. The Schedule Distance Algorithm.

		χ_5	χ_4		χ_3			χ_2	χ_1	
P_1	t_{top}	15	14	12	11	8	6	4	2	t_{bot}
P_2	–		10	13	–	9	7	5	3	1	–

| time | 1 | 2 | 3 | 4 | 5 | 6 | 7 | 8 | 9 | 10 |

Figure 3. This is an LMJ schedule for the graph in Figure 1; each χ_i is boxed.

Examine how the schedule distance algorithm determines the schedule distance between an arbitrary pair of tasks, t and t'. Let $\chi_1, \chi_2, \ldots, \chi_h$ be a set of χ_i's for $G_{t'}^t$, $\chi_{h+1} = \{t\}$, and $\chi_0 = \{t'\}$. After the first iteration of the outer loop, the distance computed between any task in χ_i and one in χ_{i-2} is at least $\lceil |\chi_{i-1}|/2 \rceil$. After the second iteration, the distance computed between any task in χ_i and any task in χ_{i-4} is at least $\lceil |\chi_{i-1}|/2 \rceil + \lceil |\chi_{i-2}|/2 \rceil + \lceil |\chi_{i-3}|/2 \rceil$. This is an easy consequence of Lemma 3.1 with $S = \chi_{i-2}$, $k = \lceil |\chi_{i-1}|/2 \rceil$, and $l = \lceil |\chi_{i-3}|/2 \rceil$. In each iteration we double the number of χ_i's accounted for. After $\log h$ iterations, the computed distance between t and t' is at least the length of any optimal schedule for $G_{t'}^t$, and thus at least $SD(t, t')$. Also note that the estimates computed by the schedule distance algorithm are never too big as can be seen from an easy induction on i, the index of the outer loop in the algorithm.

Since G contains n tasks, each $G_{t'}^t$ has at most $n - 2$ χ_i's. Therefore, after $\lceil \log n \rceil$ iterations the algorithm has converged to the schedule distances for each pair of tasks. $\square$

The distance algorithm can be used to compute the length of optimal schedules for a graph. Augment the graph with two dummy tasks, t_{top} and t_{bot}, which are a predecessor and successor (respectively) of all other tasks in G. Now $SD(t_{top}, t_{bot})$ is the length of G's optimal schedules, and can be found using the schedule distance algorithm.

The method for converting the schedule distance algorithm into one which finds an optimal schedule involves several constructions. For the sake of brevity this paper contains only an outline of our method. Interested readers may consult [14] for a more detailed presentation.

The search for an optimal schedule can be restricted to the class of *Lexicographically Maximal Jump* (LMJ) schedules. Each task t in the precedence graph is assigned a *level* equal to the number of tasks in a longest path from t to a sink. A *level schedule* gives preference to tasks on higher levels. More precisely, suppose levels $L, \ldots, l + 1$ have already been scheduled and there are k unscheduled tasks remaining on level l. If k is even a level schedule puts the k tasks in pairs, and there is no *jump* from level l. If k is odd, a level schedule pairs $k - 1$ of the tasks with each other and pairs the remaining task with a task from a lower level $l' < l$. In this case, level l *jumps* to level l'. We assume that there is a sufficiently large number of dummy tasks on level 0 which can be paired with any other task. The *jump sequence* of a level schedule is the sequence of levels jumped *to*, listed in the order in which the jumps occur (see Figure 1). The *Lexicographically Maximum Jump* (*LMJ*) sequence is the jump sequence (resulting from some level schedule) that is lexicographically greater than any other jump sequence resulting from a level schedule. An *LMJ schedule* is a level schedule whose jump sequence is the LMJ sequence. Note that our definition of LMJ is similar to the definition of *highest level first* in [7] and [27]. The following theorem establishes the importance of LMJ schedules.

Theorem 4 [7] *Every LMJ schedule is optimal.* □

Our two processor algorithm uses the schedule distance algorithm to find the LMJ sequence and which jump (if any) a pair of tasks can be used for. In general, there will be many possible pairs for each jump. A path doubling computation finds a consistent set of task pairs for the jumps. The remaining tasks are paired up within levels. Since there are never precedence constraints between any two tasks on the same level, this pairing can be done arbitrarily. An LMJ schedule is obtained by sorting the resulting set of task pairs (both for jumps and within levels). We refer the reader to [14] for a complete description of the technically more involved parts of this construction.

4. Transitive Orientation

The transitive orientation problem is nontrivial because some edges cannot be oriented independently. If the edges (a, b) and (b, c) are in the graph to be oriented, but the edge (a, c) is not, then the edges $(a, b), (b, c)$ cannot be oriented independently. If we choose the arc $\langle a, b \rangle$ then we are forced to include the arc $\langle c, b \rangle$ in the transitive orientation (see Figure 4). The binary relation r reflects this simple kind of forcing [24]. Given $G = (V, E)$, we say that $\langle a, b \rangle r \langle a, c \rangle$ and $\langle b, a \rangle r \langle c, a \rangle$ whenever $(a, b) \in E$, $(a, c) \in E$ and $(b, c) \notin E$.

The reflexive, transitive closure r^* of r is an equivalence relation on the possible orientations of edges in E. For obvious reasons, we call these equivalence classes *implication classes*. If $\vec{A}$ is a set of arcs (*e.g.* an implication class) then A denotes the set of undirected edges $\{(a, b): \langle a, b \rangle \in \vec{A} \lor \langle b, a \rangle \in \vec{A}\}$, and $\vec{A}^{-1}$ is the set of arcs $\{\langle b, a \rangle: \langle a, b \rangle \in \vec{A}\}$. A set of arcs $\vec{A}$ is *consistent* if $\vec{A} \cap \vec{A}^{-1} = \varnothing$, and is *inconsistent* when $\vec{A} \cap \vec{A}^{-1} \neq \varnothing$.

Implication classes have been studied by M. C. Golumbic and many of the lemmas in this section have originally been shown in [10] or [11].

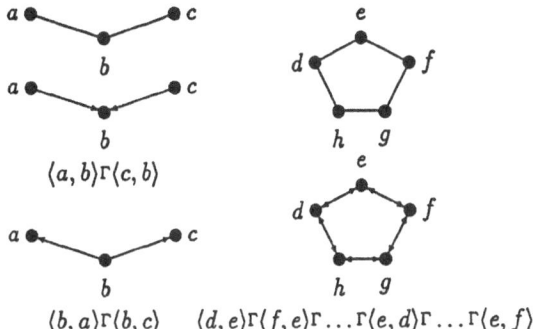

Figure 4. Graphs and Implication Classes

Lemma 4.1 *If $\vec{A} \neq \vec{B}$ are implication classes of G then either $\vec{A} = \vec{B}^{-1}$ or $A \cap B = \varnothing$.*

Proof: Assume that $(a, b) \in A \cap B$. Without loss of generality, let $\langle a, b \rangle \in \vec{A}$. If $\langle a, b \rangle \in \vec{B}$ then $\vec{B} = \vec{A}$ since implication classes are equivalence classes. Therefore $\langle b, a \rangle \in \vec{B}$, and $\langle b, a \rangle \notin \vec{A}$. By definition, if $\langle a, b \rangle \Gamma \langle a', b' \rangle$ then $\langle b, a \rangle \Gamma \langle b', a' \rangle$. Thus some $\langle c, d \rangle \Gamma^* \langle a, b \rangle$ if and only if $\langle d, c \rangle \Gamma^* \langle b, a \rangle$, so $\vec{A} = \vec{B}^{-1}$. □

Given an undirected graph $G_1 = (V, E)$ pick any implication class $\vec{B}_1$, delete B_1 from G_1, forming $G_2 = (V, E - B_1)$. Next form G_3 by removing the underlying set B_2 of some implication class $\vec{B}_2$ of G_2. Continue the process until removing B_k from G_k results in a graph with no edges. The sequence of implication classes removed, $\vec{B}_1$, $\vec{B}_2, \ldots, \vec{B}_k$, is called a *$\Gamma$-decomposition* of G. The following theorem points out the usefulness of Γ-decompositions.

Theorem 5 (TRO Theorem [10]) *Let $\vec{B}_1, \vec{B}_2, \ldots, \vec{B}_k$ be a Γ-decomposition of an undirected graph G. The following statements are equivalent:*

  i. *G is a comparability graph.*
 ii. *Every implication class of G is consistent.*
iii. *Each $\vec{B}_i$ in the Γ-decomposition is consistent.*

Furthermore, when these conditions hold, $\vec{B}_1 \cup \vec{B}_2 \cup \cdots \cup \vec{B}_k$ is a transitive orientation of G.

Proof: The proof of this theorem requires several technical lemmas, and thus is beyond the scope of this paper. The interested reader is referred to [10, 11]. □

Let $\vec{A}$ be any implication class of the graph G. Then we call the underlying set of edges A its *color class*. The TRO theorem suggests a sequential algorithm for finding transitive orientations of comparability graphs. One can take any edge, orient it arbitrarily, find the associated implication class, add the implication class to the transitive orientation and remove its color class from the comparability graph. Repeating this procedure yields a Γ-decomposition of the comparability graph and therefore a transitive orientation. This is essentially the algorithm in [24].

If we are dealing with a comparability graph it is sufficient to consider color classes instead of implication classes, since every color class A represents an implication class $\vec{A}$ and its inverse $\vec{A}^{-1}$. When talking about color classes we always assume that the corresponding implication classes are consistent.

In order to parallelize the sequential algorithm above it is necessary to understand how color classes change during a Γ-decomposition. We will see below that the changes are very simple: color classes are either merged with other color classes or remain unchanged.

Lemma 4.2 *Let B be a color class of $G = (V, E)$. Every implication class of $G' = (V, E - B)$ is the union of color classes of G.*

Proof: The Γ relation for G', restricted to $E - B$, contains the corresponding restriction of the Γ relation for G. $\square$

The three edges of a triangle in the undirected graph G form a *tricolored* triangle if they belong to three distinct color classes. We say that two color classes A and B are *triangle related*, written $A \vartriangle B$, if there is a tricolored triangle in G with one edge in A and another edge in B.

Lemma 4.3 *Let A and B be two distinct color classes in $G = (V, E)$. A is not an implication class of $G' = (V, E - B)$ iff $A \vartriangle B$.*

Proof: The proof is a simple consequence of the definition of the Γ relation. It will be omitted here. $\square$

An immediate implication of Lemma 4.3 is

Lemma 4.4 *Let the color classes $B_1, \ldots, B_k$ of $G = (V, E)$ be an independent set under the $\vartriangle$ relation. Then in $G' = (V, E - B_1)$, the collection $\{B_2, \ldots, B_k\}$ is an independent set under $\vartriangle$.*

Corollary 4.4.1 *If color classes $B_1, \ldots, B_k$ of G form an independent set under the $\vartriangle$ relation, then they are the first k color classes for some Γ-decomposition of G.*

Proof: Follows from the definition of independent set. $\square$

Lemma 4.5 *Let $B_1, \ldots, B_k$ be a maximal independent set under the $\vartriangle$ relation for some graph $G_1 = (V, E)$. Every color class of $G_{k+1} = (V, E - B_1 - B_2 - \cdots - B_k)$ is the union of at least two color classes of G_1.*

Corollary 4.5.1 *The number of color classes for G_{k+1} is at most half the number of color classes for G.*

Proof: Since the B_i form a maximal independent set under $\vartriangle$ every color class of G which is not one of the B_i must be adjacent to one of the B_i. Because of Lemma 4.3 it will be merged with some other color class. $\square$

The input to our algorithm is an undirected graph $G = (V, E)$. The ouptut is either $\vec{G}$, a transitive orientation of G, or an indication that G has no transitive orientation. With G_1 initialized to be G, and $\vec{G}$ initially equal to $(V, \varnothing)$, if no inconsistent implication class is found in the first iteration, the algorithm proceeds in iterations as long as the set of color classes is non-empty.

Each iteration consists of the following four steps:

1. Determine the color classes of G_i. This can be done using standard parallel techniques such as solving 2-SAT formulae or finding connected components [25].
2. Determine the Δ relation on color classes.
3. Use a maximal independent set subroutine [17, 19] to obtain a maximal independent set M of color classes.
4. In parallel, for each B_j in M, delete B_j from G_i, and add $\vec{B}_j$ or $\vec{B}_j^{-1}$ to $\vec{G}$.

Step 3 is the most expensive of these steps, requiring $O(\log^2 n)$ time and n^4 processors. The $\log n$ iterations can therefore be done in $O(\log^3 n)$ time on n^4 processors.

5. Maximum Matching

The two processor scheduling and transitive orientation algorithms can be used to find maximum matchings on co-comparability graphs. To find a maximum matching on the co-comparability graph $G = (V, E)$, first create the comparability graph $\bar{G}$, the complement of G. Applying the transitive orientation routine converts $\bar{G}$ into a precedence graph. An optimal two processor schedule can be found for the precedence graph using our scheduling algorithm. We will see below that the pairs of tasks scheduled together form a maximum cardinality matching of G.

Let S be any optimal two processor schedule for $\vec{\bar{G}}$. A *task-pair* of S is a pair of tasks mapped to the same timestep by S. Since there are no precedence relationships between tasks in a task-pair, the set of task-pairs of S forms a matching in G. Because S is an optimal schedule, no schedule has more task-pairs.

A task is *available* at some time step in a schedule if it could be executed in the next step without violating the precedence constraints.

Lemma 5.1 *If a co-comparability graph G has a perfect matching then $\vec{\bar{G}}$ has a schedule where every task is in a task-pair.*

Proof: We say a pair of tasks is *mated* if the pair is in the perfect matching. Construct a schedule (and modify the "mated" relationship) iteratively as follows:

> If two mated tasks are both available, schedule one such mated pair. Otherwise find two mated pairs, (t, t') and (s, s'), such that t and s are available and there is no precedence relationship between t' and s'. Schedule t with s and mate t' with s'.

Note that there are never precedence constraints between a pair of mated tasks. This method clearly takes two tasks each timestep and does not violate the precedence constraints. What we want to show is that it always constructs an optimal schedule for $\vec{\bar{G}}$. For this it suffices to prove that every time step contains two tasks.

Assume to the contrary that at some point the above routine does not find a pair of tasks to schedule. Let U be the set of available tasks and U' be the set of tasks which are mated to tasks in U. Since the method fails, $U \cap U' = \varnothing$ and, by assumption, there is a precedence relationship between every pair of tasks in U' (i.e. U' is totally ordered). Let t' be the task in U' which precedes all other tasks in U'. Since $t' \notin U$, there must be some $t \in U$ such that $t \prec t'$. However, by the transitivity of precedence, t also precedes its mate—contradiction. $\square$

Lemma 5.2 Let $\vec{G} = (V, \prec)$ be a precedence graph and S a two processor schedule for $\vec{G'} = (V - \{t\}, \prec)$. A single timestep containing t can be inserted into S yielding a schedule for $\vec{G}$.

Proof: Let t' be the last predecessor of t in S. Insert task t immediately after the timestep containing t'. Obviously there are no precedence conflicts between t and its predecessors. Since S is a valid schedule, there are no precedence conflicts between tasks in $V - \{t\}$. Therefore any precedence conflict would be of the form $t \prec \hat{t}$. By transitivity t' also precedes $\hat{t}$, so $\hat{t}$ comes strictly after t' in S. Since t is inserted in the step immediately after t', task t appears before $\hat{t}$ in the modified schedule. $\square$

Let M be the tasks in a maximum matching on G. The above Lemmas suggest a way to obtain a schedule, S, for $\vec{G} = (V, \prec)$ where the paired tasks of S are precisely the tasks in M. Start by finding an optimal schedule, S' for the subgraph of $\vec{G}$ induced by M and add the tasks in $V - M$ one at a time. One $\mathcal{NC}$ implementation of this algorithm involves bucket sorting the tasks in $V - M$ based on which task-pair of S' they follow. By topologically sorting the tasks within each bucket we can quickly determine where each task should be inserted.

Theorem 6 The task-pairs of any optimal schedule for $\vec{G}$ form a maximum cardinality matching on G.

Proof: Let M be the tasks in some maximum cardinality matching of G. Let S be an optimal schedule for the subgraph of $\vec{G}$ induced by M. By Lemma 5.1, the task-pairs of S form a maximum cardinality matching on G. By Lemma 5.2 we can insert the other tasks of $\vec{G}$ one at a time without disturbing the task-pairs. Therefore, the task-pairs of the resulting schedule for $\vec{G}$ form a maximum matching on G. Since every optimal schedule has the same number of task-pairs and the task-pairs of every schedule form a matching, the task-pairs of any optimal schedule for $\vec{G}$ form a maximum matching on G. $\square$

If $\bar{G}$ is not transitively orientable it may still be possible to find a maximum matching in $G = (V, E)$. Assume we are given a set U, consisting of $O(\log n)$ edges, such that $\bar{G} \cup U$ is transitively orientable. The following method finds a maximum matching in G.

For each $S' \subseteq U$ such that S' is a matching find (in parallel) a maximum matching in $G' = (V - \{v: (v, v') \in S'\}, E - U)$. Since G' is a vertex induced subgraph of the graph $G'' = (V, E - U)$, $\bar{G'}$ is transitively orientable since $\bar{G''}$ is. A maximum cardinality matching for G occurs whenever the cardinality of the maximum matching for G' plus $|S'|$ is maximal.

A graph G is a k-nearly comparability graph when:

—G has at most $k \log n$ inconsistent implication classes and
—each inconsistent implication class of G is split into consistent implication classes by the addition of at most k edges.

A k-nearly co-comparability graph is the complement of a k-nearly comparability graph.

Theorem 7 *Let G be a k-nearly co-comparability graph, for some constant k. Then there is an $\mathcal{NC}$ algorithm to find a maximum cardinality matching for G.*

Proof: In parallel examine each set, T, of at most k edges not in $\bar{G}$. Determine which inconsistent implication classes are split when T is added to $\bar{G}$. For each inconsistent implication class $\vec{A}$, pick any set of at most k edges which splits $\vec{A}$ into consistent implication classes. At most $k^2 \log n$ edges are picked, so the method described above can now be used to find a maximum cardinality matching for G. □

6. Conclusion

Although the algebraic approach was used to obtain the first parallel matching algorithms [16, 21], these are randomized algorithms. It is interesting to note that we can obtain deterministic matching algorithms for some wide classes of graphs using a purely combinatorial approach. We may speculate whether the combinatorial approach will yield deterministic algorithms for matching on other classes of graphs as well.

With regard to the two processor scheduling algorithm it was surprising to us how much more difficult computing the actual schedule was than simply computing its length (the details are given in [14]). In higher complexity classes such as $\mathcal{P}$ and $\mathcal{NP}$ it is often easy to go from the decision problem to computing an actual solution, because of self-reducibility. However this does not necessarily seem to be the case for parallel complexity classes. To support this observation we note that the random $\mathcal{NC}$ algorithm for finding the cardinality of a maximum matching is much simpler than the random $\mathcal{NC}$ algorithm for determining an actual maximum cardinality matching [15].
There are several open problems related to parallel scheduling algorithms. We are attempting to extend our two processor result to the case when the tasks have nonuniform start times and/or deadlines. When the precedence constraints are restricted to in-trees or out-trees there are parallel algorithms for generating schedules on an arbitrary number of processors [4, 12]. It is an open problem whether interval-ordered tasks [22] can be scheduled quickly in parallel.

One variant of the two processor problem that we know to be $\mathcal{NP}$-complete (by reduction from the clique problem) allows incompatibility edges as well as precedence constraints. When there is an incompatibility constraint between two tasks they can be executed in either order, but not concurrently. Incompatibility constraints arise naturally when two or more tasks need the same resource, such as special purpose hardware or a database file. An interesting question is to find restricted versions with feasible (parallel) solutions.

References

[1] A. Aho, J. Hopcroft, and J. Ullman. The Design and Analysis of Computer Algorithms. Addison-Wesley, New York, 1974.

[2] E. Coffman, editor. Computer and Job/Shop Scheduling Theory. Wiley, 1976.

[3] E. Coffman, Jr., and R. Graham. Optimal scheduling for two processor systems. Acta Informatica, 1 : 200–213, 1972.

[4] D. Dolev, E. Upfal, and M. Warmuth. Scheduling trees in parallel. In Bertolazzi, P., Luccio, F. (eds.): VLSI: Algorithms and Architectures. Proceedings of the International Workshop on Parallel Computing and VLSI, pages 1–30, North-Holland, 1985.

[5] S. Fortune and J. Wyllie. Parallelism in random access machines. In Proceedings of the 10th Ann. ACM Symp. on Theory of Computing (San Diego, CA), pages 114–118, 1978.

[6] M. Fujii, T. Kasami, and K. Ninamiya. Optimal sequencing of two equivalent processors. SIAM J. Appl. Math., 17(4) : 784–789, 1969.

[7] H. Gabow. An almost-linear algorithm for two-processor scheduling. JACM, 29(3) : 766–780, 1982.

[8] H. Gabow and R. Tarjan. A linear time algorithm for a special case of disjoint set union. JCSS, 30 : 209–221, 1985.

[9] P. Gilmore and A. Hoffman. A characterization of comparability graphs and of interval graphs. Canad. J. Math, 16, 1964.

[10] M. Golumbic. Algorithmic Graph Theory and Perfect Graphs. Academic Press, New York, 1980.

[11] M. Golumbic. Comparability graphs and a new matroid. J. Combinatorial Theory (B), 22(1) : 68–90, 1977.

[12] D. Helmbold and E. Mayr. Fast scheduling algorithms on parallel computers. In Preparata, F. P. (ed.): Advances in Computing Research 4: Parallel and Distributed Computing, pages 39–68, JAI Press, 1987.

[13] D. Helmbold and E. Mayr. Perfect graphs and parallel algorithms. In Proceedings of the IEEE 1986 International Conference on Parallel Processing, pages 853–860, August 1986.

[14] D. Helmbold and E. Mayr. Two processor scheduling is in $\mathcal{NC}$. SIAM J. on Comput., 16 : 747–759, 1987.

[15] R. Karp, E. Upfal, and A. Wigderson. Are search and decision problems computationally equivalent? In Proceedings of the 17th Ann. ACM Symp. on Theory of Computing (Providence, RI), pages 465–475, 1985.

[16] R. Karp, E. Upfal, and A. Wigderson. Constructing a perfect matching is in random $\mathcal{NC}$. Combinatorica, 6 : 35–48, 1986.

[17] R. Karp and A. Wigderson. A fast parallel algorithm for the maximal independent set problem. J. ACM, 32(4) : 762–773, 1985.

[18] D. Kozen, U. Vazirani, and V. Vazirani. $\mathcal{NC}$ algorithms for comparability graphs, interval graphs, and testing for unique perfect matching. In 5th Conf. Found. of Software Tech. and Theor. Comp. Sci. (New Dehli), 1985.

[19] M. Luby. A simple parallel algorithm for the maximal independent set problem. SIAM J. Comput., 15 : 1036–1042, 1986.

[20] E. Mayr. Well Structured Programs Are Not Easier to Schedule. Technical Report STAN-CS-81-880, Department of Computer Science, Stanford University, September 1981.

[21] K. Mulmuley, U. Vazirani, and V. Vazirani. Matching is as easy as matrix inversion. Combinatorica, 7 : 105–120, 1987.

[22] C. Papadimitriou and M. Yannakakis. Scheduling interval-ordered tasks. SIAM J. Computing, 8(3) : 405–409, 1979.

[23] N. Pippenger. On simultaneous resource bounds. In Proceedings of the 20th IEEE Symp. on Foundations of Computer Science, pages 307–311, 1979.

[24] A. Pnueli, A. Lempel, and S. Even. Transitive orientation of graphs and identification of permutation graphs. Can. J. Math., 23(1) : 160–175, 1971.

[25] Y. Shiloach and U. Vishkin. An $O(\log n)$ parallel connectivity algorithm. *J.* Algorithms, *3*(1) : 57–63, 1982.
[26] J. Ullman. $\mathcal{NP}$-complete scheduling problems. JCSS *10*(3) : 384–393, 1975.
[27] U. Vazirani and V. Vazirani. The two-processor scheduling problem is in $\mathcal{RNC}$. In Proceedings of the 17th Ann. ACM Symp. on Theory of Computing (Providence, RI), pages 11–21, 1985.

David Helmbold
University of California
Santa Cruz, California
U.S.A.

Ernst W. Mayr
Fachbereich Informatik
Johann Wolfgang Goethe Universität
D-6000 Frankfurt a. M.
Federal Republic of Germany

Computing Suppl. 7, 109–124 (1990)

Orders and Graphs

Ulrich Faigle** and Rainer Schrader*, Bonn

Abstract — Zusammerfassung

Orders and Graphs. This paper surveys the relationship between graphtheoretic and ordertheoretic questions. In the first part, we discuss recent results which answer ordertheoretic questions in a more general graphtheoretic framework. In the second part we address ordertheoretic approaches to graph-theoretic problems.

AMS Subject Classifications: 05C20, 05C25, 06A10.

Key words: partial orders, graphs, order invariants, monotone graph properties, polyhedral combinatorics, antimatroids.

Ordnungen und Graphen. Die Arbeit gibt einen Überblick über die Beziehungen zwischen ordnungs- und graphentheoretischen Fragestellungen. Im ersten Teil werden neuere Resultate vorgestellt, die ordnungs-theoretische Fragestellungen in einem allgemeineren graphentheoretischen Rahmen beantworten. Im zweiten Teil werden umgekehrt Probleme auf Graphen diskutiert, die mit ordnungstheoretischen An-sätzen erfolgreich gelöst werden können.

Introduction

Orders and graphs may be viewed as two faces of the same coin. While undirected graphs may be analyzed within the context of directed graphs (replace each edge by a pair of oppositely directed arcs), also (partially) ordered sets fit into this framework in the sense that they correspond to transitively oriented graphs. On the other hand, we may associate with each directed graph D an ordered set $P(D)$ in a standard way: we replace each vertex by two vertices v, v' and impose an order on the augmented set of vertices by letting $v < w'$ if (v, w) is an arc in D. Thus, D may be studied in terms of $P(D)$.

Often, however, ordertheoretic aspects of graphtheoretic problems come up in a more direct way. Similarly, many ordertheoretic problems allow a more general approach within an appropriate graphtheoretic formulation.

This survey addresses such relationships between graphtheoretic and ordertheoretic problems. It is not intended as an exhaustive and comprehensive treatment of the

* Institut für Operations Research, Universität Bonn. Supported by Sonderforschungsbereich 303 (DFG)
** Faculty of Applied Mathematics, University of Twente

subject. Generally, our approach is more ordertheoretic. It concentrates on recent developments and reflects to large extent also the research interests of the authors.

The survey consists of two sections. The first section emphasizes graphtheoretic techniques for ordertheoretic problems while in the second section ordertheoretic techniques are in the foreground.

1. Graphtheoretic Approaches to Ordertheoretic Problems

Throughout this paper we are concerned with structures defined on finite ground sets. Recall that the pair $P = (E, \leq)$ is a (partially) ordered set if for all $x, y, z \in E$

(1.1) $x \leq x$
(1.2) $x \leq y, y \leq x$ implies $x = y$
(1.3) $x \leq y, y \leq z$ implies $x \leq z$.

Note that $P = (E, \leq)$ can also be interpreted as a directed graph $D = D(P)$ defined on the set E of vertices and with edges of the form (x, y) whenever $x \leq y$. The **comparability graph** $G = G(P)$ is the undirected version of $D(P)$, i.e. has vertex set E and edges $\{x, y\}$ whenever $x \leq y$. (We remark that sometimes $D(P)$ and $G(P)$ are defined without loops, i.e., relative to the strict order relation.)

We say that y covers x if $x < y$ and there is no $z \in E$ with $x < z < y$. Hence the complete information about P is contained in the **Hasse diagram** $H = H(P)$, namely the subgraph of $D(P)$ retaining just the covering edges. (Usually $H(P)$ is drawn as an undirected graph with the understanding that the orientation is from "bottom" to "top".) Comparability graphs of ordered sets are, in particular, perfect. Let us quickly review some basic properties of perfect graphs.

In an arbitrary graph G on the vertex set V we consider two types of subsets of vertices. $C \subseteq V$ is a **clique** if there is an edge between every pair of vertices in C. $S \subseteq V$ is a **stable set** if no pair is linked by an edge. The **clique covering number** $\kappa(G)$ is the smallest number of cliques needed to cover V. The **stability number** $\alpha(G)$ is the size of the largest stable set of G. G is called **perfect** if for each vertex induced subgraph G' of G, the following equality is true

$$\alpha(G') = \kappa(G').$$

Lovász [1972] proved the **weak perfect graph conjecture:** G is perfect if and only if its complement $\bar{G}$ is perfect. There is also a **strong perfect graph conjecture**, which has been found to be true for many classes of graphs (for example planar graphs (Tucker [1973]) or claw-free graphs (Parthasarathy and Ravindra [1976])) but generally is open: the graph G is perfect if and only if neither G nor $\bar{G}$ contain an induced odd cycle of size at least five.

Computing $\alpha(G)$ and $\kappa(G)$ is generally NP-complete (cf. Garey and Johnson [1979]). In the case of perfect graphs, however, polynomial algorithms are available via the ellipsoid method (see Grötschel, Lovász, Schrijver [1988]). Let us take a look at comparability graphs. A clique in a comparability graph $G(P)$ corresponds to a

chain in P, i.e. a set of pairwise comparable elements, while a stable set of $G(P)$ is an **antichain** in P. Dilworth [1950] proved for ordered sets P:

$$\alpha(G(P)) = \kappa(G(P)).$$

Since vertex induced subgraphs of comparability graphs are comparability graphs, Dilworth's theorem implies that comparability graphs are perfect. Hence the result of Lovász [1972] yields a special case of Greene's theorem [1976]: the size of the longest chain of an order P equals the smallest number of antichains needed to cover P.

Fulkerson [1956] reduced Dilworth's theorem to König's matching theorem: the maximal number of pairwise disjoint edges in a bipartite graph G equals the minimum number of vertices needed to cover all edges of G. (Here an edge cover means a set of vertices which contains from each edge at least one endpoint). This reduction allows to determine a maximal antichain efficiently using matching algorithms (see Lovász and Plummer [1986]) or network flow techniques (cf. Lawler [1976]).

Greene and Kleitman have achieved a substantial generalization of Dilworth's result. A k-**antichain** in an ordered set P is a subset A such that $|A \cap C| \leq k$ for every chain C. (Thus, the 1-antichains are exactly the antichains in Dilworth's theorem.) The k-**weight** of a chain C is

$$w(C) = \min\{k, |C|\}.$$

Greene and Kleitman proved that the maximum size of a k-antichain equals the minimum k-weight of a chain covering.

Also in the situation considered by Greene and Kleitman it is possible to efficiently determine the numerical quantities involved by formulating the problem as a weighted matching problem (Hoffman [1982]) or min-cost flow problem (Frank [1980]). Standard duality results for network flows then yield the equality statement in the Greene–Kleitman theorem.

Frank's [1980] graphtheoretic setting yields a constructive approach for the general form of Greene's [1976] theorem. One is interested in the size of a largest subset of P containing no antichain of size $k + 1$ (or equivalently, by Dilworth's theorem, being coverable by at most k chains). Weighting antichains A with $w(A) = \min\{k, |A|\}$, Greene's theorem says that the largest size achievable equals the minimum k-weight of an antichain cover of P.

An interesting point, however, should be raised. Although the constructive approaches to the theorems of Greene and Greene–Kleitman are graphtheoretic and involve only comparability graphs, the analogous statements are false for general perfect graphs.

Greene's theorem has a direct application for a machine scheduling problem within the context of so-called loss systems. n jobs arrive at known points in time $t_1, \ldots,$ t_n. k identical machines are available, each requiring processing time p_i for job i. Assuming that a job i is lost if it is not processed at time t_i, the problem consists in processing as many jobs as possible. Note that the set of jobs carries an **interval order** P in a natural way:

$$i < j \text{ iff } t_i + p_i < t_j \quad \text{for all } i, j.$$

Clearly, the problem now is equivalent to finding a largest subset of P that can be covered by at most k chains of P.

Nawijn [1989] extends this model to the case where job i may have different processing times on different machines. This yields the following ordertheoretic problem: given orders $P_1, \ldots, P_k$ on the same ground set, find subsets $C_1, \ldots, C_k$ such that C_i is a chain in P_i and $|C_1 \cup \cdots \cup C_k|$ is as large as possible.

Assuming that the orders $P_1, \ldots, P_k$ are compatible in the sense that $x < y$ in P_i and $y < x$ in P_j cannot occur simultaneously, Nawijn's solution associates with the problem an acyclic directed graph in which an optimal solution corresponds to a longest path. Since this graph has n^k vertices the problem can be solved in polynomial time for fixed k.

A different type of k-machine scheduling problem is notorious (see, e.g., Poguntke [1986]). n jobs with unit processing times are ordered by precedence constraints P and have to be processed on at most k machines so that job i cannot be processed while one of its predecessors is still unfinished. The aim is a feasible schedule with the last job finishing as early as possible.

While an efficient solution is not known for $k = 3$, the first polynomial algorithm for $k = 2$ by Fujii et al. [1969] was based on graphtheoretic concepts. Noting that each feasible schedule corresponds to a matching in the complement $\bar{G}(P)$ of the comparability graph $G(P)$, they show that, in fact, every maximal matching can be arranged into an optimal schedule.

Observe that two non-isomorphic ordered sets may have the same comparability graph. We call a function f defined on ordered sets a **comparability invariant** if

$$f(P) = f(Q) \quad \text{whenever} \quad G(P) = G(Q).$$

Obviously, the width and the size of a longest chain only depend on the comparability graph. The aforementioned result by Fujii et al. in particular shows that the optimal value for the 2-machine scheduling is also a comparability invariant.

A further, nontrivial example of a comparability invariant was independently described by Gysin [1976] and Trotter et al. [1976]: The **intersection** of two ordered sets P and Q defined on the same ground set is the ordered set $P \cap Q$ with $v_i < v_j$ in $P \cap Q$ if and only if $v_i < v_j$ both in P and in Q. A **linear extension** of P is a linear order $v_1, \ldots, v_n$ of the vertices such that $v_i < v_j$ in P implies $i < j$. The (linear) **order dimension** $dimP$ of P now is the minimum number of linear extensions whose intersection is P. Equivalently, $dimP$ may be thought of as the smallest number k such that P can be embedded in $\mathbb{R}^k$ with the componentwise ordering.

Computing the dimension of an ordered set is NP-hard in general. More precisely, testing whether $dimP = k$ is NP-complete for any fixed $k \geq 3$. The case $k \leq 2$ is well-solved: The 2-dimensional orders (a.k.a **permutation orders**) are the orders P for which both the comparability graph $G(P)$ and its complement $\bar{G}(P)$ are transitively orientable. So checking if $dim P = 2$ can easily be done by applying a transitive orientation routine. For more details on the order dimension we refer the interested reader to the survey article by Kelly and Trotter [1982].

Faigle and Schrader [1986] derive a proof that the order dimension is a comparability invariant by constructing a canonical bijection between the sets of linear extensions of two orders with the same comparability graph. In addition, this bijection preserves the setups of a linear extension, i.e. pairs (v_i, v_{i+1}) in the linear order $L = v_1 \ldots v_n$ such that $v_i \not\leq v_{i+1}$ in P. This approach unifies and extends earlier results that the number of linear extensions and the **setup number** (the minimum number of setups in a linear extension) are comparability invariants (see also Habib [1984] and Faigle and Schrader [1985] for more comparability invariants).

The notion of order dimension can be extended in several ways. Instead of viewing P as the intersection of linear orders, we may allow more general classes of orders to form the intersection. One such class, which has been investigated in the literature, is the class of **interval orders**. Recall that P is an interval order if the elements of P can be represented by closed intervals $I_1, \ldots, I_n$ on the real line with the ordering

$$I_j < I_k \text{ if } I_j \text{ is completely to the left of } I_k.$$

(Equivalently, P is an interval order if and only if its cocomparability graph $\bar{G}(P)$ is chordal, see Section 2). The **interval dimension** $idim P$ of P is the minimum number of interval orders whose intersection is P.

The interval dimension also turns out to be a comparability invariant (Habib et al. [1988]). As is the case for linear orders, testing whether $idim P \leq 3$ is NP-complete (Yannakakis [1982]). Orders with interval dimension at most two can be recognized in polynomial time (cf. Habib and Möhring [1988]). The recognition algorithm is based on the following equivalent characterization of ordered sets P with $idim P \leq 2$. Given two parallel lines and a set $T_1, \ldots, T_n$ of trapezoids with vertices on the two lines. In the **trapezoid graph** we associate with every vertex v_i a trapezoid T_i and introduce an edge between v_i and v_j if $T_i \cap T_j \neq \emptyset$. An order P has interval dimension at most two if and only if its cocomparability graph is a trapezoid graph.

The concept of order dimension and interval dimension has a natural extension to directed graphs. Since interval orders are complements of chordal graphs, they are characterized by the fact that the successor sets (and similarly the predecessor sets) of the elements are linearly ordered by inclusion. A directed graph with this property is called a **Ferrers digraph**.

Observe that the complete symmetric directed graph with one arc left out is a Ferrers digraph. Hence we may obtain any directed graph as a suitable intersection of Ferrers digraphs. We can therefore speak of the **Ferrers dimension** $fdim D$ of a directed graph D as the minimum number of Ferrers digraphs whose intersection is D.

The Ferrers dimension has an interesting relation to the order dimension. Cogis [1982] proved that if $D = D(P)$ is the directed comparability graph of an ordered set P, then the Ferrers digraphs whose intersection is D may be taken to be reflexive, antisymmetric, transitive and complete. So $fdimD(P) = dimP$. In particular, testing if $fdim \leq 3$ is NP-complete, the case $k = 2$ being well-characterized and polynomially solvable. For a rather comprehensive survey on order invariants, see West [1985].

Recently, antimatroids have received a lot of attention in connection with the investigation of greedy-type algorithms (see, e.g., Korte, Lovász and Schrader [1989].) An **antimatroid** is a collection $\mathscr{A}$ of subsets of the set E such that

(1.4) $E \in \mathscr{A}$
(1.5) $A \cup B \in \mathscr{A}$ for all $A, B \in \mathscr{A}$
(1.6) $A \setminus x \in \mathscr{A}$ for all $A \in \mathscr{A}$ and some $x \in A$.

(Equivalently, antimatroids may be defined as the collection of complements of closed sets relative to a closure operator enjoying a matroid-like antiexchange property, see also below).

Special examples of antimatroids arise as **poset antimatroids** from ordered sets by taking their systems of order ideals. (Recall that an **order ideal** of P is a subset $I \subseteq E$ so that $x \in I$ implies $y \in I$ for all $y \leq x$ in P.)

Poset antimatroids are in many respects the simplest antimatroids. Yet they may serve as a canonical representation of all antimatroids in the following way. A **path with endpoint** x is a set $A \in \mathscr{A}$ such that $x \in A$ is the only element with $A \setminus x \in \mathscr{A}$. (For example, the paths in poset antimatroids correspond to principal ideals).

We construct a (labeled) poset by ordering the paths by inclusion and labeling the elements of this **path poset** with their endpoints. It turns out that any antimatroid can be considered to be the poset antimatroid of its path poset (for more details, see, e.g., Korte et al. [1989]).

Antimatroids allow a meaningful extension of Dilworth's theorem. For our purposes it is now useful to think of an **antimatroid** as a collection $\mathscr{P}$ of permutations of some ground set for which the following is true. If we define the language $\mathscr{L}(\mathscr{P})$ as the set of all initial segments of members of $\mathscr{P}$, then $\mathscr{L}(\mathscr{P})$ has the **augmentation property** for all $\alpha, \beta \in \mathscr{L}(\mathscr{P})$:

If not all letters of β occur in α, then there is a letter x in β such that $\alpha x \in \mathscr{L}(\mathscr{P})$.

The collection of all linear extensions of an order, for instance, is an antimatroid; but also the collection of all simplicial decomposition sequences of a chordal graph (see Section 2) has this property (cf., e.g., Korte et al. [1989]). It should not be difficult to see that sets underlying $\mathscr{L}(\mathscr{P})$ indeed satisfy the condition (1.4)–(1.6) and that conversely the elements of any antimatroid can be ordered as described above.

Given a permutation $L = v_1 \ldots v_n \in \mathscr{P}$, we can consider the poset antimatroid $\mathscr{A}_L$ on the linear order $v_1 < v_2 < \cdots < v_n$. Then, clearly,

$$\mathscr{A} = \bigcup \{\mathscr{A}_L : L \in \mathscr{P}\}.$$

Let the **convex dimension** $cdim\mathscr{A}$ of $\mathscr{A}$ be the minimum number of permutations $L \in \mathscr{P}$ whose union $\bigcup \mathscr{A}_L$ is $\mathscr{A}$. Edelman and Saks [1988] show that $cdim\mathscr{A}$ has a direct combinatorial interpretation in the spirit of a chain covering. The convex dimension of an antimatroid is equal to the width of its path poset.

Given weights $c(e)$ on the elements of the ground set E the intersection problem for matroids is to find a subset of E of largest weight which is independent in both of two given matroids. This problem, which generalizes the bipartite matching problem, is well-solved (cf., e.g. Lawler [1976]). The corresponding problem for antimatroids is NP-hard in general.

It is, however, easy to see that the problem of finding a maximum weighted subset which is an ideal relative to two given orders P and Q on E, is polynomially solvable.

Associating with each $i \subset E$ a variable x_i consider the linear programm

$$\max \sum_{i \in E} c_i x_i$$

(LP)   s.t. $x_j - x_i \leq 0$ if $i < j$ in P or $i < j$ in Q

$$0 \leq x_i \leq 1$$

Observe that the feasible $0 - 1$ solutions of (LP) correspond exactly to the ideals common to P and Q. Because linear programming is polynomial it therefore suffices to show that (LP) has an integral optimum solution. The constraint matrix of (LP) is the transpose of the vertex-edge incidence matrix of the directed graph $D(P \cup Q)$. As it is well-known that incidence matrices of directed graphs are totally unimodular, also the constraint matrix of (LP) is totally unimodular, which yields the desired result.

The previous proof is one of the few results in order theory which are based on a polyhedral approach. Linear programming and polyhedral combinatorics seem to have been neglected in this field for a long time. Before closing this section we mention two more recent results making use of polyhedral arguments, both modelling machine scheduling problems.

Consider two ordered sets $P_1 = (E_1, \leq_1)$, $P_2 = (E_2, \leq_2)$ and a cost function c_{ij} for assigning element $i \in E_1$ to element $j \in E_2$. The **order preserving matching problem** consists in finding a **strict order preserving injection** of minimum weight, i.e. a mapping $h: E_1 \rightarrow E_2$ such that

$$i <_1 j \text{ in } E_1 \text{ implies } h(i) <_2 h(j) \text{ in } E_2$$

and $\sum_{i \in E_1} c_{i,h(i)}$ is minimal.

This problem, which models the optimal assignment of jobs to machines, is NP-complete in general, while special cases are polynomially solvable (see Chang and Edmonds [1985]). For the case where P_1 is linear and P_2 is arbitrary, Margot et al. [1988] present a dynamic programming algorithm and a polyhedral characterization of the set of strict order preserving injections.

For an antichain A let

$$I(A) = \{x: x \le y \text{ for some } y \in A\}$$

be the **ideal generated by** A and $I'(A) = I(A)\backslash A$ be the corresponding **open ideal**. Margot et al. show that if P_1 is a linear order, the incidence vectors x_{ij} of strict order preserving injections are the extreme points of the following polytope:

$$\sum_{j \in E_2} x_{ij} = 1 \qquad i \in E_1$$

$$\sum_{j \in I(A)} x_{ij} - \sum_{j \in I'(A)} x_{i-1,j} \le 0 \qquad i \in E_1, A \text{ antichain in } E_2$$

$$x_{ij} \ge 0.$$

For the second result we associate with every linear extension π of P an **incidence vector** $x(\pi) = (\pi(1), \ldots, \pi(n)) \in \mathbb{R}^n$. The **permutahedron** $Perm(P)$ is the convex hull of all incidence vectors of linear extensions, i.e.

$$Perm(P) = \text{conv}\{x(\pi): \pi \text{ a linear extension of } P\}.$$

The linear programming problem

$$\max\{cx: x \in Perm(P)\}$$

is equivalent to the following 1-machine scheduling problem: given processing times c_i, the **completion time** of job i is c_i plus the sum of all processing times of jobs processed before i. The **mean finish time problem** is to find a schedule of the jobs so that the average completion time is minimal. This scheduling problem is treated in Sidney [1975], where a polynomial algorithm for series-parallel orders but no polyhedral description is given.

Rado [1952] considered the case where P is an antichain and gave a linear description of $Perm(P)$ via the following inequalities

$$x(S) \ge \binom{|S| + 1}{2} \qquad \text{for all } S \subseteq E$$

$$x(P) = n(n + 1)/2.$$

In fact, this linear description is irredundant since all of the inequalities are facets (cf, e.g. Gaiha and Gupta [1977]).

In v. Arnim et al. [1989] this result is extended to the case where P is a series-parallel order. Recall that an ordered set is **series-parallel** if it does not contain the following order as induced suborder

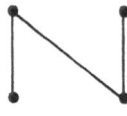

Figure 1

We call a pair (A, B) of subsets **series-reducible** if $a < b$ for all $a \in A$ and $b \in B$. If P is series-parallel then the permutahedron $Perm(P)$ is given by

$$x(I) \geq \binom{|I| + 1}{2} \qquad \text{for all ideals } I \subseteq E$$

$$|A|x(B) - |B|x(A) \geq \frac{1}{2}|A||B|(|A| + |B|) \qquad \text{for all series-reducible pairs } (A, B)$$

$$x(P) = n(n + 1)/2.$$

Not all of these inequalities induce facets. However, the facets among them are well-characterized and violated inequalities can be detected efficiently by a polynomial separation algorithm.

Closing this section, let us mention a combinatorial object which so far has received surprisingly little attention in general. It is well-known, for example, that the set of permutations can be generated via adjacent transpositions starting from one specified permutation (see, e.g., Chap. I of Even [1973]). In other words, the graph having as vertices all permutations and the neighboring relation defined via adjacent transpositions admits a hamiltonian path and, in particular, is connected. Note that this graph can be viewed as the skeleton of the unrestricted permutahedron (cf. Rado [1952], see also above).

As before, let us think of the maximal feasible sets of an antimatroid as a collection $\mathcal{P}$ of permutations. The **basis graph** $\mathcal{B}(\mathcal{P})$ of the antimatroid $\mathcal{P}$ has $\mathcal{P}$ as its set of vertices with the neighboring relation as in the general permutation graph defined above. We will simply write $\mathcal{B}(P)$ if $\mathcal{P}$ consists of the linear extensions of the order P.

It is not hard to prove by induction that $\mathcal{B}(\mathcal{P})$ always is connected. Even for orders P, however, $\mathcal{B}(P)$ need not be hamiltonian. An interesting unsolved case, for example, is given by the question whether $\mathcal{B}(P)$ has a hamiltonian path whenever P is an interval order (cf. Ruskey [1988]).

Note that also the setup number problem can be formulated for basis graphs. Indeed, the setup number of the order P is exactly the minimal vertex degree of the graph $\mathcal{B}(P)$. This formulation suggests that it might be meaningful to explore the setup number for other classes of antimatroids as well.

2. Ordertheoretic Techniques

Many combinatorial optimization problems require to determine an "independent" set which is maximal relative to some weight function. Here the notion of "independence" implies that subsets of independent sets are also independent. Algorithms to solve such problems often impose a linear order on the elements of the ground set in question and then scan through the set in that order before building up the desired object.

A well-known example is Kruskal's algorithm to find a maximal edge-weighted spanning tree in a graph. The linear order on the edge set here lists the edges according to decreasing weights. Scanning through the edge set, one then builds up a tree "greedily" (see Welsh [1976] for matroid-theoretic ramifications).

The **maximum stable set problem** in a vertex-weighted graph G asks for a maximum weight set of vertices that are pairwise non-incident. The problem is known to the NP-complete in general (see Garey and Johnson [1979]). Frank [1976] gives an efficient algorithm provided that the vertices of G can be linearly ordered $v_1, v_2, \ldots,$ v_n such that v_{i+1} is **simplicial** in the induced graph $G_i = G \setminus \{v_1, \ldots, v_i\}$ $(i = 1, \ldots, n)$, i.e. such that the neighbors of v_{i+1} form a clique in G_i. The graphs which admit such an ordering are exactly the **chordal** graphs (a.k.a **triangulated**) graphs, where, by definition, each circuit of size at least 4 has a chord (see Golumbic [1980]).

Hoffman et al. [1985] consider a class of linear programs with $(0, 1)$-constraint matrices that can be solved by a certain greedy algorithm. The crucial point of their algorithm consists in the observation that each constraint matrix in the class gives rise to a **strongly chordal graph** G, whose vertices can be ordered $v_1, \ldots, v_n$ such that v_{i+1} is **simple** in the induced graph $G_i = G \setminus \{v_1, \ldots, v_i\}$ (see Anstee and Farber [1984]). To explain this terminology, we denote by $N(v)$ the **(open)** set of neighbors of the vertex v in a graph G and let $N[v] = N(v) \cup \{v\}$ be the **closed** neighborhood set. The vertex v is then said to be simple in G if the collection $\{N[u] : u \in N(v)\}$ is linearly ordered by set-inclusion. (Note that, in particular, each simple vertex is simplicial).

Let us turn to a standard heuristic algorithm for the **graph coloring problem**, where the vertices of a graph G are to be colored with as few colors as possible such that adjacent vertices carry different colors. The heuristic scans the vertices $v_1, v_2, \ldots,$ v_n in some linear order and assigns colors by a "first fit" method: give the vertex v_j the smallest positive integer assigned to no neighbor v_i $(i < j)$ of v_j. Chvátal [1984] calls a graph G **perfectly orderable** if G admits an ordering which makes the heuristic produce an optimal coloring when applied to G or any induced subgraph F of G. Chvátal shows that a perfectly orderable graph is, in particular, **strongly perfect** (and hence perfect), i.e. each induced subgraph F has a stable set that meets all maximal cliques of F. Moreover, he characterizes perfect orderings as those orderings which admit no induced chordless path P_4 on four vertices $abcd$ with $a < b$ and $c > d$.

Taking the simplicial ordering, one sees that chordal graphs are perfectly orderable. Another class of examples is formed by comparability graphs. To see this, we orient a comparability graph according to a partial order P and choose a linear extension of P. (For more examples, see Chvátal [1989] and the references cited there). It is, however, interesting for our purposes that also graphs with **Dilworth number** at most three are perfectly orderable (Chvátal et al. [1987]).

The Dilworth number of a graph G was introduced by Foldes and Hammer [1978], who define the **vicinal preorder** on the vertices of G as follows:

$$x \leq y \text{ if and only if } N(x) \subseteq N[y].$$

The vicinal preorder is reflexive and transitive (though not necessarily antisymmetric). The Dilworth number of G now is the minimal number of chains with respect to this preorder needed to cover all vertices of G.

The graphs with Dilworth number 1 are exactly the **threshold graphs** (Chvátal and Hammer [1977]). Recall that such graphs are characterized by the possibility to assign real numbers $a_1, a_2, \ldots, a_n$ to the vertices so that a number S exists with the property

$$i \text{ and } j \text{ are adjacent if and only if } a_i + a_j \geq S.$$

Graphs with Dilworth number 2 correspond to so-called **threshold signed graphs**. They can be thought of as being constructed from pairs of disjoint threshold graphs by inserting new edges according to Galois connections relative to the respective vicinal preorders (Benzaken et al. [1985]).

Since the concept of a Galois connection, which goes back to Ore [1944], is also of importance in other contexts, we give a formal definition:

A **Galois connection** between two (partially) ordered sets P and Q is a pair (σ, τ) of maps $\sigma: P \to Q$ and $\tau: Q \to P$ such that

(2.1) $\sigma(p) \geq \sigma(p')$ for all $p \leq p'$ in P
(2.2) $\tau(q) \geq \tau(q')$ for all $q \leq q'$ in Q
(2.3) $p \leq \tau(\sigma(p))$ for all $p \in P$
(2.4) $q \leq \sigma(\tau(q))$ for all $q \in Q$.

To illustrate this concept, consider a bipartite graph G with a partition $S \cup T$ of its vertex set such that each edge has one endpoint in S and the other endpoint in T. We choose P and Q to be the collections of all subsets of S and T respectively, ordered by inclusion. For each $A \subseteq S$ and $B \subseteq T$, we define a Galois connection via

$$\sigma(A) = \bigcap_{s \in A} N(s)$$

$$\tau(B) = \bigcap_{t \in B} N(t)$$

(with the understanding that the intersection of the empty set equals the ground set).

Wille [1985] suggests the following interpretation in the language of data analysis. S is a set of "objects" and T a set of "properties". An edge between t and s in G signifies that s has property t. In this sense, the bipartite graph G can be viewed as a **context**. A **concept** in this context is determined as a collection of all those objects that have a set $B \subseteq T$ of properties in common. In other words, concepts correspond to sets of objects of the form $\tau(B)$. (Since the role of objects and properties is completely symmetric, one may equivalently view concepts as sets of properties of the form $\sigma(A)$). Hence the **concept lattice**

$$\mathscr{L}(G; S) = \{\tau(B) : B \subseteq T\}$$

(or equivalently $\mathscr{L}(G; T)$) contains the complete information about the concepts that can be distinguished within the data structure G (see also Section 9 in Bock [1988].)

Orders that are induced via set inclusion are also central in the study of monotone graph properties. We consider the class Γ of all simple graphs on the vertices $1, 2, \ldots, n$. For $x, y \in \Gamma$, we write $x \leq y$ if the edge set of x is contained in the edge set of y. A **monotone property** of Γ is a subset $F \subseteq \Gamma$ such that for all $x, y \in \Gamma$,

$$x \in F \text{ and } x \leq y \text{ implies } y \in F.$$

(Usually, monotone graph properties are additionally assumed to be isomorphism-invariant). One fundamental fact is that any two monotone properties F_1 and F_2 are **positively correlated**, i.e.

$$\frac{|F_1 \cap F_2|}{|\Gamma|} \geq \frac{|F_1|}{|\Gamma|} \frac{|F_2|}{|\Gamma|}.$$

(This means that F_1, say, is at least as likely to hold if F_2 is known to hold than it is if nothing is known about F_2).

The proof of this fact is an application of the so-called **FKG-inequality** for distributive lattices. To formulate it, we consider a finite family $\mathcal{D}$ of sets which is closed under union and intersection. We assume to be given two monotone functions f, $g: \mathcal{D} \to \mathbb{R}$ and a nonnegative function $\mu: \mathcal{D} \to \mathbb{R}_+$ satisfying

$$\mu(x \cap y)\mu(x \cup y) \geq \mu(x)\mu(y) \qquad \text{for all } x, y \in \mathcal{D}.$$

The FKG-inequality then concludes

$$\sum_{x \in \mathcal{D}} f(x)g(x)\mu(x) \cdot \sum_{x \in \mathcal{D}} \mu(x) \geq \sum_{x \in \mathcal{D}} f(x)\mu(x) \cdot \sum_{x \in \mathcal{D}} g(x)\mu(x).$$

In our application above we take $\mathcal{D} = \Gamma$, $\mu \equiv 1$ and let f and g be the indicator functions of F_1 and F_2. We just remark that the FKG-inequality in turn is a consequence of the even deeper Ahlswede-Daykin inequality for nonnegative functions on distributive lattices. We will omit the details and refer instead to the comprehensive introduction into the subject by Graham [1982].

One of the most outstanding problems about Γ is the question whether there exists a monotone isomorphism-invariant graph property $F \neq \Gamma$ which is **non-evasive**. The question refers to the following setting. We want to determine whether the graph $x \in \Gamma$, which is unknown to us at the outset, belongs to F. We may query an oracle to find out if a given edge belongs to x. The minimal number $c(F)$ needed to settle the problem is the **complexity** of F. F is **evasive** if

$$c(F) = \binom{n}{2},$$

i.e. if we have to ask all possible edges. Rivest and Vuillemin [1976] proved the **Anderaa–Rosenberg conjecture**: If F is a non-trivial isomorphism-invariant monotone graph property in Γ, then

$$c(F) > \frac{n^2}{16}.$$

Currently still open is the **Karp conjecture**: Each non-trivial isomorphism-invariant monotone graph property is evasive.

Many interesting partial results towards the Karp conjecture have been obtained so far. A very readable account therof can be found in Aigner [1988].

An analogue of the recognition complexity of graph properties for ordered structures is investigated in Faigle and Turán [1988]. Here an **ordered set property** is taken to be a class F of orders which is isomorphism-invariant. The oracle model gives as answers to queries the comparability status of two elements x and y presented ("$x < y$" or "$x > y$" or "$x \| y$"). There is an obviously evasive ordered set property, namely, the property of being an antichain. Equally, the property of having exactly one comparable pair is evasive. Are there more evasive properties?

In view of the Karp conjecture for graphs, it is curious that evasive ordered set properties seem to be hard to find. In contrast, there are many "easy" ordered set properties (for example: having a unique maximal element or being a linear order).

Clearly related to the recognition problem for linear orders is the **sorting problem**, which in its classical formulation consists in identifying an a priori unknown linear order on the ground set by repeatedly asking the comparability status of pairs of elements. It is not the place here to discuss the various sorting techniques, which can be found in the standard literature (e.g. Knuth [1973], see also Bollobás and Hell [1985] for a more graphtheoretically oriented survey).

Let us look at a natural generalization of the sorting problem, which is due to Schönhage [1976]. Suppose we only want to identify the 3rd-largest element of the unknown linear order. Then it clearly suffices to stop asking comparability questions when the answers sofar have produced a partially ordered set P of Figure 2 into which the order can be embedded in an order-preserving manner.

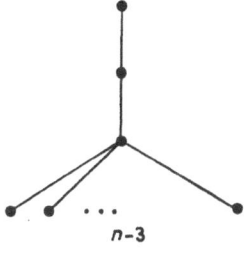

n-3

Figure 2

Generally, the **order production problem** is based on a linear order, which is only known to a "comparison oracle", and on a partial order P on the same ground set, which is known to us. We are to identify a partial order Q into which P can be embedded with as few calls to the oracle as possible.

Equivalently, we want to find a permutation σ of the ground set such that $x < y$ in P implies $\sigma(x) < \sigma(y)$ in the underlying order. From this, Schönhage [1976] deduced by an information-theoretic argument the lower bound

$$c(P) \geq \log_2\left(\frac{n!}{e(P)}\right)$$

for the number of oracle calls, where $e(P)$ is the number of linear extensions of P. Saks [1985] conjectured that this lower bound can be achieved asymptotically in the sense that

$$c(P) = O\left(\log_2\left(\frac{n!}{e(P)} + n\right)\right).$$

In a remarkable paper, Yao [1988] recently proved the conjecture to be correct.

Another generalization of the sorting problem goes back to Fredman [1976]: Suppose that some of the relations of the unknown linear order have been determined and give rise to the partial order P. How many comparisons are still needed for completely identifying the underlying linear order? Or, equivalently: How many comparisons are needed in order to determine a fixed (but a priori unknown) linear extension of a given partial order P?

The complexity $c(P)$ in this case can be bounded from below by the information-theoretic bound

$$c(P) \geq \log_2 e(P).$$

A standard application of the binary search principle would establish $c(P) = O(\log_2 e(P))$ if there existed a universal constant $0 < \delta < \frac{1}{2}$ with the property: Each ordered set P which is not a linear order contains two elements x and y such that the fraction $P(x < y)$ of linear extensions where x precedes y satisfies $\delta < P(x < y) < 1 - \delta$.

Denote by $\delta(P)$ the best such δ relative to the fixed order P. Kahn and Saks [1984] have shown

$$\delta(P) \geq \frac{3}{11} \qquad \text{for all non-linear orders } P,$$

which yields $c(P) \leq 2.2 \log_2 e(P)$. The 3-element order P_3 with exactly one nontrivial comparability has $\delta(P_3) = \frac{1}{3}$. Nevertheless, a challenging conjecture of Kahn and Saks claims

$$\lim_{w(P) \to \infty} \delta(P) = \frac{1}{2},$$

where $w(P)$ is the width of P.

4. References

M. Aigner [1988]: Combinatorial Search. Chichester: Wiley-Stuttgart: Teubner, 1988.

R. P. Anstee and M. Farber [1984]: Characterization of totally balanced matrices. Journal of Algorithms 5 215–230, (1984).

A. v. Arnim, U. Faigle, and R. Schrader [1989]: The permutahedron of series-parallel posets. to appear in: Discr. Appl. Math..

C. Benzaken, P. L. Hammer and D. de Werra [1985]: Threshold characterization of graphs with Dilworth number two. Journ. of Graph Theory 9, 245–267, (1985).

H. H. Bock (ed.) [1988]: Classification and Related Methods of Data Analysis. Amsterdam: North Holland, 1988.

B. Bollobás and P. Hell [1985]: Sorting and graphs. In: Graphs and Order (I. Rival, ed.), Reidel, 169–184, 1985.

G. J. Chang and J. Edmonds [1985]: The poset scheduling problem. Order 2, 113–118, (1985).

V. Chvátal [1984]: Perfectly orderable graphs. In: Topics on Perfect Graphs (C. Berge and V. Chvátal eds.), North Holland, Amsterdam, 63–65, 1984.

V. Chvátal [1989]: A class of perfectly orderable graphs. Report No. 89573-OR, Institute of Operations Research, Universität Bonn.

V. Chvátal and P. L. Hammer [1977]: Aggregation of inequalities in integer programming. Ann. Discr. Math 1, 145–162, (1977).

V. Chvátal, C. T. Hoàng, N. V. R. Mahadev and D. de Werra [1987]: Four classes of perfectly orderable graphs. Journ. of Graph Theory 11, 481–495, (1987).

O. Cogis [1982]: On the Ferrers dimension of a digraph. Discr. Math. 38, 47–52, (1982).

P. H. Edelman and M. Saks [1988]: A combinatorial representation and convex dimension of convex geometries. Order 5, 23–32, (1988).

S. Even [1973]: Algorithmic Combinatorics. Macmillan, New York, 1973.

U. Faigle and R. Schrader [1985]: Comparability graphs and order invariants. In: U. Pape (ed.): Graptheoretic Concepts in Computer Science, Trauner Verlag, Linz, 136–145, 1985.

U. Faigle and R. Schrader [1986]: A combinatorial bijection between linear extensions of equivalent orders. Discr. Math. 58, 295–301, (1986).

U. Faigle and Gy. Turán [1988]: Sorting and recognition problems for ordered sets. SIAM Journ Comput. 17, 100–113, (1988).

S. Foldes and P. L. Hammer [1978]: The Dilworth number of a graph. Ann. Discr. Math. 2, 211–219, (1978).

A. Frank [1976]: Some polynomial algorithms for certain graphs and hypergraphs. In: C.St.J.A. Nash-Williams (ed.): Proceedings of the 5th British Combinatorial Conference (1975), Congressus Numerantium 15, Utilitas Mathematica 211–226, (1976).

A. Frank [1980]: On chain and antichain families of a partially ordered set. J. Comb. Th. B 29, 176–184, (1980).

M. Fredman [1976]: How good is the information theory bound for sorting. Theoretical Comp. Sci. 1, 355–361, (1976).

M. Fujii, T. Kasami and K. Ninomiya [1969]: Optimal sequencing of two equivalent processors. SIAM Journ. Appl. Math. 17 (1969), 784–789 (Erratum 20, 141, (1977)).

D. R. Fulkerson [1956]: Note on Dilworth's decomposition theorem for partially ordered sets. Proc. Amer. Math. Soc. 7, 701–702, (1956).

P. Gaiha and S. K. Gupta [1977]: Adjacent vertices on a permutahedron. SIAM J. Appl. Math. 32, 323–327, (1977).

M. R. Garey and D. S. Johnson [1979]: Computers and Intractability. Freeman and Co., San Francisco, 1979.

M. C. Golumbic [1980]: Algorithmic Graph Theory and Perfect Graphs. Academic Press, New York, London, San Francisco, 1980.

R. L. Graham [1982]: Applications of the FKG-inequality and its relatives. In: Mathematical Programming–The State of the Art (A. Bachem et al. (eds.), Heidelberg-New York-Berlin: Springer, 115–131, (1982).

C. Greene [1976]: Some partitions associated with a partially ordered sets Journ. Comb. Th. A 20, 69–79, (1976).

C. Greene and D. J. Kleitman [1976]: Strong versions of Sperner's theorem. Journ. Comb. Th. A 20, 80–88, (1976).

M. Grötschel, L. Lovász and A. Schrijver [1988]: Geometric Algorithms and Combinatorial Optimization. Heidelberg-New York-Berlin: Springer, 1988.

R. Gysin [1977]: Dimension transitiv orientierbarer Graphen. Acta Math. Acad. Sci Hungar. 29, 313–316, (1977).

M. Habib [1984]: Comparability invariants. Ann. Discr. Math 23, 371–386, (1984).

M. Habib, D. Kelly and R. H. Möhring [1988]: Interval dimension is a comparability invariant. Preprint, Technische Universität Berlin.

M. Habib and R. H. Möhring [1988]: A fast algorithm for recognizing trapezoid graphs and partial orders of interval dimension 2. Preprint, Technische Universität Berlin.

A. J. Hoffman [1982]: Ordered sets and linear programming. In: Ordered Sets (I. Rival, ed.), Reidel, Dordrecht, 619–654, 1982.

A. J. Hoffman, A. W. J. Kolen and M. Sakarovitch [1985]: Totally balanced and greedy matrices. SIAM Journ. Alg. Disc. Meth. 6, 721–730, (1985).

J. Kahn and M. Saks [1984]: Balancing poset extensions. Order 1, 113–126, (1984).

D. Kelly and W. T. Trotter [1982]: Dimension Theory for ordered sets. In: Ordered Sets (I. Rival, ed.), Reidel, Dordrecht, 171–212, 1982.

D. E. Knuth [1973]: The Art of Computer Programming, Vol. 3, Sorting and Searching. Addison-Wesley, Reading, 1973.

B. Korte, L. Lovász and R. Schrader [1989]: Greedoid Theory. Berlin-Heidelberg-New York-Tokyo-Hong Kong: Springer, 1989.

E. L. Lawler [1976]: Combinatorial Optimization: Networks and Matroids. Holt, Rinehart and New York; Winston, 1976.

L. Lovász [1972]: Normal hypergraphs and the perfect graph conjecture. Discr. Math. 2, 253–267, (1972).

L. Lovász and M. Plummer [1986]: Matching Theory. North-Holland Mathematics Studies Vol. 121, 1986.

F. Margot, A. Prodon and Th.M. Liebling [1988]: A note on order preseving matchings. Preprint, École Polytechnique Fédérale de Lausanne.

W. Nawijn [1989]: Minimum loss scheduling. In: U. Faigle and C. Hoede, eds.: Twente Workshop on Graphs and Combinatorial Optimization, Research Memorandum No. 787, Universiteit Twente.

O. Ore [1944]: Galois connexions. Trans. Am. Math. Soc. 55, 493–513, (1944).

K. R. Parthasarathy and G. Ravindra [1976]: The strong perfect graph conjecture is true for $K_{1,3}$-free graphs. Journ. Comb. Th. B 21, 212–233, (1976).

W. Poguntke [1986]: Order-theoretic aspects of scheduling. Contemporary Mathematics (Amer. Math. Soc.) 57, 1–32, (1986).

R. Rado [1952]: An inequality. Journ. London Math. Soc. 27, (1952).

R. L. Rivest and J. Vuillemin [1976]: On recognizing graph properties from adjacency matrices. Theor. Comp. Sci. 3 371–382, (1976).

F. Ruskey [1988]: Research Problem 90/91 Discr. Math. 70, 111–112, (1988).

M. Saks [1985]: The information-theoretic bound for problems on ordered sets and graphs. In: Graphs and Order (I. Rival, ed.), Reidel, Dordrecht, 137–168, 1985.

A. Schönhage [1976]: The production of partial orders. Astérisque 38–39, 229–246, (1976).

J. B. Sidney [1975]: Decomposition algorithms for single-machine sequencing with precedence relations and deferral costs. Oper. Res. 23, 283–298, (1975).

W. T. Trotter, J. I. Moore and D. P. Sumner [1976]: The dimension of a comparability graph. Proc. Amer. Math. Soc. 60, 35–38, (1976).

A. Tucker [1973]: The strong perfect graph conjecture for planar graphs. Canad. Journ. Math. 25, 103–114, (1973).

D. Welsh [1976]: Matroid Theory. Academic Press, London, 1976.

D. West [1985]: Parameters of partial orders and graphs: packing, covering and representation. In: Graphs and Order (I. Rival, ed.), Reidel, Dordrecht, 267–350, 1985.

R. Wille [1985]: Restructuring lattice theory: an approach based on hierarchies of concepts. In: Graphs and Order (I. Rival, ed.), Reidel, Dordrecht, 445–470, 1985.

M. Yannakakis [1982]: The complexity of the partial order dimension problem. SIAM J. Ald. Discr. Meth. 3, 351–358, (1982).

A. C. Yao [1988]: On the complexity of partial order productions. Preprint, Dept. of Computer Science, Princeton University.

Rainer Schrader
Forschungsinstitut für Diskrete Mathematik,
Rhein.
Friedrich-Wilhelms-Universität Bonn,
Nassestrasse 2
D-5300 Bonn
West-Germany

Ulrich Faigle
Department of Applied Mathematics
Universiteit Twente
Postbus 217
NL-7500 AE Enschede
The Netherlands

Computing Suppl. 7, 125–139 (1990)

Computing
© by Springer-Verlag 1990

Dynamic Partial Orders and Generalized Heaps

Hartmut Noltemeier, Würzburg

Abstract — Zusammenfassung

Dynamic Partial Orders and Generalized Heaps. Classical and recent results are surveyed in the development of efficient representations of dynamic partial orders by heaps and their generalizations.

AMS Subject Classifications: 68B15, 06A10.

Key words: partial order, heap, implicit data structure, double-ended priority queue, interval heap, heap ordered tree, Fibonacci heap, binomial queue, priority search tree.

Dynamische Partialordnungen und verallgemeinerte Heaps. Die Möglichkeiten und Probleme der Repräsentation von dynamischen Partialordnungen durch Heaps und ihrer Verallgemeinerungen werden diskutiert; klassische und neueste Resultate werden überblicksmäßig vorgestellt.

1. Introduction

1.1. Basic notations

Let V be a finite set of objects, drawn from a possibly infinite set U (universe), on which a partial order "$\prec$" is given.

A partial order is a binary relation $PO \subset U \times U$ on U (resp. V)—an element $(v_1, v_2) \in PO$ is denoted by $v_1 \prec v_2$—, which is

reflexive: $v \prec v$ for all $v \in U$ (resp. V),

antisymmetric: $(v_1 \prec v_2$ and $v_2 \prec v_1) \Rightarrow (v_1 = v_2)$

and transitive: $(v_1 \prec v_2$ and $v_2 \prec v_3) \Rightarrow (v_1 \prec v_3)$.

An element $\underline{v} \in V$ is *minimal* in V iff

$$v \prec \underline{v} \Rightarrow v = \underline{v},$$

an element $\bar{v} \in V$ is *maximal* in V iff

$$\bar{v} \prec v \Rightarrow v = \bar{v}.$$

A *chain* is a nonempty sequence of pairwise different elements of V

$$w = (v_{i_1}, v_{i_2}, \dots, v_{i_k}) \qquad (k \geq 1),$$

which are totally ordered:

$$v_{i_j} \prec v_{i_{j+1}} \qquad (j = 1, \ldots, k-1).$$

k is the length $l(w)$ of the chain w; a chain of length 1 is called a trivial chain. Any partial order PO can be represented by a directed acyclic graph (DAG) $G = (V, E)$ with $E := \{(v, v')/v \neq v'$ and $v \prec v'\}$; its (unique) *transitively irreducible kernel* (transitive reduction) is called the Hasse diagram of PO.

But notice:

The number of arcs of the Hasse diagram may have at most $\left\lfloor \dfrac{n^2}{4} \right\rfloor$ arcs, where n denotes the cardinality of V; this $O(n^2)$-upper bound is sharp (f.e. in the complete bipartite graph $K_{n/2, n/2}$) ([18], [19]).

1.2. Basic problems

As V ("the actual set") may change in time we are concerned with the following problems:

(1) *represent* V with respect to some given objectives (f.e. support special questions on V efficiently: report all maximal elements, give the Dilworth number of PO, etc.)
(2) *maintain the representation* when V changes in time, especially
    (a)   if a "new element" $v \in U - V$ has to be *inserted*
    (b)   if an element $v \in V$ has to be *deleted*
    (b') if a minimal (maximal) element has to be deleted
    (b'') if a minimal (maximal) element has to be deleted and a new element to be inserted
(3) *divide* $V := V_1 \cup V_2$, that means split the representation of V into two representations of V_1 and V_2 respectively $(V_1 \cap V_2 = \emptyset)$
(4) *merge* $V := V_1 \cup V_2$, that means given representations of V_1 resp. V_2 construct a representation of $V_1 \cup V_2$ (especially if $V_1 \cap V_2 = \emptyset$).

Some more special operations are given in forthcoming chapters.

2. Heaps

2.1. The classical min-Heap

Originally the concept of a heap (Williams [27]) was as follows: A real-valued array $A[1 \ldots n]$ is a *heap* iff

(*) $A[\lfloor i/2 \rfloor] \leq A[i]$ for $i = 2, \ldots, n$.

These $n - 1$ conditions pose a special partial order on the components of array A illustrated in figure 1 (for simplicity let a_i denote $A[i]$):

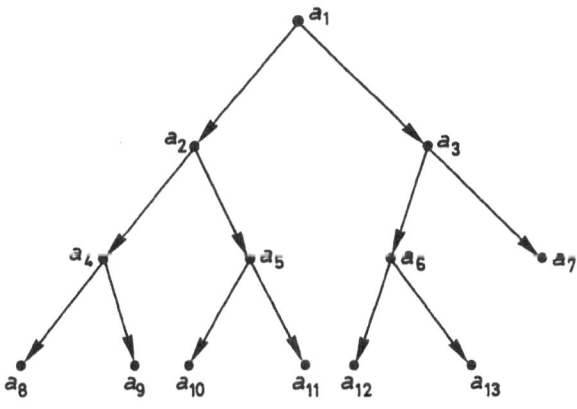

Figure 1

A heap was designed as a data structure for the following situation: let $k: U \to \mathbb{R}$ be a real-valued function (not necessarily an injective one); $k(v)$ is called the *key* of object v.

A heap $A[1 \ldots n]$ *represents* a set of objects V *endogenously*, if A has pairwise different components and V is the set of keys itself; it represents a set of objects V *exogenously*, if there is a further array $P[1 \ldots n]$, where $P[i]$ points to the object associated with key $A[i]$.

For simplicity let us restrict ourselves to the endogenous case.

It is worthwhile to mention that the original concept of an endogenous heap is a *pointerless implementation* (*implicit data structure*) of the binary tree given in figure 1.

To be more independent from implementation techniques let us define a heap in a more general way.

Definition: A (min-)*Heap* is an ordered* binary tree H with the following properties:

(1) H is *heap-ordered*, i.e. a key in any node is not less than the key of its father (if there is a father)
(2) H has a *heap-shape*, i.e. is a *left complete binary* tree (that means: all levels except the last one are complete; the leaves in the last level are as far to the left as possible; see figure 2).

---

* ordered means: every son is uniquely characterized as a left son resp. right son

Heap-shape of H:

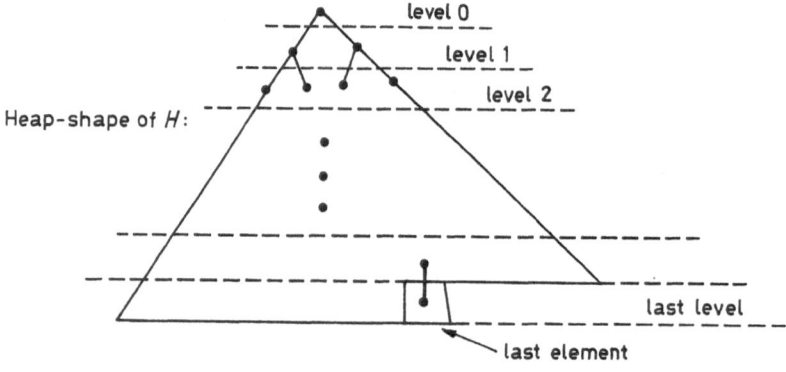

Figure 2

Remarks:

1.) The root contains the minimal element.
2.) The height of H is $h(H) = \lfloor \log_2 n \rfloor$.
3.) A breadth-first-search of H (starting at the root and respecting the order of the sons: left son before right son) results in a heap as defined originally. The rank of an element in this order is called its *position*.

(min-)Heaps support the following basic operations, measured in the number of comparisons as well as data movements:

Operation	Description	Complexity
FindMin	find element with minimum key	$O(1)$
Insert	insert a new element and restore the heap-property	$O(\log n)$
DeleteMin	remove element with minimum key and restore the heap-property	$O(\log n)$
Create	construct a heap with n elements	$O(n)$
Sort	sort all keys (in decreasing order)	$O(n \log n)$

To give a rough idea how these operations can work, let us first look at *Insert*:

1.) *place* the new element (the $(n + 1)$th) "just behind" the last one (the n-th element), where the shortest binary coding of the number $n + 1$ $(d_{i_k} d_{i_{k-1}} \ldots d_{i_1} d_{i_0})$ immediately gives us the path from the root to the correct position: read 0 as "go left" and 1 as "go right" and start at the root with $d_{i_{k-1}}$, continue with $d_{i_{k-2}}$ etc.
2.) if father (new) $\leq$ new then STOP else

    bubble up (new)   {exchange father (new) and new recursively as long as the if-condition is not satisfied}

Notice: The place-routine is independent of the value of the new element. In a pointerless implementation using consecutive addresses $(1, 2, \ldots, n, n + 1, \ldots)$ the father of i is given by $\lfloor i/2 \rfloor$, which can be realized by shifting one bit in the binary coding of i; to find the k-th ancestor one has to shift just k bits! This allows to reduce the number of comparisions even to $O(\log(\log n))$ by binary search on the path of ancestors of the $(n + 1)$-th position (Gonnet, Munro [10]). Although the number of data movements again may be of order $O(\log n)$ this idea leads to an $O(\log(\log n))$-time INSERT-algorithms on an $O(\log n/\log \log n)$-processor parallel CREW-RAM.

To *delete the minimal element* and to restore the remaining $n - 1$ elements in a heap, put the old n-th element in position 1 (root) and

> *trickle_down* (v) {exchange the actual element v and the smaller of its
> sons as long as this son is smaller than the
> actual father recursively}, starting with the root v.

To *create a heap* with n elements, at first place the elements arbitrarily in positions 1, ..., n; then (iteratively) trickle_down $(\lfloor n/2 \rfloor)$, trickle_down $(\lfloor n/2 \rfloor - 1)$, ..., trickle_down (1).

To *sort* all elements first create a heap, then (iteratively) exchange the first (minimal) element and the element on position k and trickle_down (1) in the remaining set of the $k - 1$ first positions ($k = n, n - 1, ..., 2$) ("Heapsort", Williams [27]; Floyd [6]; for an improved version using the remark above see Carlsson [3]).

2.2. *Variants of Heaps*

The restriction to ordered binary trees is not essential. We can take ordered *d-ary trees* as well ($d \geq 2$).

A *d-heap* is an ordered d-ary tree, which is left complete and heap-ordered. Analogous to the case $d = 2$, the d-heap operations have running time $O(1)$ for FindMin, $O(\log_d n)$ for Insert and $O(d \log_d n)$ for DeleteMin, since a left complete d-ary tree has height at most $\lfloor \log_d n \rfloor + 1$.

The parameter d can be chosen adequately with respect to the relative frequencies of the operations DeleteMin and Insert in a given application (for example to speed up shortest-path-algorithms): as the proportion of deletions decreases, one can increase d, saving time on insertions (see Tarjan [24]).

Again we do not need any pointer: if we search the tree in breadth-first-search-manner (and left to right), the resulting positions are as follows:

$$\text{father } (i) = \lceil (i - 1)/d \rceil \quad \text{and}$$

$$\text{sons } (i) \quad = [d \cdot i - d + 2, \min\{d \cdot i + 1, n\}].$$

The concept of (min-)Heap can easily turned to max-Heaps.

A *max-Heap* is an ordered binary tree, which has the heap-shape and is *max-heap-ordered*, i.e. a key in any node is *not greater* than the key of its father. max-Heaps obviously can be reduced to min-Heaps: any max-Heap is a min-Heap with respect to the inverse order of keys and vice versa.

Thus max-Heaps support FindMax, Insert, DeleteMax, Create and SORT, in a similar way with analogous complexities.

Sometimes it is desirable, to support

> FindMin, DeleteMin, Insert and Create as well as
> FindMax, DeleteMax simultaneously.

Any data structure, which supports these operations, is called a *double-ended priority queue (DEPQ)*.

A simple collection of a min-Heap and a max-Heap (for the same set V) doesn't give a reasonable solution: besides the doubling of space requirement it implies some bad worst-case-time complexities: a DeleteMin need $O(n)$ time in a max-Heap, a Deletemax similarly $O(n)$ time in min-Heaps.

Some more sophisticated solutions were given in [1], where the elements are divided into even levels (which form a min-Heap) and odd levels (which form two max-Heaps) in the following way:

a *min-max-Heap* is an ordered binary tree, which has the Heap-shape and additionaly is *minmax-ordered*, i.e. all elements on even levels are less than or equal to all of their descendants, while elements on odd levels are greater than or equal to their descendants (if any).

This concept even can be enlarged to *minmaxMedian-Heaps*, which support

> FindMin, FindMax, FindMedian   in $O(1)$ time
> DeleteMin, DeleteMax, DeleteMedian   in $O(\log n)$ time
> Insert   in $O(\log n)$ time
> Create   in $O(n)$ time

using only n storage cells for data ([1]).

The most interesting and elegant approach to get an efficient implementation of a DEPQ is the

> *INTERVAL-HEAP*, due to J. v. Leeuwen and D. Wood [13].

Let $U = \mathbb{R}$ (real numbers) with the total order "$\leq$" on it,

> $0 := \{[a,b]/a, b \in U \text{ and } a \leq b\}$ the set of closed intervals on $\mathbb{R}$ and
> $[a,b] \prec [c,d]$ iff $[a,b] \subset [c,d]$
> the partial order on 0 induced by inclusion.

Definition: An *Interval-Heap IH* is an ordered binary tree, which obeys the following properties:

1) *IH* has the *heap-sheap*, i.e. is left complete
2) for each node $v \in V$ (except the last node l) $I(v)$ *is an interval* from 0, assigned to v; to the last node l there is assigned either an interval or a single value $a \in \mathbb{R}$ (which of course can be interpreted as $[a,a]$).
3) *IH* is *max-heap-ordered*, i.e.

$$I(v) \prec I \text{ (father } (v))$$

Figure 3 gives an example of an interval-heap.

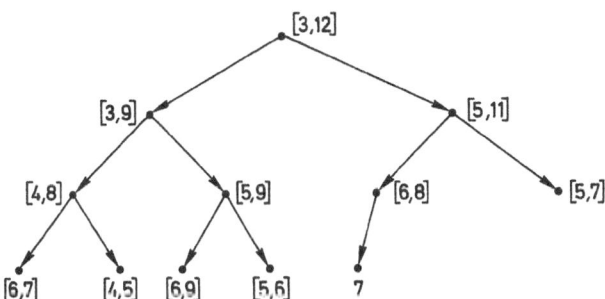

Figure 3. Interval-Heap

As an immediate consequence of 3) we have the following:

let IH be an interval-heap and $P(IH) =: P$ the *multiset* of *endpoints* of intervals represented by IH, and let $[a, b]$ be the interval assigned to the root, then $a = \min\{p/p \in P\}$ and $b = \max\{p/p \in P\}$.

The question arises: given a finite *set* $P \subset \mathbb{R}$; is there an interval-heap IH such that $P(IH) = P$ and–if any–how much effort is needed to construct IH?

Lemma: *Interval-heaps allow Insert (in P) in* $O(\log|V|)$ *time.*

The idea, to solve this problem, is very similar to the Insert in ordinary max-heaps. Roughly we look for the last node l:

α) if $I(l) = a$ is a single value, then do $I(l) := [a, \text{new\_}p]$ if a is less than the new point new_p (and $[\text{new\_}p, a]$ in the other case);

  Bubble up (l) {the actual interval $I(v) = [a, x]$ which was enlarged just
         before, is compared with
         I (father (v)) $= [b, c]$: if $c < x$ then do begin
         $I(v) := [a, c]$; I (father (v)) $:= [b, x]$ end
         and continue with father (v) until the tree is max-heap-
         ordered (all other, but similar cases are omitted)};

β) if $I(l) = [a, b]$, store new_p in the next (new last) position l' and compare new_p
  with I (father (l')) $= [a, b]$:

  if new_$p \in [a, b]$ then STOP
   else begin $I(l') := \text{Median } \{a, b, \text{new\_}p\}$;
      I (father (l')) $:= [\min\{a, b, \text{new\_}p\}, \max\{a, b, \text{new\_}p\}]$;
      Bubble up (father (l'))
     end.

A more detailed analysis yields the following ([13])

Theorem: *For any finite set* $V \subset \mathbb{R}$ *(with n elements) an interval-heap IH with* $P(IH) = V$ *can be constructed in* $O(n)$ *time using only n storage cells for data, which allows the following operations*

FindMin	*in* $O(1)$ *time*
FindMax	*in* $O(1)$ *time*
DeleteMix	*in* $O(\log n)$ *time*
DeleteMax	*in* $O(\log n)$ *time.*

Interval heap find applications in the field of *intersection* and *vision problems* in *computational geometry* very well ([13]).

An interesting *question* remains: can this technique be adapted to other sets of "complex objects", where the partial order is induced in a simple way from a total order on the "basic constituents"?

In chapter 4 we will give a partial answer to this question.

3. Heap-ordered trees

The "heap-shaped"-condition of heaps is very disadvantageous to support further basic operations (f.e. Divide and Merge).

To represent arbitrary partial orders we have to omit this condition too. Thus we introduce the following concept.

Definition: A *heap-ordered tree* T is an ordered finite rooted tree (not necessarily a binary tree) containing a set of items, one item in each node, with the items arranged in *heap order*:

if v is any node, then the key of the item in v is no less than the key of the item in its father, provided v has a father.

Consequently the root of T contains an item with *minimal* key. Now to merge two item-disjoint heap-ordered trees T_1 (with root r_1) and T_2 (with root r_2) into one heap-ordered tree T, we compare the roots:

if the item in r_1 has a smaller key, we make r_2 a new child of r_1, otherwise we make r_1 a child of r_2 (see figure 4):

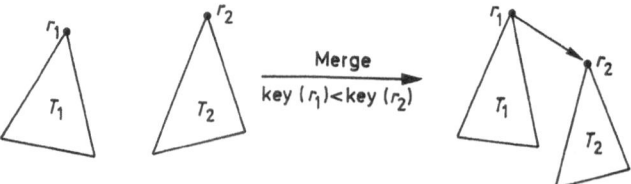

Figure 4. Merging two item-disjoint heap-ordered trees

Thus this basic operation takes $O(1)$ time if we use an appropriate tree representation.

There are two commonly used representations of heap-ordered trees:

A) *"Child sibling"* representation or *"binary tree"* representation: each node has a *left pointer* to its *first child* and a *right pointer* to its *next sibling* (or null).

The effect of the representation is to convert a heap-ordered tree T into a half-ordered binary tree T' with empty right subtree, where by half-ordered we mean that the key of any node is at least as small as the key of any node in its *left* subtree (see figure 5).

In order to support further operations it sometimes appears to be useful, to store with each node a *third pointer* to its *father* in the binary tree T'.

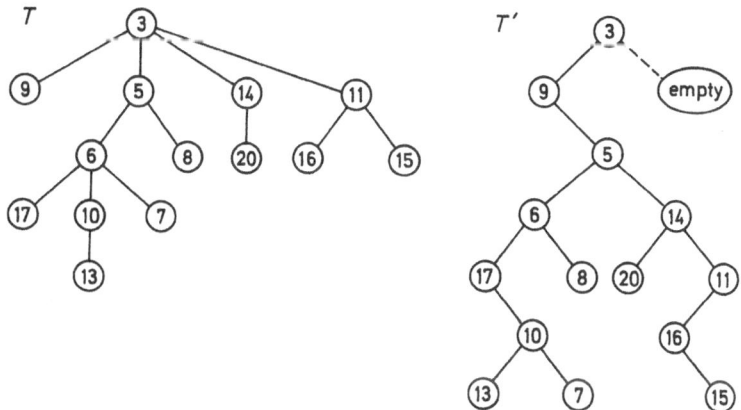

Figure 5

Remark: It is easy to realize that *merging two item-disjoint heap-ordered trees* using this representation can be implemented in $O(1)$ *time*.

Insert (x) by making x into a one-node tree and merge it with the actual tree yields an $O(1)$-*Insert-algorithm* too. The same holds for FindMin and for an important operation, which frequently appears in solving shortest path problems, assignment problems and minimum spanning tree problems:

DecreaseKey (Δ, v, h):   decrease the key of the item, associated to node v by subtracting the non-negative real number Δ.

This operation can be reduced to "merge" in the following sense:

Subtract Δ from the key in v. If v is not the root of the tree, cut the arc joining v to its father and merge the two trees as usually (figure 4).

The "child sibling" representation obviously guarantees $O(1)$-*worst-case running time* for *DecreaseKey* too.

It turns out that the *crucial operation* is $\boxed{\text{DeleteMin}}$, where we have to merge possibly a large number of subtrees (see figure 6).

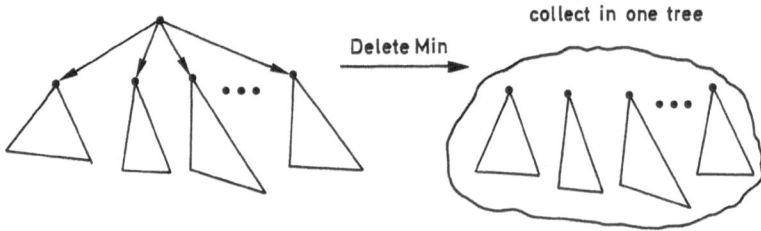

Figure 6

⌈An arbitrary Delete (v), where we know the position of v, reduces to DeleteMin by cutting the arc joining v and its father, performing a DeleteMin on the subtree with root v and merging the resulting tree with the other tree formed by the cut.⌋

Fredman, Sedgewick, Sleator, Tarjan [9] proposed a *pairing procedure*: order the children of each node in the order they were attached by merge operations, with the first (youngest) child being the one most recently attached. Then merge the first and second, the third and fourth, and so on. Afterwards merge each remaining tree to the last one, working from the next-to-last back to the first. The resulting tree is called a *pairing heap*. The authors can show the following (see also [25])

Theorem: *On pairing heaps the operation FindMin has $O(1)$ amortized time, the DeleteMin runs in $O(\log n)$ amortized time.*

But the authors conjecture $O(1)$-amortized time bounds for the operations Insert, Merge, DecreaseKey too. Jones [12] has given some experimental results which may indicate that pairing heaps are competitive in practice with all known alternatives.

The best known alternative is the *Fibonacci-Heap* (*F-heap*), which is a finite collection of item-disjoint heap-ordered trees (Fredman, Tarjan [7]). The standard representation of Fibonacci-Heaps is as follows.

B) *"F-Heap" representation*:
Each node contains a pointer to its father (or null, if it is a root), and a pointer to one of its children (if any). The children of each node are doubly linked in a circular list. Each node also contains its degree and a bit for marking purposes. The roots of all trees, which constitutes the *F*-Heap, are doubly linked in a circular list, access to the heap is done by a pointer pointing to a root with minimal key ("minimal node" of *F*-Heap).

To carry out Insert (new), one creates a new *F*-Heap consisting of one node and merges two *F*-Heaps.

The merge in general can be performed by combining the root lists into a single list and pointing to the minimal of the two minimal elements.

These operations can be implemented in $O(1)$ time.

The ⎡ DeleteMin ⎤ , can be done as follows:

remove the minimum node v from F-Heap H, then concatenate the list of children of v with the list of roots of H other than v, and repeat the following *Linking Step*: find any two trees, whose root have the *same degree*, and *merge* them as usually (figure 4).

The new tree root has degree one greater than the degrees of the old tree roots. Once there are *no* two trees with root of the same degree, we form a list of the remaining roots.

The authors could prove the following ([7])

Theorem: *If we start with an empty F-Heap and perform an arbitrary sequence of F-Heap operations FindMin, DecreaseKey, Insert, Merge and DeleteMin or Delete* (v), *then the total sequence takes at most* $O(m + n \log n)$ *time, where n denotes the number of DeleteMin and Delete-operations and m denotes the number of all remaining operations.*

Using F-Heaps the authors could improve running times for several network optimization problem like shortest path problems, assignment problems and minimum spanning tree problems, where the set of "candidates" (like in Dijkstra's shortest path algorithm) can be best represented by data structures which support the F-Heap operations efficiently [7].

The F-Heap is a "lazy merging" version of the wellknown structure, the *binomial tree (binomial queue)* ([26], [23]).

A binomial tree is defined inductively as follows:

a binomial tree of rank 0 consists of a single node; a binomial tree of rank $k > 0$ is formed by merging two binomial trees of rank $k - 1$ (see figure 7).

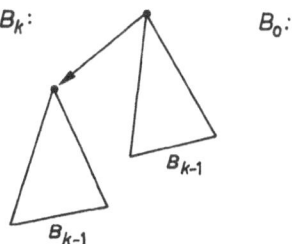

Figure 7. Binomial tree

If additionally B_k is heap-ordered, we call B_k a *binomial heap*.

A forest of binomial heaps is called a *binomial queue*.

A binomial tree B_k of rank k contains exactly 2^k nodes and its root has exactly degree k. Thus every node in an n-item binomial tree has degree at most $\log n$. If n is binary coded as $(d_k d_{k-1} \ldots d_1 d_0)$, the set of n items can be represented by the collection of binomial trees B_i (where $d_i \neq 0$). There are obviously strong connections to dynamic decomposition techniques related to number systems ([2], [21]).

Based on these techniques the following recent results can be summarized.

Theorem: *There is an implicit implementation of a binomial queue (IBQ), which allows Insert in constant time and DeleteMin in $O(\log n)$ worst-case running time (Carlsson, Munro, Poblete [4], using the redundant binary number system).*

Another recent result is due to Driscoll, Gabow, Shrairman and Tarjan [5]. The authors use *relaxed heaps*, a type of binomial queue that allows heap order to be violated on "small" parts.

Theorem: *The relaxed heap achieves worst-case-time bounds $O(1)$ for DecreaseKey and $O(\log n)$ for DeleteMin.*

Especially relaxed heaps give a processor-efficient *parallel* implementation of Dijkstra's shortest path algorithm and hence of a lot of other algorithms in network optimization (see for more details [5]).

4. Priority search trees—the concept of symbiosis

The last question of chapter 2 can be partially answered by another approach, the symbiosis of two dissimilar data structures. We will demonstrate the idea by the well known priority search tree (McCreight [14]).

Let E be the set of points $P = (x, y)$ in the (real) plane, where $\prec$ denotes the lexicographic partial order induced by

$$P_1 = (x_1, y_1) \prec P_2 = (x_2, y_2) \quad \text{iff} \quad (x_1 < x_2) \quad \text{or} \quad (x_1 = x_2 \text{ and } (y_1 \leq y_2)).$$

Let V be a finite subset of E,

$$V = \{P_i = (x_i, y_i)/i = 1, \ldots, n\} \quad \text{and} \quad V_x := \{x_i/(x_i, y_i) \in V\}.$$

For simplicity let us assume furthermore

1) $x_i \in U_x$ (a finite universe of x-coordinates)
2) $x_i \neq x_j$ for $i \neq j$

Definition: A *priority search tree* (*PST*) is a binary tree representing a finite set of points V of the plane with the following properties:

A1) PST is a leaf-oriented *search tree* for the x-coordinates, especially
    a) for each x-coordinate of V exists a leaf with "*splitvalue*" x and
    b) every non-leaf v has a *splitvalue* $s(v) \in U_x$ which is the maximum of split-values of all nodes in the left subtree of v.
A2) Each node v can store besides the splitvalue $s(v)$ a point $P(v) \in V$ (eventually unused)
B1) Every point $P_i = (x_i, y_i) \in V$ is located on the x-search path from the root to the leaf with splitvalue x_i.
B2) If node v stores a point $P \in V$ then its father does it too ($v \neq$ root (PST) of course).
B3) PST is (min-)*heap-ordered* with respect to y-coordinates.

Example: $V = \{(4, 1), (9, 4), (15, 18), (20, 3), (19, 9), (12, 4), (1, 2), (5, 6)\}$,

$U_x = \{1, 4, 5, 9, 12, 15, 19, 20\}$.

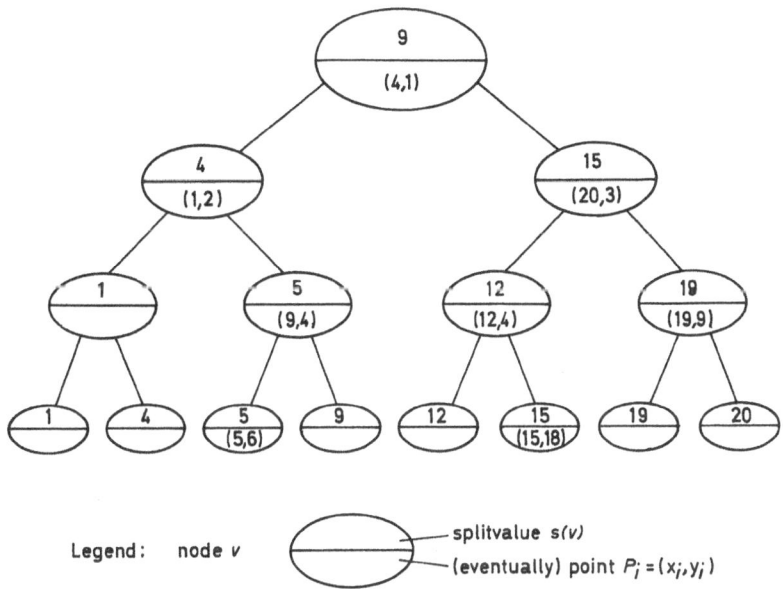

Figure 8. Priority search tree

If we assume for simplicity that a left complete skeleton tree with exactly all splitvalues from U_x was preprocessed (the "upper part" of all nodes),

Insert a new point $P = (x, y)$ is easy and runs as follows:
    if v is the "current node" (at first the root) with splitvalue $v.s$ and (eventually) a point v.point, then do
      if v.point is undefined, then store P in v
      else if v.point.$y \leq y$ then {follow $P.x$}
         begin if $v.s \geq x$ then Insert P in leftson (v)
                 else Insert P in rightson (v)
      end
      else {heap-order has to be guaranteed}
            store P in v and continue insertion with the old point v.point.

Moreover priority search trees can be *balanced* by rotations, where indeed each rotation may cause $O(h)$ time (h: height of the relevant PS-subtree).

Thus taking any class of balanced trees, which only needs a constant number of rotations for any balancing step

  – f.e. take "half balanced trees" (Olivié [20]), "red-black trees" (Guibas, Sedgewick [11]) or more general (a, b)-trees with $b \geq 2a + 2$, which additionally allow (parallel) top-down-balancing (Mehlhorn [15])–

we get the following results (McCreight [14]).

Theorem: *Let V be a dynamic set of points of the (real) plane, V_x a subset of a finite universe U_x and let n denote the cardinality of the actual set V. The following operations*

(1) *Insert (Delete) a point into (from) V*
(2) *given $x_0 \in U_x$, $x_1 \in U_x$ and $y_1 \in \mathbb{R}$, find among all points $P = (x, y)$ of V such that $x_0 \leqq x \leqq x_1$ and $y \leqq y_1$*
    a point whose x is minimal (or maximal)
(3) *given $x_0 \in U_x$, $x_1 \in U_x$ find among all points $P = (x, y)$ of V such that $x_0 \leqq x \leqq x_1$*
    a point whose y is minimal
(4) *given $x_0 \in U_x$, $x_1 \in U_x$ and $y_1 \in \mathbb{R}$, enumerate all points $P = (x, y)$ of V such that*

$$x_0 \leqq x \leqq x_1 \text{ and } y \leqq y_1$$

can be implemented by a balanced priority search tree efficiently, using $O(n)$ space and

$$O(\log n) \text{ worst-case time for operations (1), (2), (3),}$$

and at most $O(k + \log n)$ time for operation (4), where k denotes the number of points to be reported.

The last statement can be best illustrated by the following figure:

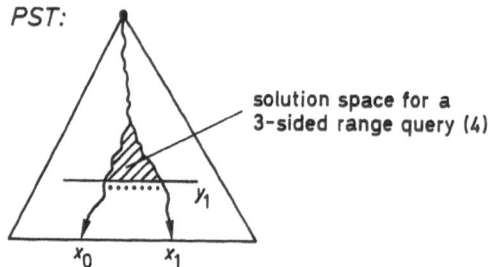

PST:

solution space for a
3-sided range query (4)

Figure 9

Remark: The efficient implementation of operations (1), (2), (3), (4) by balanced priority search trees yields important results in a wide range of applications, f.e. detecting all overlapping rectangles in a large set of axis-parallel rectangles etc.

5. Conclusion

Dynamic partial orders frequently can be represented efficiently by heaps or their generalizations, supporting a lot of important operations. These also includes some decomposition techniques, we did not mention here in detail.

Some effort was done with respect to lower bounds, too.

We finally refer to results of Fredman [8], Gonnet, Munro [10] and Noltemeier [17]: in the article mentioned last, lower complexity bounds for Find-operations as well as Delete-Find-operations in generalized heaps are given based on the well known theorem of Dilworth.

References

[1] M. D. Atkinson, J. R. Sack, N. Santoro, Th. Strothotte An efficient implicit double-ended priority queue. SCS-TR 55, Carleton Univ., Ottawa 1984.

[2] J. L. Bentley, Decomposable searching problems. Inform. Proc. Lett., 8, 244–251 (1979).

[3] S. Carlsson, A variant of heapsort with almost optimal number of comparisons. Informat. Proc. Lett., 24, 247–250 (1987).

[4] S. Carlsson, J. I. Munro, P. V. Poblete, An Implicit Binomial Queue with Constant Insertion Time in: LNCS, vol. 318, Springer: Berlin-Heidelberg-New York-Tokyo, 1–13 (1988).

[5] J. R. Driscoll, H. N. Gabow, R. Shrairman, R. E. Tarjan, Relaxed Heaps: An Alternative to Fibonacci Heaps with Applications to Parallel Computation. Comm. ACM, 31, no. 11, 1343–1354 (1988).

[6] R. W. Floyd, Algorithm 245: Treesort 3. Comm. ACM, 7, 701 (1964).

[7] M. L. Fredman, R. E. Tarjan, Fibonacci Heaps and their Uses in improved Network Optimization Algorithms. Journal ACM, 34, no. 3, 596–615 (1987).

[8] M. L. Fredman, Refined Complexity Analysis for Heap Operations. Journal of Computer and System Sciences, 35, 269–284 (1987).

[9] M. L. Fredman, R. Sedgewick, D. D. Sleator, R. E. Tarjan, The pairing Heap: A New Form of Self-Adjusting Heap. Algorithmica, 1, 111–119 (1986).

[10] G. H. Gonnet, J. I. Munro, Heaps on Heaps. SIAM J. Comput., 15, no. 4, 964–971 (1986).

[11] L. J. Guibas, R. Sedgewick, A dichromatic framework for balanced trees. Proc. 19th IEEE-FOCS, 8–21 (1978).

[12] D. W. Jones, An empirical comparison of priority queues and event set algorithms. Comm. ACM, 29, no. 4, 300–311 (1986).

[13] J. v. Leeuwen, D. Wood, Interval Heaps. Techn. Report, Dec. 1987.

[14] E. M. McCreight, Priority Search Trees. SIAM J. Comput., 14, no. 2, 257–276 (1985).

[15] K. Mehlhorn, Datenstrukturen und effiziente Algorithmen, Band 1: Sortieren und Suchen Stuttgart: Teubner, 1986.

[16] H. Noltemeier, Informatik III–Einführung in Datenstrukturen. 2. edit., München-Wien: Carl Hanser, 1988.

[17] H. Noltemeier, On a generalization of heaps, in: Graphtheoretic Concepts in Computer Science (WG'80), (ed. H. Noltemeier) LNCS, vol. 100, Berlin-Heidelberg-New York: Springer, 127–136 (1981).

[18] H. Noltemeier, Reduction of directed graphs to irreducible kernels. Techn. Rep., Göttingen 1974.

[19] H. Noltemeier, Graphentheorie mit Algorithmen und Anwendungen. Berlin-New York: de Gruyter, 1976.

[20] H. J. Olivié, A new class of balanced search trees: half-balanced binary search trees. RAIRO Theor. Inform., 16, 51–71 (1982).

[21] M. H. Overmars, J. v. Leeuwen, Two general methods for dynamizing decomposable searching problems. Computing, 26, 155–166 (1981).

[22] D. D. Sleator, R. E. Tarjan, Self-adjusting heaps, SIAM J. Comput., 15, 1, 52–69 (1986).

[23] Th. Strothotte, J. R. Sack, Heaps in heaps. SCS-TR-67, Carleton Univ., Ottawa 1985.

[24] R. E. Tarjan, Data Structures and Network Algorithms. SIAM, Reg. Conf. Series in Appl. Math., Philadelphia, 1983.

[25] R. E. Tarjan, Amortized computational complexity. SIAM, J. Algebraic Discrete Methods 6, 2, 306–318 (1985).

[26] J. Vuillemin, A data structure for manipulating priority queues. Comm. ACM, 21, 4, 309–315 (1978).

[27] J. W. J. Williams, Algorithm 232–Heapsort. Comm. ACM, 7, 347–348 (1964).

Prof. Dr. Hartmut Noltemeier,
Institut für Informatik I,
University of Würzburg, Am Hubland,
D-8700 Würzburg, Federal Republic of Germany
email: noltemei @ uniwue.uucp

Computing Suppl. 7, 141–153 (1990)

Communication Complexity

Ulrich Faigle, Enschedel and **György Turán,** Chicago, Ill.

Abstract — Zusammenfassung

Communication Complexity. In this introductory survey, the general communication complexity problem is discussed from an ordertheoretic point of view. In particular, results about special classes of ordered sets are reported. Furthermore, open problems and related ordertheoretic questions are mentioned.

AMS Subject Classification: 68C25.

Key words: communication complexity, communication protocol, order, linear extension, rank.

Daskommunikationskomplexitätsproblem. In dieser einführenden Übersicht wird das Kommunikationskomplexitätsproblem von einem ordnungstheoretischen Standpunkt aus diskutiert. Insbesondere werden Resultate über spezielle Klassen geordneter Mengen vorgestellt. Außerdem wird auf offene Probleme und verwandte ordnungstheoretische Fragen eingegangen.

1. Introduction

A basic technique for proving lower bounds on the complexity of VLSI layouts relates the size of a cut through a prospective chip with the information flow across it (Thompson [1979], see also Lipton and Sedgewick [1981]). Roughly, the argument goes as follows. Suppose a chip of area A is to compute the value of the Boolean function $f(z_1, \ldots, z_k)$ in T time units. Also suppose we can separate the input variables into two groups $x = (z_1, \ldots, z_m)$ and $y = (z_{m+1}, \ldots, z_k)$ via a cut through the chip that cuts through $\sqrt{A}$ horizontal wires (Fig. 1).

During the computation of $f(x, y)$ then a total of not more than $T \cdot \sqrt{A}$ bits are exchanged between part I and part II of the processor. Hence, if the computation of f requires the exchange of a least J bits of information, we obtain the lower bound

$$J^2 \le AT^2$$

on the layout complexity of the proposed chip. The communication complexity tries to obtain a bound on this number J.

A mathematical model for the communication complexity is due to Yao [1979]. There are two finite sets E_1 and E_2 associated with two players (or "processors") I

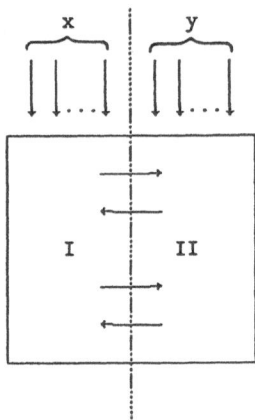

Figure 1

and II respectively and a function

$$f: E_1 \times E_2 \rightarrow \{0,1\}.$$

We assume that f is completely known to both players. Now I chooses an "input" $x \in E_1$ and II an "input" $y \in E_2$. In a cooperative effort, I and II try to answer the query

$$? \, f(x,y) = 1 \, ?$$

by exchanging as few bits of information as possible. The communication between I and II goes in "rounds" and is governed by a "protocol" (for a formal definition, see Section 2): one player sends some bits of information to the other player. Based on the information available to him so far, the other player responds by sending some bits of information back etc.. The game continues until at least one player has sufficient information to answer the query.

Thus, for example, the communication exchange can be done using at most $\lceil \log |E_2| \rceil$ bits via the **trivial** protocol: II sends the "name" of his element $y \in E_2$ to I. Since I has complete knowledge of f, he can then determine $f(x,y)$.

There are different interpretations of the Boolean function f possible. We may think of f as the indicator function of some binary relation between E_1 and E_2. Equivalently, we may associate with f its **incidence matrix** $M = M(f)$ with $(0,1)$-entries so that E_1 and E_2 represent the rows and columns respectively of M. In this sense, each $(0,1)$-matrix can be viewed as arising from some communication problem and hence is a **communication matrix**.

Another aspect of the communication problem takes the columns of the $(0,1)$-matrix M as incidence vectors of subsets of rows. Thus one could formulate the game relative to a finite set E and a family $\mathscr{F}$ of subsets of E: player I chooses an element $x \in E$ and player II a subset $Y \in \mathscr{F}$. The relevant query is

$$? \, x \in Y \, ?$$

Finally, players I and II may select elements $x, y \in P$, where P is a (partially) ordered set completely known to both players, and ask the query

$$? \, x < y \, ?$$

(Note that we consider here the **strict** order relation of P as the "interesting" binary relation because the trivial protocol turns out to be already optimal for the query $? \, x \leq y \,$? (see Section 3)). This seemingly special case contains the original model: we may order $E_1 \cup E_2$ via the only non-trivial relations for $e_1 \in E_1$, $e_2 \in E_2$,

$$e_1 < e_2 \quad \text{if} \quad f(e_1, e_2) = 1.$$

Relative to this order, we then have

$$f(x, y) = 1 \qquad \text{if and only if} \quad x < y.$$

In this introductory survey on the communication complexity problem, we take the ordertheoretic point of view (for a comprehensive general survey see Lovász [1988]). In Section 2, we define the deterministic and the nondeterministic complexity of the communication problem for Boolean functions (or, equivalently, for binary relations). Lower and upper bounds are presented in Sections 3 and 4. We then look at the comunication problem for special classes of ordered sets and finally discuss further ordertheoretic ramifications and open problems.

2. Communication Complexity

We will now describe the model for communication complexity as formulated by Lovász and Saks [1988].

Given $f : E_1 \times E_2 \to \{0, 1\}$, we consider the binary relation $Q = f^{-1}(1)$. The decision problem to solve is whether a given input (x, y) satisfies $f(x, y) = 1$, i.e., $(x, y) \in Q$.

A **deterministic communication protocol** for recognizing Q is a decision tree T whose nodes are of two types. An internal node of type i ($i = 1, 2$) is labeled by a function from E_i to the set of children of that node. A leaf of type i is labeled by a function from E_i to the set $\{\text{YES}, \text{NO}\}$. Each input $(x, y) \in E_1 \times E_2$ specifies a unique path from the root to a leaf of T in such a way that Q consists exactly of those inputs (x, y) yielding the outcome YES.

T is a **one-way** protocol if T has depth 1. The cost of an internal node in T equals the logarithm (here always assumed relative to base 2) of the number of its children, i.e., the number of bits needed to specify a child. The cost $c(P)$ of a path P from the root to a leaf in T is the sum of the costs of its internal nodes. Thus the complexity $c(T)$ of the protocol T can be defined as

$$c(T) = \max\{c(P) : P \text{ path in } T\}.$$

The **deterministic (communication) complexity** $cc(Q)$ of the binary relation Q is given by

$$cc(Q) = \min\{c(T) : T \text{ protocol for } Q\}.$$

We say that Q **is one-way optimal** if there exists an optimal one-way protocol for Q.

Let the $(m \times n)$-matrix M be the $(0, 1)$-incidence matrix of Q, and denote by m^* and n^* the number of pairwise different rows and columns respectively. The clearly

$$cc(Q) \leq \min\{\lceil \log m^* \rceil, \lceil \log n^* \rceil\}.$$

In fact, Q is one-way optimal if and only if $cc(Q) = \min\{\lceil \log m^* \rceil, \lceil \log n^* \rceil\}$. Another general upper bound follows from the rank of the matrix M:

$$cc(Q) \leq rk\, M.$$

To see this, assume that the first r rows M_r of M form a row basis. Then M_r (and hence $M!$) has at most 2^r-different columns. Thus $n^* \leq 2^r$.

A **proof scheme** for the relation $Q \subseteq E_1 \times E_2$ consists of a set $\mathscr{P}$ of **proofs** together with two **verification relations** $V_1 \subseteq E_1 \times \mathscr{P}$ and $V_2 \subseteq E_2 \times \mathscr{P}$ such that $(x, y) \in Q$ if and only if there exists a proof $p \in \mathscr{P}$ with the property $(x, p) \in V_1$ and $(y, p) \in V_2$. The **nondeterministic (communication) complexity** $cc^*(Q)$ is the number

$$cc^*(Q) = \min\{\lceil \log |\mathscr{P}| \rceil : \mathscr{P} \text{ proof scheme for } Q\}.$$

Say that $R \subseteq Q$ is a **1-rectangle** of Q if there are subsets $F_1 \subseteq E_1$ and $F_2 \subseteq E_2$ such that $R = F_1 \times F_2$. With this terminology, Lipton and Sedgewick [1981] have observed that a proof scheme of Q may equivalently be defined as a set of 1-rectangles whose union equals Q. Indeed, for each proof p of the proof scheme $\mathscr{P}$, the set

$$R(p) = \{(x, y) \in Q : (x, p) \in V_1 \text{ and } (y, p) \in V_2\}$$

is a 1-rectangle in Q.

An important parameter, therefore, is the minimal number $\kappa_1 = \kappa_1(Q)$ of 1-rectangles needed to cover the relation Q. Introducing $\bar{\kappa}_1 = \bar{\kappa}_1(Q)$ as the minimal number of **disjoint** 1-rectangles needed to cover Q, we have

$$\kappa_1 \leq \bar{\kappa}_1.$$

Our original problem, of course, could also have been phrased relative to the query

$$?\, f(x, y) = 0\,?$$

i.e., relative to the complementary relation $Q^c = (E_1 \times E_2) \backslash Q$. This leads to the analogous parameters $\kappa_0 = \kappa_1(Q^c)$ and $\bar{\kappa}_0 = \bar{\kappa}_1(Q^c)$ associated with 0-**rectangles** of Q.

The equality $cc(Q) = cc(Q^c)$ certainly holds. On the other hand, $cc^*(Q)$ and $cc^*(Q^c)$ may differ greatly. Consider, for examle, the strict order relation P with the following Hasse diagram:

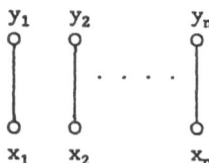

Here we have $\kappa_1(P) = n$ while the 0's of the associated incidence matrix can be covered with $O(\log n)$ 0-rectangles.

As outlined in the Introduction, we may associate with any $(0, 1)$-matrix M an ordered set, which we denote by $P(M)$, via a natural construction. $P(M)$ is **bipartite** in the sense that its Hasse diagram is a bipartite graph. If M is the (strict) incidence matrix of the order P, we write $PP = P(M)$ for this **bipartite reduction** of P.

It is clear that the communication problem is the same for the order P and its bipartite reduction PP. In particular, we have $\kappa_1(P) = \kappa_1(PP)$ etc.

Let us interprete a 0-rectangle of M in the order $P(M)$. We have a subset $X \subseteq E_1$ and a subset $Y \subseteq E_2$ such that there are no order relations between X and Y. We consider

$$\mathrm{fil}(X) = \{e \in E_1 \cup E_2 : e \geq x \text{ for some } x \in X\}$$

$$\mathrm{id}(Y) = \{e \in E_1 \cup E_2 : e \leq y \text{ for some } y \in Y\}.$$

Then $\mathrm{fil}(X) \cap \mathrm{id}(Y) = \phi$. Moreover $\mathrm{id}(Y)$ is an **ideal** in P, i.e., closed with respect to smaller elements, while $\mathrm{fil}(X)$ is a **filter**, i.e., the settheoretic complement of some ideal. In other words, the 0-rectangles in M correspond to the pairs of disjoint ideals and filters in $P(M)$.

3. Lower Bounds

To prove lower bounds for the communication complexity, we consider the binary relation under investigation to be given via its $(0, 1)$-incidence matrix.

Let $g(M)$ be some nonnegative integervalued function which is defined for $(0, 1)$-matrices M and satisfies for all disjoint row (or column) partitions (M_0, M_1) of M the inequality

$$g(M_0) + g(M_1) \geq g(M).$$

Theorem 1: $\lceil \log g(M) \rceil \leq cc(M)$. $\square$

We sketch the proof of Theorem 1: Let T be an optimal protocol which starts with player II, say, who has selected a column of M. M_0 is the submatrix of M consisting of those columns whose choice would lead to a message starting with "0". M_1 is defined analogously. By hypothesis, we may assume $g(M_0) \geq g(M)/2$ or $g(M_1) \geq g(M)/2$. Observe now that $cc(M) - 1$ is an upper bound for both $cc(M_0)$ and $cc(M_1)$. Hence, by induction on the size of M, we have

$$\log(g(M)/2) \leq cc(M) - 1.$$

which yields the bound.

Examples for functions satisfying our hypothesis are the parameters $\kappa_1(M), \bar{\kappa}_1(M), \kappa_0(M)$, and $\bar{\kappa}_0(M)$. They are strong enough to demonstrate that for "almost all" communication problems the trivial protocol is optimal. In fact, Yao [1979] proved

Theorem 2: *Let M be a random $(n \times n)$-matrix with $(0, 1)$-entries. Then*

$$cc(M) \geq \log n - 2$$

with probability at least $1 - 2^{-n^2/2}$. $\square$

An interesting choice is $g(M) = rk(M)$ (Mehlhorn and Schmidt [1982]). Interest in this **rank lower bound** arises from the curious fact that many researchers feel this bound to be possibly far from optimal. Yet, no class of examples is known for which lower bounds of higher order than $O(\log(rk\, M))$ could be proved.

The rank lower bound quickly exhibits communication problems for ordered sets relative to the "$\leq$"-relation to be trivial: the corresponding incidence matrix is easily seen to be of full rank, which implies that the trivial protocol is optimal.

A further example is the **positive rank** $g(M) = rk^+(M)$ (Yannakakis [1988]). Formally, $rk^+(M)$ is the minimal number p such that there are nonnegative matrices A (with p columns) and B (with p rows) with the property $M = AB$. From a geometrical point of view is $rk^+(M)$ the smallest number of nonegative vectors needed to generate a cone that contains all column vectors of M.

Obviously $rk(M) \leq rk^+(M)$ holds. But it is already not known whether $rk^+(M) = O(rk\, M)$ is true. It is easy to see that

$$rk(M) \leq \bar{\kappa}_1(\overline{M})$$

and one can also verify

$$\kappa_1(M) \leq rk^+(M).$$

The last inequality is a consequence of the observation that

$$M = Z_1 + Z_2 + \cdots + Z_p,$$

where $Z_i = a_i b_i^T$ with $a_i = i$-th column of A and $b_i^T = i$-th row of B. The desired cover of 1-rectangles is obtained from the supports of the matrices Z_i.

Let us briefly discuss two lower bounds which are implied by the rank lower bound. We consider the communication problem for the order P (as always, relative to the strict order relation). A **linear extension** of P is an arrangement $L = x_1 x_2 \cdots x_n$ of the ground set underlying P such that $x_i < x_j$ in P implies $i < j$ in L. The **lineality** of L is the number

$$\ell(L) = |\{(x_i, x_{i+1}) : x_i < x_{i+1} \text{ in } P\}|,$$

i.e., the number of adjacent comparabilities in L. The **lineality** of P is

$$\ell(P) = \max\{\ell(L) : L \text{ linear extension of } P\}.$$

Gierz and Poguntke [1983] have observed that

$$\ell(P) \leq rk(P) \leq n - w(P),$$

where $w(P)$ is the **width** of P, namely the maximal number of pairwise incomparable elements of P. Since $n - w(P)$ actually yields an upper bound on the communication complexity of P (see Section 4), we note

$$\lceil \log \ell(P) \rceil \leq \lceil \log rk(P) \rceil \leq cc(P) \leq \lceil \log(n - w(P)) \rceil + 1.$$

It is furthermore straightforward to see that

$$\ell(P) \leq \kappa_1(P).$$

One only has to write down the incidence matrix of P according to an optimal linear extension. Then exactly $\ell(P)$ 1's will appear on the side-diagonal. We illustrate the use of these inequalities for the Boolean algebra $P = B_n$ with 2^n elements. Füredi and Reuter [1989] have shown that

$$\ell(B_n) = 2^{n-1}.$$

Hence we obtain

$$n - 1 \leq cc^*(B_n) \leq cc(B_n) \leq n.$$

Comparing the bounds obtained from P with those obtained from its bipartite reduction, we have $rk(P) = rk(PP)$, while $\ell(P) < \ell(PP)$ may be possible. Such an improvement, however, can never yield more than 1 bit as Reuter [1988] has observed:

$$\tfrac{1}{2}\ell(PP) \leq \ell(P) \leq \ell(PP).$$

4. Upper Bounds

A general technique for proving upper bounds on the communication complexity was exhibited by Lovász and Saks (see Lovász [1988]). We describe it in its ordertheoretic setting.

Let $h(P)$ be a nonnegative integer-valued function which is defined for all bipartite orders P and satisfies the two properties:

(i) $h(P) = 0$ if and only if P is an antichain
(ii) $h(A) + h(P \setminus A) \leq h(P)$ for all ideals $A \subseteq P$.

Denoting by $P = P(M)$ the bipartite order associated with the communication matrix M, we obtain

Theorem 4: $cc(P) \leq \lceil \log h(P) \rceil \cdot (1 + \lceil \log \kappa_0(P) \rceil)$. $\square$

We sketch the proof of Theorem 4 by describing an appropriate protocol recursively. Its validity can be established by induction on $h(P)$.

The first thing to observe is that for each ideal $A \subseteq P$, we have $h(A) \leq h(P)/2$ or $h(P \setminus A) \leq h(P)/2$. We now choose κ_0 0-rectangles that cover all 0's of M and let $A_1, \ldots, A_k, A_{k+1}, \ldots, A_{\kappa_0}$ be the associated ideals in $P(M)$ (see Section 2). We may assume that

$$h(A_i) \leq h(P)/2 \quad \text{if} \quad i \leq k$$

$$h(P \setminus A_i) \leq h(P)/2 \quad \text{if} \quad i > k.$$

Player II now tries to find an ideal A_i $(i \leq k)$ containing his chosen element $y \in E$. If he is successful, he sends "0" and the index i with a total of at most $1 + \lceil \log \kappa_0 \rceil$ bits. Otherwise he sends "1".

Player I in turn tries to find a filter $P \backslash A_j$ $(j > k)$ containing his chosen element $x \in E$. If he is successful, he sends "0" and the index j of that filter $P \backslash A_j$.

If neither player is successful $x < y$ must hold (because we started out with a 0-cover) and the game ends. Otherwise the game continues with either A_i or $P \backslash A_j$ and h-value at most $h(P)/2$.

Applying Theorem 4 with the lineality $h(P) = \ell(P)$, one gets the original result of Lovász and Saks:

$$cc(M) \leq \lceil \log \ell(P(M)) \rceil (1 + \lceil \log \kappa_0(M) \rceil).$$

In view of the discussion in Section 3, this upper bound has a number of consequences:

(a) $cc(M) \leq \lceil \log rk(M) \rceil (1 + \lceil \log \kappa_0(M) \rceil)$

(b) $cc(M) \leq \lceil \log \kappa_1(M) \rceil (1 + \lceil \log \kappa_0(M) \rceil)$

(c) $cc(M) \leq \lceil \log \bar{\kappa}_1(M) \rceil \cdot \lceil \log \bar{\kappa}_1(M) \rceil)$

(d) $cc(M) \leq \lceil \log rk(M) \rceil \cdot \lceil \log rk^+(M) \rceil)$

(b) is the well-known upper bound of Aho et al. [1983]. (c) and (d) are due to Yannakakis [1988]. It is a challenging open problem to decide, for example, whether $rk^+(M)$ can be replaced by $rk(M)$ in (d).

Also note that for general ordered sets P, the Lovász-Saks bound implies

$$cc(P) \leq (1 + \lceil \log \ell(P) \rceil)(1 + \lceil \log \kappa_0(P) \rceil).$$

As already mentioned in the previous section, for general ordered sets P on n elements the bound

$$cc(P) \leq 1 + \lceil \log(n - w(P)) \rceil$$

is valid. This can be established with a two-way protocol as follows. To set up the game, both players agree on a fixed antichain $W \subseteq P$ of size $|W| = w(P)$. After the choice of their elements $x, y \in P$, the first player sends a "0" to the second player if his element x lies in W. Otherwise he sends "1" and the name of his element. If player II receives "0" and his element y also lies in W, $x < y$ cannot hold. If y does not lie in W, player II sends the name of y to the other player with at most $\lceil \log(n - w(P)) \rceil$ bits.

An application of the **width upper bound** allows to determine the communication complexity of cycle-free orders up to one bit. Recall that an order P is **cycle-free** if the comparability graph $G(P)$ of P is **chordal**, i.e., contains no vertex-induced circuits of length 4 or more. Duffus et al. [1982] have shown that the equality

$$\ell(P) = n - w(P)$$

is valid for cycle-free orders P. Hence such an order P satisfies

$$\lceil \log(n - w(P)) \rceil \le cc^*(P) \le cc(P) \le 1 + \lceil \log(n - w(P)) \rceil.$$

5. Some Classes of Orders

We will now briefly discuss results and problems concerning special classes of orders. The details can be found in Faigle and Turán [1989].

Let us begin with a general result and consider a class $\mathscr{P}$ of orders which is closed under taking suborders.

Theorem 5: *If there exists at least one bipartite order Q such that $Q \notin \mathscr{P}$, then for all $P \in \mathscr{P}$,*

$$cc(P) = O(\log(rk\,P)). \quad \square$$

The idea of the proof consists in showing that each order P in the class $\mathscr{P}$ gives rise to an incidence matrix M that has not more than $(rk\,P)^c$ different columns, where $c = c(Q)$ is a constant depending on Q. Theorem 5 will then be implied by the trivial protocol. The existence of such a constant $c(Q)$, however, can be argued with the help of "Sauer's Lemma" (see Lovász [1979, Problem 13.10c]):

Lemma: *Let R be some finite set and $\mathscr{F}$ a family of distinct subsets of R such that*

$$|\mathscr{F}| > \binom{|R|}{0} + \binom{|R|}{1} + \cdots + \binom{|R|}{k}.$$

Then, there exists a subset $R' \subseteq R$ with $|R'| = k + 1$ such that $\{F \cap R' : F \in \mathscr{F}\}$ comprises all subsets of R'. $\square$

The Lemma is applied as follows. If the incidence matrix M had "too many" distinct columns, M would contain the characteristic vectors of some power set large enough to exhibit Q as an induced suborder of P.

Unfortunately, the requirement that the order Q in Theorem 5 be bipartite turns out to be essential in the proof. Whether an analogous statement is true for, say, Q equal to a 3-element chain, is not known (note that an affirmative answer would imply the communication complexity for arbitrary $(0, 1)$-matrices M to be $O(\log(rk\,M))$ since then $\mathscr{P}$ could be taken to be the class of **all** bipartite orders).

As examples for Theorem 5, we could choose $\mathscr{P}$ as the class of orders P not containing

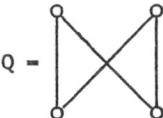

$$Q =$$

as a suborder. In particular, $\mathscr{P}$ could be the class of all cycle-free orders. By taking

$$Q = $$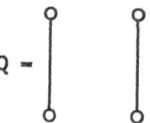

we, furthermore, could select $\mathscr{P}$ as the class of interval orders (see Fishburn [1970]).

We should make clear, however, that a direct analysis often yields a sharper bound on the communication complexity. For cycle-free orders, we have seen this in the previous section. For so-called generalized interval orders P,

$$cc^*(P) = cc(P) = \lceil \log(rk\, P) \rceil$$

can be proved (cf. Faigle et al. [1988]).

What about classes of orders that cannot be characterized by forbidden induced suborders? Recall that an order is said to be N-**free** if its Hasse diagram (!) does not contain

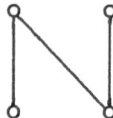

as an induced subconfiguration. For each N-free order P, one can show

$$\lceil \log(rk\, P) \rceil = cc^*(P) = cc(P).$$

Our proofs for the rank bound yielding the exact communication complexity relies on the notion of rank optimality. Thereby an order P is said to be **rank-optimal** if

$$\ell(P) = rk(P).$$

N-free orders are rank-optimal. While non-bipartite generalized interval orders need not be rank optimal, their bipartite reductions always are. We do not know whether the rank bound is sharp for all rank-optimal orders. An unsolved test case is presented by cycle-free orders, which are known to be rankoptimal. The best we can prove is that here the rank bound is "nearly optimal".

Similarly, no counterexample to the conjecture that

$$\kappa_1(P) \geq rk(P)$$

be true for all rank-optimal orders is known to us. Again, already the case of cycle-free orders is unsolved.

We finally mention a generalization of the notion of N-freeness. Say that an order is M-**free** if its Hasse diagram does not admit

as an induced subconfiguration. (Note that the class of M-free orders is **not** closed under ordertheoretic duality). In general, an M-free order need not be rank-optimal. Yet, one can prove that for each M-free order P,

$$cc(P) \leq 1 + \lceil \log(rk\,P) \rceil,$$

where the upper bound can be achieved with a one-way protocol. The nondeterministic complexity $cc^*(P)$ for M-free orders is unclear.

6. Remarks

In our ordertheoretic formulation of the communication problem, we have considered the query

(i) "Is x (strictly) smaller than y?"

We could similarly have investigated the query

(ii) "is x a lower neighbor of y?"

Within the framework of communication complexity, which of the two queries is easier to decide? Are they equally difficult? (An affirmative answer to the last question can be given for, e.g., N-free orders or interval orders).

Observe that the rank lower bounds for (i) and for (ii) may be different as shown by the order with the following Hasse diagram (Lovász and Zádori [1988]):

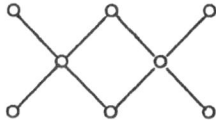

Here (i) leads to rank = 5, while (ii) yields only rank = 4.

In view of the observed strength of the rank lower bound for the communication complexity problem, it appears necessary to develop a better combinatorial understanding of the rank parameter for binary relations. Determining the rank of types of binary relations is generally a non-trivial problem. It might therefore be interesting to sketch a powerful technique due to Lovász and Saks [1988] for computing the rank of certain communication matrices.

The following communication problem is considered. Player I chooses an element x and player II chooses an element y in a **lattice** L (i.e., an ordered set L with maximal element 1 and minimal element 0 and the property that any two elements $a, b \in L$ have a unique maximum $a \vee b \in L$ and a unique minimum $a \wedge b \in L$). The query to be decided is now

(iii) "Is $x \wedge y = 0$?"

(Note that (iii) is different from (i)!). This problem is termed the **meet problem**. Let C be the communication matrix for the meet problem relative to L. In order to determine $rk(C)$, one considers the matrix $Z = (\zeta_{xy})$ associated with L:

$$\zeta_{xy} = \begin{cases} 1 & \text{if } x \leq y \\ 0 & \text{otherwise} \end{cases}$$

The combinatorial identity of Wilf allows to express $C = Z^T D Z$, where D is the diagonal matrix defined via the Moebius function:

$$(D)_{xx} = \mu(0, x).$$

(Recall that the **Moebius function** $\mu(x, y)$ of L, by definition, is given via the entries of the inverse matrix Z^{-1} (see Rota [1964]). Hence

Theorem 6: $rk\,C = |\{x \in L : \mu(0, x) \neq 0\}|$. $\square$

Theorem 6 has far-reaching consequences. In Hajnal et al. [1988], for example, it is used to show that the communication complexity is $\Omega(n \log n)$ for the following problem: player I and II want to decide if a graph G on n vertices is connected. G is unknown to I and II. But for one half of all possible edges player I knows which one's belong to G. Player II similarly supervises the other half of he possible edges. (For more examples, see Lovász [1988]).

References

A. V. Aho, J. D. Ullman, and M. Yannakakis (1983): On notions of information transfer in VLSI circuits. Proc. 15th ACM STOC, 133–139

D. Duffus, I. Rival, and P. Winkler (1982): Mimimizing setups for cycle-free ordered sets. Proc. Amer. Math. Soc. 85, 509–513

U. Faigle and Gy. Turán (1989): On the communication complexity of ordered sets. Working paper.

U. Faigle, R. Schrader, and Gy. Turán (1988): The communication complexity of generalized interval orders. Memorandum No. 745, Fac. of Applied Math., Universiteit Twente

P. C. Fishburn (1970): Intransitive indifference with unequal indifference intervals. Journ. Math. Psychol. 7, 144–149

Z. Füredi and K. Reuter (1989): The jump number of suborders of the power set order. In Memorandum No. 787 (U. Faigle and C. Hoede eds.), Faculty of Applied Mathematics, Universiteit Twente, 57–59 To appear in: ORDER

G. Gierz and W. Poguntke (1983): Mimimizing setups for ordered sets: a linear algebraic approach. SIAM Journ. Algebr. Discr. Methods 4, 132–144

A. Hajnal, W. Maass, and Gy. Turán (1988): On the communication complexity of graph properties. Proc. 20th ACM STOC, 186–191

R. J. Lipton and R. Sedgewick (1981): Lower bounds for VLSI. Proc. 13th ACM STOC, 300–307

L. Lovász and M. Saks (1988): Lattices, Möbius functions and communication complexity. Preprint, Department of Computer Science, Eötvös Loránd University, Budapest, Hungary

L. Lovász (1988): Communication complexity: A survey. Report No. 89555-OR, Institut für Operations Research, Universität Bonn

L. Lovász (1979): Combinatorial Problems and Exercises. North-Holland, Amsterdam

L. Lovász and L. Zádori (1988): personal communication.

K. Mehlhorn and E. M. Schmidt (1982): Las Vegas is better than determimism in VLSI and distributed computing. Proc. 14th ACM STOC, 330–337

K. Reuter (1988): The jump number and the lattice of maximal antichains. Preprint, FB Mathematik, TH Darmstadt

G.-C. Rota (1964): On the foundations of combinatorial theory I. Theory of Möbius functions. Z. Wahrscheinlichkeitstheorie 2, 340–368

C. D. Thompson (1979): Area-time complexity for VLSI. Proc. 11th ACM STOC, 81–88
M. Yannakakis (1988): Expressing combinatorial optimization problems by linear programs. Preprint
A. C.-C. Yao (1979): Some complexity questions related to distributive computing. Proc. 11th ACM
   STOC, 209–213.

Ulrich Faigle, Faculty of Applied Mathematics,
Universiteit Twente, NL-7500 AE Enschede,
The Netherlands
György Turán, Department of Mathematics and
Computer Science University of Illinois, Chicago,
Ill. 60637, U.S.A. and Automata Theory Research
Group, Hungarian Academy of Sciences, Szeged,
Hungary

Computing Suppl. 7, 155–189

Path Problems in Graphs*

Günter Rote, Graz

Abstract — Zusammenfassung

Path Problems in Graphs. A large variety of problems in computer science can be viewed from a common viewpoint as instances of "algebraic" path problems. Among them are of course path problems in graphs such as the shortest path problem or problems of finding optimal paths with respect to more generally defined objective functions; but also graph problems whose formulations do not directly involve the concept of a path, such as finding all bridges and articulation points of a graph. Moreover, there are even problems which seemingly have nothing to do with graphs, such as the solution of systems of linear equations, partial differentiation, or the determination of the regular expression describing the language accepted by a finite automaton.
We describe the relation among these problems and their common algebraic foundation.
We survey algorithms for solving them: vertex elimination algorithms such as Gauß-Jordan elimination; and iterative algorithms such as the "classical" Jacobi and Gauß-Seidel iteration.

AMS 1980 mathematics subject classification (1985 revision): 68-01, (68E10, 68R10, 68Q, 05C, 65-01, 65F05, 65F10, 16A78, 90C35, 90C50)
CR categories and subject descriptors (1987 version): F.2.1. [**Analysis of algorithms**]: Numerical algorithms and problems—*computations on matrices*; G.1.0. [**Numerical analysis**]: *Numerical algorithms*; G.1.3. [**Numerical analysis**]: Numerical linear algebra—*linear systems (direct and iterative methods), matrix inversion*; G.2.2. [**Discrete mathematics**]: Graph theory—*network problems, path and circuit problems*; I.1.2. [**Algebraic manipulation**]: Algorithms—*algebraic algorithms*

Additional keywords and phrases: algebraic path problem, iteration equation, Gauß-Jordan elimination, block decomposition, shortest paths, optimal paths, automatic differentiation, finite automata, regular expression
General terms: algorithms, theory
IAOR categories: computational analysis, graphs, networks, network programming.

Wegeprobleme in Graphen. Es gibt eine Vielfalt von Problemen, die sich als "algebraische" Wege-Probleme interpretieren lassen. Dazu gehören natürlich Wege-Probleme auf Graphen wie das gewöhnliche kürzeste-Wege-Problem oder das Bestimmen bester Wege unter allgemeineren Optimalitätskriterien, aber auch Probleme, deren Definition nur indirekt mit Wegen zu tun hat, wie das Bestimmen aller Brücken und Artikulationsknoten eines Graphen. Sogar einige Probleme, die anscheinend überhaupt nichts mit Graphen zu tun haben, lassen sich als algebraische Wege-Probleme behandeln: Man kann z. B. lineare Gleichungssysteme lösen, auf schnellem Weg alle partiellen Ableitungen eines Ausdrucks berechnen, oder einen regulären Ausdruck für die formale Sprache bestimmen, die ein endlicher Automat akzeptiert.
In dieser Überblicksarbeit wird einerseits dargestellt, wie man alle diese Problem unter einen Hut bringt, indem man eine gemeinsame algebraische Formulierung für sie findet; andererseits werden verschiedene Lösungsalgorithmen besprochen: Kneneliminations-Algorithmen (z. B. Gauß-Jordan-Elimination) und iterative Algorithmen (wie die klassischen Iterationsverfahren von Jacobi und Gauß-Seidel).

* This work was written while the author was at the Freie Universität Berlin, Fachbereich Mathematik. It was partially supported by the ESPRIT II Basic Research Action Program of the EC under contract no. 3075 (project ALCOM).

Contents

1. Introduction

Path problems can be seen as a unified framework for a lot of problems from different fields. Solution procedures for these problems were initially discovered independently of each other. When the connection between these solution methods became apparent, various attempts have been made to lay a common theoretical basis for them. Also, new applications of the method were explored.

It would be difficult to give a complete account of the area of path problems. A complete bibliography including all applications would fill many pages. There have been several good accounts in textbooks and treatises, like Gondran and Minoux [6], chapter 3; Zimmermann [17], chapter 8; Carré [4], chapters 3 and 4.

The purpose of this exposition is to give an introduction to this area and an overview of some of the more interesting applications and interpretations of path problems, and to give a relatively small glimpse of the theory which has been established in this field. We shall do this in a very elementary way.

We shall not deal with specialized algorithms for the shortest path problem in particular. Also, algorithms which use special properties of the underlying graphs will only be mentioned.

The reader who wants to know more about path problems in general or about specific applications should consult the above-mentioned references. References to the literature about various applications are almost completely omitted from this survey unless they appeared recently.

2. Two example problems

2.1. Example 1: The shortest path problem

2.1.1. Description of the problem—a numerical example

Consider the directed graph shown in figure 1. It has $n = 4$ vertices and ten arcs, which are labeled with weights. A *path* in a graph is a sequence of $l + 1$ vertices $(v_0, v_1, \ldots, v_l)$ such that (v_i, v_{i+1}) is an arc of the graph, for $i = 0, 1, \ldots, l - 1$. It is called a path *from v_0 to v_l*. For example, $p = (1, 3, 4, 4, 1, 1, 3, 2)$ is a path from 1 to 2. Note that we allow repetition of vertices and of arcs in a path. With every path, we may associate its *weight*, which is the sum of the weights of its arcs. The weight of the example path p is thus $7 + 3 + 2 + 2 + (-5) + 7 + (-1) = 15$. Note that we must distinguish between an *empty* path without arcs, like the path $q = (1)$ from 1 to 1, and the path $r = (1, 1)$, which contains one arc (a loop). The weight of the empty path is assumed to be zero.

The weights can be interpreted as lengths of the arcs, and then the weight of the path is simply its total length. Or the weights could be the time taken to traverse an arc; or the money that one has to pay (or that one gains) for traversing an arc.

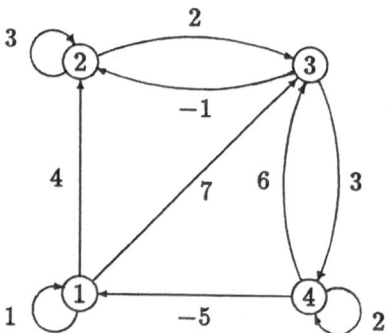

Figure 1. A network

The last interpretation is one for which arcs of negative weight—as in the example—make sense.

The (all-pairs) shortest path problem is the following:

For each pair (i, j) of vertices, compute the weight x_{ij} of the *shortest path* (i.e., the path of smallest weight) from i to j.

2.1.2. A system of equations

With the graph G, we may associate its weighted adjacency matrix

$$A = \begin{pmatrix} 1 & 4 & 7 & \infty \\ \infty & 3 & 2 & \infty \\ \infty & -1 & \infty & 3 \\ -5 & \infty & 6 & 2 \end{pmatrix}.$$

The element a_{ij} is the weight of the arc (i, j), if this arc exists. Artificial weights of ∞ have been inserted in the places where no arc exists. These artificial arcs will do no harm, because a path using such an arc has weight ∞; thus it will certainly not affect the shortest path.

Now we are going to set up a system of equations which the desired quantities x_{ij} will fulfill. Consider a shortest path p from i to j. If $i \neq j$, this path must contain at least one arc, i.e., it is of the form $(i = v_0, v_1, \ldots, v_l = j)$, with $l \geq 1$. If it is a shortest path, then the subpath $p' = (v_1, v_2, \ldots, v_l = j)$, must be a shortest path from v_1 to j. Thus $x_{ij} = a_{ik} + x_{kj}$, for some $k = v_1$. On the other hand, the expression $a_{ik} + x_{kj}$, for any k, is the length of some path from i to j, namely the path starting with the arc (i, k) and continuing along the shortest path from k to j. Thus, we have

$$x_{ij} = \min_{1 \leq k \leq n} (a_{ik} + x_{kj}), \qquad \text{for } i \neq j. \tag{1}$$

For $i = j$, the above considerations apply with one change: The empty path from i

to i without arcs is an additional candidate for the shortest path, and thus we have to extend the above equation:

$$x_{ii} = \min\left\{ \min_{1 \le k \le n} (a_{jk} + x_{kj}), 0 \right\}. \tag{1'}$$

In the above example,

$$X = \begin{pmatrix} 0 & 4 & 6 & 9 \\ 0 & 0 & 2 & 5 \\ -2 & -1 & 0 & 3 \\ -5 & -1 & 1 & 0 \end{pmatrix}$$

is the unique solution of this system, and it represents the lengths of the shortest paths.

2.2. Example 2: The language accepted by a finite automaton

A finite automaton is a machine which reads words (sequences of symbols over some alphabet Σ) and either accepts them or rejects them. It can be specified by its *transition diagram*, which is a finite directed graph (see figure 2). The vertices of the graph are the *states* of the automaton. One of the vertices (vertex 1 in our case) is designated as the *start state*, and a subset of the vertices is designated as the *final states*. The arcs are labeled by subsets of letters from Σ. ($\Sigma = \{f, g, h\}$ in our example.) The automaton starts in the designated start state and reads the symbols of an input word one by one. A label z on an arc (i, j) means the following: If the automaton is in state i and the next symbol which it reads is z, it may go to state j. When the automaton is in state i and there is no arc labeled z which leaves i, the automaton cannot continue and stops. When there is at most one choice of an arc for each state and each input letter, the automaton is called a deterministic automaton; otherwise it is a non-deterministic automaton, but this difference does not concern us here.

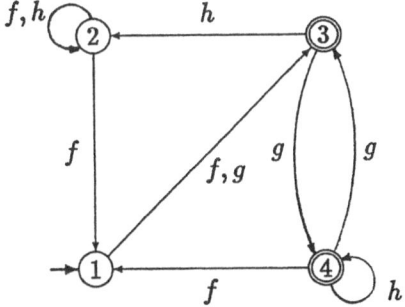

Figure 2. The transition diagram of a finite automaton. The initial state is state 1. The final states are marked by double circles.

We say that the automaton *accepts* a word, if there is a sequence of state transitions leading from the start state to a final state while reading this word. To put it differently, let $p = (v_0, v_1, \ldots, v_l)$ be a path from the start state v_0 to some final state v_l. If z_i is a label of the edge (v_{i-1}, v_i), for $1 \leq i \leq l$, then the word $z_1 z_2 \ldots z_n$ is accepted by the automaton. For example, the automaton shown in figure 2 accepts the word *fgghhfhfff* because it leads from state 1 to state 3 via the path $(1, 3, 4, 3, 2, 2, 2, 2, 2, 1, 3)$. Thus, the automaton defines a subset of words (a formal language) which it accepts.

Thus our problem is now the following:

> For each final state j, determine the set x_{1j} of words which lead from the initial state 1 to state j.

In order to solve this problem, we must introduce a few notations. We are working with words (finite sequences) over some alphabet Σ, including the *empty word* ε, which contains no symbols. We write the concatenation of two words a and b as $a \cdot b$ or simply as ab. If A and B are sets of words, then $A \cdot B$ denotes the set $\{ab \mid a \in A, b \in B\}$.

As above, we can set up a matrix (a_{ij}), where a_{ij} denotes the set of labels of the arc (i, j). Let x_{ij} denote the set of all words by which the automaton can be lead from state i to state j. We shall solve the more general problem of computing x_{ij} for all pairs of states i and j.

As in the case of the shortest path problem, we shall set up a system of equations. Consider the set x_{ij}. When the automaton is started in state i, the first state transition must lead to some state k. In order to go to k the automaton must read a symbol from a_{ik}. Then it must eventually go to j. The possible words which lead from k to j are collected in the set x_{kj}. Thus, the words which lead from i to j via k as the first vertex are exactly the set $a_{ik} \cdot x_{kj}$. This is also true if there is no arc from i to k, because then $a_{ik} = \emptyset$. Now we just have to take the union over all possible states k, and we get an equation for x_{ij}. Again, if $i = j$, we have to consider the additional possibility that the automaton reads nothing and stays in state i, and thus we have to adjoin the empty word.

$$x_{ij} = \bigcup_{k=1}^{n} (a_{ik} \cdot x_{kj}), \qquad \text{for } i \neq j, \text{ and}$$

$$x_{jj} = \bigcup_{k=1}^{n} (a_{jk} \cdot x_{kj}) \cup \{\varepsilon\}$$

(2)

2.3. Summary

In this section, we have described two examples of path problems. In both cases, we have stated the problem, and we have derived a set of equations which the solutions have to fulfill. It is, however, not the case that every solution of the equations is a solution of the respective problem that we started with. We will say more about the relation between the solution of equations and the original formulation of path problems in sections 4.2 and 7.

In the next section we will exhibit the common algebraic structure of our two sample problems.

3. An algebraic framework

3.1. *Semirings—the algebraic path problem*

The two systems of equations (1)–(1') and (2) have a similar structure:

$$x_{ij} = \bigoplus_{k=1}^{n} (a_{ik} \otimes x_{kj}), \qquad \text{for } i \neq j, \text{ and}$$

$$x_{jj} = \bigoplus_{k=1}^{n} (a_{jk} \otimes x_{kj}) \oplus ① \tag{3}$$

In the case of the shortest path problem, $\oplus$ denotes max, $\otimes$ denotes $+$, and $①$ denotes 0, and in the second example problem, $\oplus$ denotes $\cup$, $\otimes$ denotes product (concatenation), and $①$ means $\{\varepsilon\}$. "$\bigoplus_{k=1}^{n}$" is a notation for the $\oplus$-sum of a sequence of elements, analogous to $\sum_{k=1}^{n}$.

The algebraic structure which is behind these two operations is a *semiring* $(S, \oplus, \otimes)$, i.e., a set S with two binary operations $\oplus$ and $\otimes$, which fulfills the following axioms:

(A$_1$) $(S, \oplus)$ is a commutative semigroup with neutral element $⓪$:

$$a \oplus b = b \oplus a,$$

$$(a \oplus b) \oplus c = a \oplus (b \oplus c),$$

$$a \oplus ⓪ = a.$$

(A$_2$) $(S, \otimes)$ is a semigroup with neutral element $①$, and $⓪$ as an absorbing element:

$$(a \otimes b) \otimes c = a \otimes (b \otimes c)$$

$$a \otimes ① = ① \otimes a = a,$$

$$a \otimes ⓪ = ⓪ \otimes a = ⓪.$$

(A$_3$) $\otimes$ is distributive over $\oplus$:

$$(a \oplus b) \otimes c = (a \otimes c) \oplus (b \otimes c),$$

$$a \otimes (b \oplus c) = (a \otimes b) \oplus (a \otimes c).$$

We shall now discuss why these axioms are natural assumptions for any path problem. $\oplus$ must be commutative and associative, because the sum $\bigoplus_{k=1}^{n}$ in equation (3) must be independent of the order of the operands. $\otimes$ is the operation by which the weight of a path is computed from the weights of its arcs, and we require the operation to be associative.

$$w((v_0, v_1, \ldots, v_l)) = a_{v_0 v_1} \otimes a_{v_1 v_2} \otimes \cdots \otimes a_{v_{l-1} v_l}.$$

① is the weight of the empty path. What we want to compute is, in terms of the semiring, the ⊕-sum of the weights of all paths from i to j:

$$x_{ij} = \bigoplus_{\substack{p \text{ is a path} \\ \text{from } i \text{ to } j}} w(p) \tag{4}$$

In this formulation, the problem is called the *algebraic path problem*. However, this formulation contains an infinite sum. This raises questions of "convergence", which fall outside the realm of classical algebra. Thus, we shall mainly stick to the formulation as a system of equations (3). Later, in section 4.3, we shall also work with the interpretation of x_{ij} as a sum of paths.

We have implicitly used (left) distributivity in the derivation of the equations (1), (2), and (3), when we have expressed the sum of the paths from i to j whose first arc is (i, k) as $a_{ik} \otimes x_{kj}$.

The axioms regarding ⓪ are not essential, since a semiring without ⓪ can always be extended by adding a new zero element according to the axioms, like the element ∞ in the shortest path problem. Thus, we shall not insist that there is always a zero element. (The axioms regarding the existence of ① could also be omitted w. l. o. g., but it requires a trickier construction to show this.) We shall denote the product of an element with itself by the power notation

$$a^k = a \otimes a \otimes \cdots \otimes a \qquad (k \text{ times})$$

with the usual convention $a^0 = ①$. Also, for better readability, we shall often omit the multiplication sign ⊗, from now on.

3.2. Types of semirings, ordered semirings

The examples of semirings which we will encounter belong mostly to three main groups:

1. $(S, \otimes, \leq)$ is a linearly or partially ordered semigroup (with neutral element ①), and ⊕ is the supremum or infimum operation (the maximum or minimum operation, in case of a linearly ordered semigroup).

   An ordered semigroup is a semigroup with an order relation which is monotone with respect to the semigroup multiplication:

   $$a \leq b \text{ and } a' \leq b' \Rightarrow a \otimes a' \leq b \otimes b'.$$

   When ⊕ is defined in this way, it is clearly an associative and commutative operation. The above monotonicity property translates into distributivity. If necessary, we must add a smallest (or largest, resp.) element ⓪.

   In the example of shortest paths, the order $\leq$ was just the usual order for real numbers, and ⊕ was the minimum operation; in the second example, the order relation is set inclusion, and ⊕ is the supremum (least upper bound) with respect to this order.

Another possibility to characterize this class of semirings is that the idempotent law holds for ⊕:

$$a \oplus a = a.$$

For this class of *idempotent semirings*, the relation defined by

$$a \le b \Leftrightarrow a \oplus b = b \tag{5}$$

is a partial order. Thus, we can either start with an ordered semigroup and define ⊕ as the supremum operation (if the supremum exists always), or we can start with an idempotent semiring and define the partial order by (5). In both cases we get the same kind of algebraic structure.

2. $(S, \oplus, \otimes)$ is a ring, or a subset of a ring. Examples are the field of real numbers $(\mathbb{R}, +, \cdot)$ with ordinary addition and multiplication, or any subsemiring of the reals, like the natural numbers. For these cases, equation (3) has a closer connection to conventional linear algebra.

3. The elements of S are sets of paths, of path weights, or the like. An example which we have already encountered is the set of words which leads a finite automaton from one state to another. Here, what we deal with are not sets of paths, but sets of label sequences that correspond to paths. Usually, ⊕ is the union operation, and thus these semirings fall also under the first category, since they are ordered by the set inclusion relation.

A semiring $(S, \oplus, \otimes)$ with a partial order relation ≤ which is monotone with respect to both operations is called an *ordered semiring* $(S, \oplus, \otimes, \le)$:

$$(a \le b \quad \text{and} \quad a' \le b') \Rightarrow a \oplus a' \le b \oplus b' \quad \text{and} \quad a \otimes a' \le b \otimes b'.$$

All semirings of the first type are ordered semirings, but there are also several examples from the second class, like the non-negative reals $(\mathbb{R}_+, +, \cdot, \le)$ with the usual order.

We say that $(S, \oplus, \otimes, \le)$ is ordered by the *difference relation*, or *naturally ordered*, if

$$\text{for all } a, b \in S: \quad (a \le b \Leftrightarrow \text{there is a } z \in S \text{ such that } a \oplus z = b). \tag{6}$$

When ⊕ is the min or inf operation of an ordered semigroup, the relation ≤ must simply be reversed in order that this definition makes sense. With this proviso, all natural examples of ordered semirings that arise in applications are ordered by the difference relation.

3.3. Matrices

The $(n \times n)$-matrices over a semiring S form another semiring if matrix addition and matrix multiplication are defined just as usual in linear algebra: If $A = (a_{ij})$ and $B = (b_{ij})$ then $A \oplus B = C$ and $A \otimes B = D$, where

$$c_{ij} = a_{ij} \oplus b_{ij}$$

and

$$d_{ij} = \bigoplus_{k=1}^{n} a_{ik} b_{kj}.$$

$(S^{n \times n}, \oplus, \otimes)$ is a semiring. The zero matrix is the matrix whose entries are all $\bigcirc$, and the unity element is the unit matrix I with $\textcircled{1}$'s on the main diagonal and $\bigcirc$'s otherwise.

Thus, we may rewrite equation (3) in matrix form as follows:

$$X = I \oplus AX. \tag{7}$$

A symmetric variation of this equation can also be derived by splitting the possible paths from i to j according to their last arc:

$$X = I \oplus XA. \tag{7'}$$

4. Direct solution procedures (elimination algorithms)

We are looking for a solution to the matrix equation (7). If we hope to find a solution for $(n \times n)$-matrices we must surely be able to solve the case of (1×1)-matrices, i.e., of scalars. Thus, we look at the following equation, the so-called *iteration* equation:

$$x = \textcircled{1} \oplus ax. \tag{8}$$

Let us consider what this equation amounts to in the two examples that we have dealt with in the beginning.

$$x = \min\{0, a + x\}$$

For $a > 0$, there is a unique solution $x = 0$. For $a = 0$ the solution of this equation is not unique: any $x \le 0$ is a solution. For $a < 0$, there is no solution.

Correspondingly, the system of equations (1)–(1') need not have a unique solution, or it can have no solution at all. For example, if we add an arc $(2, 4)$ of length 1, then the column vector $(x_{13}, x_{23}, x_{33}, x_{43})$ of the updated matrix X of shortest distances can be changed to $(-100, -104, -105, -105)$, and we still get a solution. It can be shown that this ambiguity of the solution occurs exactly if the graph contains a cycle of weight 0. In our case, this is the cycle $(1, 2, 4, 1)$.

If we reduce the length of the new arc $(2, 4)$ to 0, then no solution fulfills (1)–(1'). The reason is that the graph contains a cycle of negative weight, and hence the shortest paths are undefined. We can remedy this situation by adding a new element $-\infty$ to the semiring. This element solves (8) for $a < 0$. The result $x_{ij} = -\infty$ means then that there are arbitrarily short paths from i to j.

In the semiring of formal languages, we get

$$x = \{\varepsilon\} \cup a \cdot x.$$

This equation always has a solution, namely the set

$$a^* = \{\varepsilon\} \cup a \cup a^2 \cup a^3 \cup \cdots,$$

which consists of all words which are concatenations $w_1 w_2 \ldots w_l$ of an arbitrary number of words $w_i \in a$.

In general, we denote the solution (or some solution) of (8) by a^*, and correspondingly, we denote the solution of the matrix equation (7) by A^*. Semirings in which a^* always exists are called *closed* semirings.

If we repeatedly substitute the expression for x in (8) into the right-hand side, starting with (8), we get

$$x = ① \oplus ax$$
$$= ① \oplus a(① \oplus ax) = \cdots = ① \oplus a \oplus a^2 \oplus a^3 \oplus a^4 \oplus \cdots$$

If this sequence remains stable after a finite number of iterations, then the sum is a solution of (8).

By multiplying (8) from the right side with any element $b \in S$, we obtain that if $x = a^*$ solves (8) then $y = a^*b$ is a solution of the more general equation

$$y = b \oplus ay. \tag{9}$$

4.1. An elimination procedure—Gauß-Jordan elimination

In this section we shall derive a solution of (3) or (7) by purely algebraic means, namely by successive elimination of variables, very much like in solving ordinary systems of linear equations. Since the intuition for what is actually going on during the solution process may get lost when we write the procedure in full generality, we will first illustrate the method with a specific example. Later, in section 4.3, we will see that the coefficients that arise in the elimination process can be interpreted in a different way, namely as sums of certain subsets of path weights.

When we look at equation (3), we can see that the column index j of the unknowns x_{ij} is the same for all variables which occur in one equation. This means that the system (3) consists really of four decoupled systems of equations, one for each column of X. A column of j represents the sums of paths from all vertices to the vertex j. Similarly, an equation system for a row of X, i.e., for the paths starting from a fixed vertex i (the *single-source* path problem), can be obtained from (7').

Let us take a closer look at one specific system, say, for the third column of a (4×4)-matrix:

$$x_{13} = a_{11}x_{13} \oplus a_{12}x_{23} \oplus a_{13}x_{33} \oplus a_{14}x_{43}$$
$$x_{23} = a_{21}x_{13} \oplus a_{22}x_{23} \oplus a_{23}x_{33} \oplus a_{24}x_{43}$$
$$x_{33} = a_{31}x_{13} \oplus a_{32}x_{23} \oplus a_{33}x_{33} \oplus a_{34}x_{43} \oplus ①$$
$$x_{43} = a_{41}x_{13} \oplus a_{42}x_{23} \oplus a_{43}x_{33} \oplus a_{44}x_{43}$$

$$\tag{10.0}$$

The quantities a_{ij} are the given data, and the x_{i3} are the unknowns. This is very much like an ordinary system of equations, except that the unknowns appear on both sides: They appear explicitly on the left side, and implicitly on the right side. The iteration equation (9) is the paradigm for handling this situation in the case of one variable: Note that the first equation has the structure

$$x_{13} = ax_{13} \oplus b,$$

with $a = a_{11}$ and $b = a_{12}x_{23} \oplus a_{13}x_{33} \oplus a_{14}x_{43}$. If we assume that a^* exists, then we know that $x_{13} = a^*b$ is a solution of the above equation, and thus we get an explicit expression for x_{13}:

$$x_{13} = a_{11}^*(a_{12}x_{23} \oplus a_{13}x_{33} \oplus a_{14}x_{43})$$
$$= a_{11}^*a_{12}x_{23} \oplus a_{11}^*a_{13}x_{33} \oplus a_{11}^*a_{14}x_{43}$$

Substituting this into the other equations and collecting terms, we get a new system:

$$x_{13} = a_{12}^{(1)}x_{23} \oplus a_{13}^{(1)}x_{33} \oplus a_{14}^{(1)}x_{43}$$
$$x_{23} = a_{22}^{(1)}x_{23} \oplus a_{23}^{(1)}x_{33} \oplus a_{24}^{(1)}x_{43}$$
$$x_{33} = a_{32}^{(1)}x_{23} \oplus a_{33}^{(1)}x_{33} \oplus a_{34}^{(1)}x_{43} \oplus \text{①}$$
$$x_{43} = a_{42}^{(1)}x_{23} \oplus a_{43}^{(1)}x_{33} \oplus a_{44}^{(1)}x_{43},$$

$$(10.1)$$

where the new coefficients $a_{ij}^{(1)}$ are defined as follows:

$$a_{1j}^{(1)} = a_{11}^*a_{1j}, \qquad\qquad \text{for } j > 1,$$
$$a_{ij}^{(1)} = a_{ij} \oplus a_{i1}a_{11}^*a_{1j}, \qquad \text{for } i \neq 1, j > 1.$$

Let us summarize what we have done in order to eliminate x_{13}: First we have used the equation where x_{13} occurs on both sides for obtaining an explicit expression of x_{13} in terms of the other variables. This was done by solving the iteration equation. Then we have used this explicit expression for substituting x_{13} in all other places where it occurred.

The four equations of the last system fall in two groups: The first equation is the explicit expression for x_{13}; the remaining three equations form an implicit system for the other three variables x_{23}, x_{33}, and x_{43}, which has the same structure as the original system, but one variable less.

Thus we can repeat the elimination process in essentially the same way as we have begun it: We eliminate x_{23} from the second equation, assuming that $(a_{22}^{(1)})^*$ exists, and substitute this into the other three equations. We get a new system (10.2), which looks like (10.1) except that x_{23} does not appear on the right-hand side and the superscripts are (2) instead of (1). The elimination of x_{33} is a bit different, because of the ① on the right-hand side. We get

$$x_{33} = (a_{33}^{(2)})^*(a_{34}^{(2)}x_{43} \oplus \text{①})$$
$$= (a_{33}^{(2)})^*a_{34}^{(2)}x_{43} \oplus (a_{33}^{(2)})^*.$$

When we substitute this into the other equations, we get a constant term in all equations:

$$x_{13} = a_{13}^{(3)} \oplus a_{14}^{(3)} x_{43}$$

$$x_{23} = a_{23}^{(3)} \oplus a_{24}^{(3)} x_{43}$$

$$x_{33} = a_{33}^{(3)} \oplus a_{34}^{(3)} x_{43} \qquad (10.3)$$

$$x_{43} = a_{43}^{(3)} \oplus a_{44}^{(3)} x_{43}$$

The new coefficients are determined by the following recursions:

$$a_{33}^{(3)} = (a_{33}^{(2)})^*,$$

$$a_{i3}^{(3)} = a_{i3}^{(2)}(a_{33}^{(2)})^*, \qquad \text{for } i \neq 3,$$

$$a_{3j}^{(3)} = (a_{33}^{(2)})^* a_{3j}^{(2)}, \qquad \text{for } j > 3,$$

$$a_{ij}^{(3)} = a_{ij}^{(2)} \oplus a_{i3}^{(2)}(a_{33}^{(2)})^* a_{3j}^{(2)}, \qquad \text{for } i \neq 3, j > 3.$$

For reasons which will become clear later, we regard the constant terms as the third column of the coefficient matrix. In the remaining elimination steps (there is only one more to follow), this column will remain, whereas the remaining columns will be successively eliminated.

So we finally eliminate x_{43} from the last equation, and we are left with the explicit solution

$$x_{13} = a_{13}^{(4)}$$

$$x_{23} = a_{23}^{(4)}$$

$$x_{33} = a_{33}^{(4)} \qquad (10.4)$$

$$x_{43} = a_{43}^{(4)}$$

with

$$a_{44}^{(4)} = (a_{44}^{(3)})^*,$$

$$a_{i4}^{(4)} = a_{i4}^{(3)}(a_{44}^{(3)})^*, \qquad \text{for } i \neq 4.$$

The purpose of this calculation has been to make it clear that the solution of the matrix iteration $X = AX \oplus I$ (equation (7)) can be reduced to n solutions of the scalar iteration $x = ax \oplus \text{①}$ for the *pivot elements* $a = a_{11}, a_{22}^{(1)}, a_{33}^{(2)}, a_{44}^{(3)}$. The remaining steps in the derivation were merely substitutions of variables and applications of the semiring axioms (distributivity, etc.) which pose no problems.

Let us summarize in a general way the equations that we have obtained. In the above example, the column index of the solution was $l = 3$. The index k denotes the step number. We denote the elements of the original coefficient matrix by $a_{ij}^{(0)} = a_{ij}$.

$$x_{il} = \bigoplus_{j=k+1}^{n} a_{ij}^{(k)} x_{jl}, \qquad \text{for } 0 \leq k < l, i \neq l,$$

$$x_{ll} = \bigoplus_{j=k+1}^{n} a_{lj}^{(k)} x_{jl} \oplus \text{①}, \qquad \text{for } 0 \leq k < l, \qquad (11)$$

$$x_{il} = \bigoplus_{j=k+1}^{n} a_{ij}^{(k)} x_{jl} \oplus a_{il}^{(k)}, \qquad \text{for } l \leq k \leq n,$$

The formulas for the coefficients $a_{ij}^{(k)}$ were as follows:

$$a_{kk}^{(k)} = (a_{kk}^{(k-1)})^*,$$

$$a_{ik}^{(k)} = a_{ik}^{(k-1)}(a_{kk}^{(k-1)})^*, \qquad\qquad \text{for } i \neq k,$$

$$a_{kj}^{(k)} = (a_{kk}^{(k-1)})^* a_{kj}^{(k-1)}, \qquad\qquad \text{for } j \neq k, \qquad\qquad (12)$$

$$a_{ij}^{(k)} = a_{ij}^{(k-1)} \oplus a_{ik}^{(k-1)}(a_{kk}^{(k-1)})^* a_{kj}^{(k-1)}, \qquad \text{for } i \neq k, j \neq k.$$

When we compute only a column x_{il} of the solution, for fixed l, as in our example, we actually carry out the recursions for $a_{kk}^{(k)}$ and $a_{ik}^{(k)}$ only for $k = l$, and the recursions for $a_{kj}^{(k)}$ and $a_{ij}^{(k)}$ only for $j > k$ and for $j = l < k$. Observe, however, that the above recursions are the same *for all columns* l, as far as they overlap for different columns. Thus, when we want to determine the whole matrix X we get just the above recursions, and the final result is

$$x_{ij} = a_{ij}^{(n)}. \qquad\qquad (13)$$

We get this by setting $k = n$ in (11), whereas for $k = 0$ we get the original system (3) or (10.0). The nice thing about all these equations is that they can all be interpreted as equations between sets of paths. We will do this in section 4.3.

We can cast our recursion into an algorithm, in which we can omit the superscripts (k) from the variables. We start with the given array a_{ij} and modify this array step by step until the final solution $a_{ij}^{(n)}$ is obtained. This algorithm corresponds just to the Gauß-Jordan elimination algorithm of ordinary linear algebra, and hence it carries this name.

Gauß-Jordan elimination algorithm for
the solution of the equation $X = AX + I$

---

for k **from** 1 **to** n **do begin**
    (∗ Transformation of the matrix $A^{(k-1)}$ into $A^{(k)}$: ∗)
    $a_{kk} := (a_{kk})^*$;
    for all i **from** 1 **to** n **with** $i \neq k$ **do** $a_{ik} := a_{ik} \otimes a_{kk}$;
    for all i **from** 1 **to** n **with** $i \neq k$ **do**
        for all j **from** 1 **to** n **with** $j \neq k$ **do** $a_{ij} := a_{ij} \oplus a_{ik} \otimes a_{kj}$;
    for all j **from** 1 **to** n **with** $j \neq k$ **do** $a_{kj} := a_{kk} \otimes a_{kj}$;
end;

We get a variation of this algorithm if we do not substitute the explicit value for a variable in the equations preceding the current one, only in the following ones. The resulting system for our example would look as follows:

$$x_{13} = a_{12}^{(1)} x_{23} \oplus a_{13}^{(1)} x_{33} \oplus a_{14}^{(1)} x_{43}$$

$$x_{23} = \qquad\qquad a_{23}^{(2)} x_{33} \oplus a_{24}^{(2)} x_{43}$$

$$x_{33} = \qquad\qquad\qquad\qquad a_{34}^{(3)} x_{43} \oplus a_{33}^{(3)}$$

$$x_{43} = \qquad\qquad\qquad\qquad\qquad\qquad a_{43}^{(4)}$$

The system can now be solved in one backsubstitution pass, starting with the last equation. This method corresponds to Gaußian elimination in ordinary linear algebra.

4.2. Theorems about the solution of the elimination algorithm

We can summarize the results of the preceding section as follows:

Theorem 1. *If, for all pivot elements $a = a_{kk}^{(k-1)}$ of the algorithm, a^* is a solution of $x = \text{①} \oplus ax$, then $A^{(n)}$ is a solution of $X = I \oplus AX$.* ∎

Note, however, that the converse of this statement is not true: The Gauß-Jordan elimination algorithm may fail although a solution exists. This situation is known from ordinary matrix inversion. There, one cannot always take the next diagonal element as pivot.

We may ask under what conditions the solution of the matrix equation is unique. The following theorem, which follows readily from the elimination algorithm, gives an answer:

Theorem 2. *If, for each pivot element $a = a_{kk}^{(k-1)}$ in the algorithm, a^*b is the unique solution of $x = b \oplus ax$, for all $b \in S$, then $A^{(n)}$ is the unique solution of $X = I \oplus AX$.*

Proof. We have to review how the algorithm obtains the solution (10.4) from the original system (10.0). It does so by a sequence of transformations. In going from $(10.k - 1)$ to $(10.k)$, we solve one iteration equation $x_{k3} = a_{kk}^{(k-1)}x_{k3} \oplus b$. Under the assumption of the theorem, the resulting equation $x_{k3} = (a_{kk}^{(k-1)})^*b$ is an implication of the iteration equation. The remaining equations of $(10.k)$ are derived by substitution and rearrangement of terms and are therefore also implied by the given equations.

Thus, the final equations $(10.n)$ are implied by the original system, and therefore they represent the unique solution. ∎

Note that we must require uniqueness of the solution of $x = b \oplus ax$ for all b, since whenever we solve an equation of this form during the elimination process, a is a number that we have computed, whereas b is an expression which still involves other unknowns.

For the case of shortest paths, this means that the solution is unique as long as no $a_{kk}^{(k-1)}$ is 0. On the other hand, we have seen that, when the graph contains cycles of zero length, a solution need not be unique. In those cases, it is nevertheless desirable to get a specific solution. In the case of the shortest path problem, the *greatest* solution is the desired solution, since it can be shown that it actually represents the lengths of shortest paths. Thus, we may ask ourselves whether an analog of the above theorem holds for this case, i.e., whether we actually get the greatest (or smallest) solution of $X = I \oplus AX$, if we make sure that a^*b is always the greatest (or smallest) solution of $x = b \oplus ax$.

The following theorem shows that a somewhat weaker statement is true. We formulate it in terms of the smallest solution. One gets an analogous theorem for largest solutions by substituting "smallest" by "largest" and "$\geq$" by "$\leq$".

Theorem 3. *Assume that we have an ordered semiring. If, for each pivot element $a = a_{kk}^{(k-1)}$ in the algorithm, $y = a^*b$ is the smallest solution of $y \geq b \oplus ay$, for all $b \in S$, then $A^{(n)}$ is the smallest solution of $X = I \oplus AX$.*

Proof. We write $\geq$ instead of $=$ in all given equations (10.0) and in all equations (10.k) that are derived during the elimination process. Then, as in the proof of theorem 2, each derived inequality is an implication of the preceding inequalities: For the solution of the iteration equation, this follows from the assumptions of the theorem; for the substitution of the lower bound for this variable in the other inequalities, this follows from the monotonicity of the $\oplus$ and $\otimes$ operations. Since the final inequalities read $X \geq A^{(n)}$, we have the desired result. ∎

If the semiring is ordered by the difference relation (see (6)) we get a result completely analogous to theorem 2, where the "$\geq$" in the preceding theorem is replaced by "$=$". In fact, the following theorem strengthens theorem 2:

Theorem 4. *Assume that we have a semiring which is ordered by the difference relation. If, for each pivot element $a = a_{kk}^{(k-1)}$ in the algorithm, $x = a^*b$ is the smallest solution of $x = b \oplus ax$, for all $b \in S$, then $A^{(n)}$ is the smallest solution of $X = I \oplus AX$.*

Proof. Let a be one of the pivot elements of the algorithm. In order to reduce this theorem to the preceding one, we only have to show that a^*b is the smallest solution of $y \geq b \oplus ay$, for any b.

Let y be a solution of the inequality $y \geq b \oplus ay$. Since the semiring is ordered by the difference relation, we may write

$$y = (b \oplus z) \oplus ay,$$

for some z (see (6)). By the assumption of the theorem, $a^*(b \oplus z)$ is the smallest solution of this equation, and hence

$$y \geq a^*(b \oplus z) \geq a^*b \oplus a^*z \geq a^*b.$$

The last inequality follows again from the definition of the difference relation. ∎

Theorems 3 and 4 answer a question posed by Lehmann [9]. The requirement of theorem 4 that the semiring be ordered by the difference relation cannot be omitted completely, as can be shown by a suitable counter-example.

4.3. An interpretation with sets of paths

In this section, we shall give a different interpretation to the equations derived in section 4.1: We shall interpret the coefficients as sums of path weights. These sums are in general infinite. However, in order to avoid the technicalities which are involved in dealing with infinite sums, we shall take a naive approach and assume that all infinite sums exist. In any case, the following considerations can at least be taken as heuristic support for the equations of section 4.1.

The quantity x_{il} represents the sum of the weights of all paths from i to l. We can partition the set of all paths into disjoint subclasses according to some criterion, e.g., according to the first vertex j on the path whose number is greater than i. The paths in one subclass can be split into two subpaths in a unique way, e.g., at this vertex j. By considering all possibilities how this can be done, we get an expression for x_{il} in terms of certain sums of subpaths.

To be more specific, we define a family of sets of paths as follows: We assume that the vertices are numbered from 1 to n. For $1 \le i, j \le n$ and $0 \le k \le n$, $P_{ij}^{(k)}$ denotes the set of paths from i to j whose intermediate vertices belong to the set $\{1, 2, \ldots, k\}$. The intermediate vertices of a path $(i = v_0, v_1, \ldots, v_l = j)$ are all vertices except the first and the last one. In the case of the empty path (i) we count i as an intermediate vertex; thus, (i) is contained in $P_{ii}^{(i)}$ but not in $P_{ii}^{(i-1)}$.

We shall give the following interpretation of the coefficients $a_{ij}^{(k)}$ that arise in the elimination algorithm:

$$a_{ij}^{(k)} = \bigoplus_{p \in P_{ij}^{(k)}} w(p).$$

For $k = 0$, we get the initial values $a_{ij}^{(0)} = a_{ij}$, which is correct because $P_{ij}^{(0)}$ contains only the arc (i, j), if it is part of the graph. Starting from $k = 0$, the truth of the above expression for $a_{ij}^{(k)}$ can be verified by induction, using the recursions (12). We start with the simplest formula in (12), $a_{kk}^{(k)} = (a_{kk}^{(k-1)})^*$. A path in $P_{kk}^{(k)}$ must start at k and end at k. In the meantime, it can pass arbitrarily many times through k. When we cut the path into pieces at these intermediate vertices k, we get $l \ge 0$ partial paths which are members of $P_{kk}^{(k-1)}$. The expression $(a_{kk}^{(k-1)})^l$ is the sum of all paths which contain exactly $l + 1$ occurrences of k (including the first and the last occurrence). Thus, the expression

$$(a_{kk}^{(k-1)})^* = \text{①} \oplus a_{kk}^{(k-1)} \oplus (a_{kk}^{(k-1)})^2 \oplus \cdots \oplus (a_{kk}^{(k-1)})^l \oplus \cdots$$

accounts for every path in $P_{kk}^{(k)}$ in a unique way. On the other hand, it is easy to see that every path weight contributing to the expression on the right-hand side corresponds to the weight of some path in $P_{kk}^{(k)}$.

Now, let us consider the second equation: $a_{ik}^{(k)} = a_{ik}^{(k-1)}(a_{kk}^{(k-1)})^*$, for $i \ne k$. A path in $P_{ik}^{(k)}$ can be split, in a unique way into the initial part from i to the first occurrence of k (k must occur since it is the last vertex) and the remaining part. The first part is in $P_{ik}^{(k-1)}$, and the remaining part is accounted for by $(a_{kk}^{(k-1)})^*$. The third equation follows by a symmetric argument (splitting at the last occurrence of k instead of the first occurrence).

The last case $a_{ij}^{(k)} = a_{ij}^{(k-1)} \oplus a_{ik}^{(k-1)}(a_{kk}^{(k-1)})^* a_{kj}^{(k-1)}$, for $i, j \ne k$, is also straightforward. The difference to the previous case is, that a path in $P_{ij}^{(k)}$ need not go through k at all. This is taken into account by the term $a_{ij}^{(k-1)}$.

Using the previous arguments as inductive steps from k to $k + 1$, we finally arrive at $a_{ij}^{(n)} = x_{ij}$, because $P_{ij}^{(n)}$ is the set of all paths from i to j (cf. (13)).

Next, we shall consider the equations (11) containing the "unknowns" x_{il} and the "coefficients" $a_{ij}^{(k)}$. In this context, the difference between coefficients and unknowns is immaterial, since we interpret both as sums of path weights.

Let us interpret the first equation in (11): $x_{il} = \bigoplus_{j=k+1}^{n} a_{ij}^{(k)} x_{jl}$, for $0 \le k < l$ and $i \ne l$. The left side represents all paths from from i to l, for some $i \ne l$. Such a path must contain at least one intermediate vertex whose number is greater than k, because the last vertex l is greater than k. Let j be the first intermediate vertex along the path which is greater than k, and split the path into two parts at this vertex. The first part of the path from i to k contains no intermediate vertex greater than k, which is reflected in the superscript of the expression $a_{ij}^{(k)}$. The second part of the path can be an arbitrary path from k to l. Thus the product $a_{ij}^{(k)} x_{jl}$ is the sum of the weights of all paths from i to l whose first intermediate vertex which is greater than k is j. The vertex j can be any vertex between $k+1$ and n, and thus every path from i to l is represented in a unique way on the right-hand side.

The second equation in (11) differs from the first one only by the additional $\textcircled{1}$ on the right-hand side, which accounts for the empty path in $P_{il}^{(n)}$. The third equation: $x_{il} = \bigoplus_{j=k+1}^{n} a_{ij}^{(k)} x_{jl} \oplus a_{il}^{(k)}$, for $k \ge l$, differs from the first one in the additional term $a_{il}^{(k)}$. This term accounts for the fact that a path from i to l need not certain an intermediate vertex j whose number is greater than k: The paths which contain no intermediate vertex greater than k are exactly the paths whose weights sum to $a_{il}^{(k)}$.

4.4. Block decomposition methods

As in the case of real matrices, we can decompose a matrix into blocks and carry out the computations blockwise, as with scalar matrices. For example, when we decompose into 4 blocks, the equation $X = AX \oplus I$ becomes

$$\begin{pmatrix} X_{11} & X_{12} \\ X_{21} & X_{22} \end{pmatrix} = \begin{pmatrix} A_{11} & A_{12} \\ A_{21} & A_{22} \end{pmatrix} \otimes \begin{pmatrix} X_{11} & X_{12} \\ X_{21} & X_{22} \end{pmatrix} \oplus \begin{pmatrix} I_{11} & 0 \\ 0 & I_{22} \end{pmatrix}$$

We assume that all diagonal blocks X_{ii} and A_{ii} are square. I_{11} and I_{22} are unit matrices of the appropriate size.

We can apply the elimination algorithm for this block equation without change. The main difference is that the iteration equation $X = AX \oplus B$ which is used to eliminate a variable X_{ij} from the right-hand side is now itself a matrix equation instead of a scalar equation, and the problem of determining A^* is of the same type as the original problem, but of smaller size, however. This opens the possibility for recursive divide-and-conquer solution strategies.

Let us look at the above decomposition into 2×2 blocks and apply the Gauß-Jordan algorithm for $n = 2$.

$$
\begin{array}{ll}
1. \; A_{11}^{(1)} := (A_{11})^*; & 5. \; X_{22} = A_{22}^{(2)} := (A_{22}^{(1)})^*; \\
2. \; A_{21}^{(1)} := A_{21} A_{11}^{(1)}; & 6. \; X_{12} = A_{12}^{(2)} := A_{12}^{(1)} X_{22}; \\
3. \; A_{22}^{(1)} := A_{22} \oplus A_{21}^{(1)} A_{12}; & 7. \; X_{11} = A_{11}^{(2)} := A_{11}^{(1)} \oplus X_{12} A_{21}^{(1)}; \\
4. \; A_{12}^{(1)} := A_{11}^{(1)} A_{12}; & 8. \; X_{21} = A_{21}^{(2)} := X_{22} A_{21}^{(1)};
\end{array}
\tag{14}
$$

We will consider two opposite possibilities for the partitioning strategy: decomposition into equal-size blocks of size approximately $(n/2) \times (n/2)$, and partitioning into one block of size $(n-1) \times (n-1)$ and a scalar.

For the first choice, the above program shows that a *-operation for $(n \times n)$-matrices can be reduced to six multiplications, two additions and two *-operations on matrices of size $(n/2) \times (n/2)$. By using the reduction recursively, one can derive the result that an $O(n^c)$-time matrix multiplication algorithm for the semiring, with any fixed exponent $c \geq 2$ leads to an algorithm for computing the *-operation with the same asymptotic time complexity (cf. Aho, Hopcroft, and Ullman [2], section 5.9).

The other possibility, where A_{11} consists of the first $n-1$ rows and columns of A and A_{22} is just the element a_{nn}, corresponds to the escalator method for inverting a matrix, which adds one column and one row at a time until the whole matrix is inverted. A_{21} is the last row and A_{12} is the last column of the matrix. $(A_{11})^*$ is the matrix $(a_{ij}^{(n-1)})_{1 \leq i,j \leq n-1}$ whose elements correspond to subsets of paths in the graph with vertex n deleted. If we continue the above decomposition recursively, we get the following algorithm for computing A^*. Since the recursive step, the evaluation of A_{11}^*, comes right at the beginning of the algorithm, it is easy to write the algorithm without recursion. For easier reference, we have numbered the steps as in the algorithm above.

Escalator method for the solution of the equation $X = AX \oplus I$

for k **from** 1 **to** n **do begin**
    (* Transformation of the matrix $(b_{ij}) = (a_{ij}^{(k-1)})_{1 \leq i,j \leq k-1}$ *)
    (* into the matrix $(a_{ij}^{(k)})_{1 \leq i,j \leq k}$.   *)
    1. (* $A_{11}^{(1)}$ is already given. *)
    2. **for** j **from** 1 **to** $k-1$ **do** $b_{kj} := \bigoplus_{i=1}^{k-1} a_{ki} b_{ij}$;
    3. $b_{kk} := a_{kk} \oplus \bigoplus_{l=1}^{k-1} b_{kl} a_{lk}$;
    4. **for** i **from** 1 **to** $k-1$ **do** $b_{ik} := \bigoplus_{j=1}^{k-1} b_{ij} a_{jk}$;
    5. $b_{kk} := (b_{kk})^*$;
    6. **for** i **from** 1 **to** $k-1$ **do** $b_{ik} := b_{ik} b_{kk}$;
    7. **for** i **from** 1 **to** $k-1$ **do**
        for j **from** 1 **to** $k-1$ **do** $b_{ij} := b_{ij} \oplus b_{ik} \otimes b_{kj}$;
    8. **for** j **from** 1 **to** $k-1$ **do** $b_{kj} := b_{kk} b_{kj}$;
end;

Note we have used a different array (b_{ij}) for the result variables, because otherwise the k-th row and column would be overwritten while they are being used in steps 2 and 4. Thus the final result is contained in (b_{ij}), whereas the original data (a_{ij}) remain unchanged.

In the case of the shortest path problem, this algorithm is known as the algorithm of Dantzig. There, steps 6 and 8 can be omitted because b_{kk} is always zero, and step

5 reduces to a sign test. Moreover, since the semiring is idempotent, we do not have to differentiate between the matrices (a_{ij}) and (b_{ij}), because it does not matter if elements of A are overwritten.

The recursions of this algorithm can also be interpreted as equations between sets of paths, like in section 4.3. In order to see this, we have to add the correct superscripts. Expressions without superscripts, like a_{ki} denote the initial values of these variables: $a_{ki} = a_{ki}^{(0)}$. Since we have the explicit superscripts, we write a again instead of b:

2. $a_{kj}^{(k-1)} = \bigoplus_{i=1}^{k-1} a_{ki} a_{ij}^{(k-1)}$, for $j < k$: Since a path in $P_{kj}^{(k-1)}$ must contain at least one arc, we can partition this set of paths according to the first vertex i which comes after the start vertex k.

4. $a_{ik}^{(k-1)} = \bigoplus_{j=1}^{k-1} a_{ij}^{(k-1)} a_{jk}$, for $i < k$: This is symmetric to 2.

3. $a_{kk}^{(k-1)} = a_{kk} \oplus \bigoplus_{i=1}^{k-1} a_{ki}^{(k-1)} a_{ik}$: This is similar to the previous case, except that we have to take the single arc (k, k) into account.

The remaining recursions:

5. $a_{kk}^{(k)} = (a_{kk}^{(k-1)})^*$;

6. $a_{ik}^{(k)} = a_{ik}^{(k-1)} a_{kk}^{(k)}$, for $i < k$;

7. $a_{ij}^{(k)} = a_{ij}^{(k-1)} \oplus a_{ik}^{(k)} \otimes a_{kj}^{(k-1)}$, for $i, j < k$; and

8. $a_{kj}^{(k)} = a_{kk}^{(k)} a_{kj}^{(k-1)}$, for $j < k$;

are the same as in Gauß-Jordan elimination.

There is also a three-phase algorithm which is analogous to LU-decomposition of ordinary linear algebra (cf. Rote [13]). The top-down pass computes the matrix $L \oplus U$, where L is a strictly lower triangular matrix defined by $l_{ij} = a_{ij}^{(j)}$, for $i > j$, and $l_{ij} = \mathbb{0}$ for $i \leq j$, and U is an upper triangular matrix with $u_{ij} = a_{ij}^{(i-1)}$, for $i \leq j$, and $u_{ij} = \mathbb{0}$ for $i > j$. Then L^* and U^* are computed, with $(L^*)_{ij} = a_{ij}^{(i-1)}$, for $i > j$, and $(U^*)_{ij} = a_{ij}^{(j)}$, for $i \leq j$. Finally, U^* is multiplied with L^*, yielding the result matrix $X = A^*$. The matrices in this algorithm fulfill the following relations:

$$A \oplus LU = L \oplus U \qquad (LU\text{-decomposition})$$

$$U^*L^* = A^*$$

All of these equations can be interpreted as path equations as in section 4.3.

4.5. A graphical interpretation of vertex elimination

We can view the elimination of a variable from the right-hand side of the equations as the elimination of the corresponding vertex from the graph. This is shown in figure 3. When a vertex k is removed, we must somehow make up for the paths that have gone lost by this removal. Thus, for each pair consisting of an ingoing arc (i, k) and an outgoing arc (k, j), we add a new short-cut arc (i, j). The weight of this additional arc, which is meant to replace the piece incident to vertex k in every path passing through k, reflects the paths that were lost: $a_{ik} a_{kk}^* a_{kj}$. If the arc (i, j) is already present in the graph, we simply add this expression to its old weight.

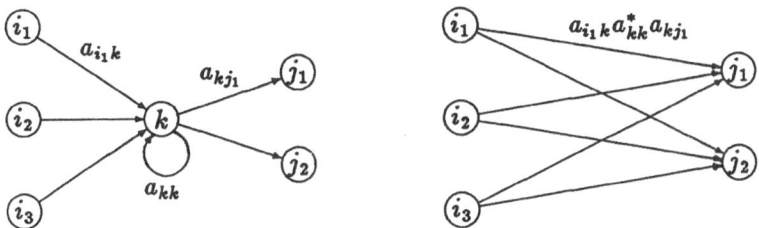

Figure 3. Elimination of the vertex k.

On certain types of sparse graphs (i.e., graphs with few arcs), Gaußian elimination can be carried out more efficiently by using a special ordering in which the variables are eliminated. For example, using a technique called *generalized nested dissection* due to Lipton and Tarjan [11], single-source path problems on *planar* graphs can be solved in $O(n^{3/2})$ steps (see also Lipton, Rose, and Tarjan [10]). Flow graphs from computer programs (cf. section 6.5) usually have a special structure: They are *reducible*. There are specialized algorithms for solving path problems on these graphs (cf. Tarjan [16]).

5. Iterative solution procedures

5.1. Matrix powers

Iterative algorithms are based on the connection between matrix powers and paths of a certain length. In particular, if $(A^l)_{ij}$ denotes the (i, j) entry of the l-th power of the matrix A, then

$$(A^l)_{ij} = \bigoplus_{\substack{p \text{ is a path} \\ \text{from } i \text{ to } j \\ \text{of length } l}} w(p),$$

and thus we get

$$(I \oplus A \oplus A^2 \oplus \cdots \oplus A^l)_{ij} = \bigoplus_{\substack{p \text{ is a path} \\ \text{from } i \text{ to } j \\ \text{of length at most } l}} w(p).$$

For many path problems, paths which are longer than some threshold play no role. For example, in case of the shortest path problem, no path of length n or longer can be a shortest path, and thus it suffices to compute $I \oplus A \oplus A^2 \oplus \cdots \oplus A^{n-1}$. When the semiring is idempotent, such a sum can be evaluated by successively squaring the matrix $(I \oplus A)$. By the idempotence law, we get

$$(I \oplus A)^l = I \oplus A \oplus A^2 \oplus \cdots \oplus A^l.$$

Thus, if we square the matrix $(I \oplus A)$ $\lceil \log_2(n - 1) \rceil$ times we get a matrix power $(I \oplus A)^l$ with $l \geq n - 1$, and thus this is the matrix of shortest distances.

5.2. Jacobi iteration and Gauß-Seidel iteration

When we want to compute only one row or one column of the matrix X, (i.e., we want to solve the single-source path problem), we can simply look at this row of the system (7′) or at this column of the system (7). For column j the corresponding system reads:

$$x = e_j \oplus Ax. \tag{15}$$

e_j denotes the j-th column of I, i.e., the j-th unit vector. One can view this equation, which defines the vector x in terms of an expression involving x, as a recursion which defines a sequence $x^{[0]}$, $x^{[1]}$, $x^{[2]}$, ... of successive approximations of x:

$$\begin{aligned} x^{[0]} &:= e_j \\ x^{[l]} &:= e_j \oplus Ax^{[l-1]}, \qquad \text{for } l \ge 1. \end{aligned} \tag{16}$$

Any fixed point of this iteration is a solution of (15). By induction one can show that

$$x^{[l]} = (I \oplus A \oplus A^2 \oplus \cdots \oplus A^l)e_j,$$

i.e., $x^{[l]}$ is the j-th column of the matrix on the right-hand side. In the case of the shortest path problem, this means that the elements $x^{[l]}$ are the lengths of shortest paths among the paths which contain at most l arcs. By the results of the previous subsection we conclude that, in case of the shortest path problem, $x^{[n-1]}$ is the j-th column of the shortest path matrix X.

We can simply iterate the recursion (16) until it converges, i.e., until two successive vectors are equal. If the iteration does not converge after n steps, we know that there is a negative cycle. If the iteration converges and the semiring is ordered by the difference relation, the resulting solution is the smallest solution (the *least fixed point*) of the iteration (cf. theorem 4).

This algorithm corresponds to the Jacobi iteration of numerical linear algebra. Gauß-Seidel iteration is a variation of this method. There, when the elements of the vector $x^{[l]}$ are computed one after the other, they are not computed from the old values of $x^{[l-1]}$, as in (16), but the new elements of $x^{[l]}$ replace the corresponding entries as soon as they are computed. It can be shown that, in the case of idempotent semirings, this modification preserves correctness of the algorithm, and, moreover, the Gauß-Seidel algorithm never needs more iterations than the Jacobi algorithm.

In contrast to elimination algorithms, these iterative algorithms do not require both (left and right) distributive laws. For example, for the column iteration (16) described above, only the left distributive law $a(b \oplus c) = ab \oplus ac$ is required. An example of a semiring where only one of the distributive laws holds occurs in the computation of least-cost paths in networks with losses and gains (cf. Gondran and Minoux [6], section 3.7).

5.3. *Acyclic graphs*

When the graph $G = (V, A)$ on which we want to solve our path problem is acyclic, one can order the vertices in such a way that an arc (i, j) can only exist if $i < j$. In this case, the matrix entries, a_{ij} are zero for $i \geq j$ and the matrix A is (strictly) upper triangular. Then one can solve the system (3) in one pass by computing the solution x_{ij} in the order of decreasing row indices i. Thus, one column of X can be computed in $O(|V| + |A|)$ time. Similarly, one can compute one row of X in linear time by considering the system (7').

5.4. *The Dijkstra algorithm*

In some cases, specialized algorithms can solve path problems more efficiently. The single-source shortest path problem in graphs with non-negative arc lengths can be solved efficiently by the algorithm of Dijkstra. This algorithm can be generalized to semirings which come from a linearly ordered semigroup in which ① is the largest element (see the examples in section 6.4.1). The algorithm works by a clever choice of the vertex to be eliminated next. This is somehow analogous to elimination algorithms in linear algebra which use pivoting.

6. Further applications

In this last section, we shall present a selection of examples from different areas which can be interpreted and solved as path problems.

In the first three parts of this section, we shall consider problems which involve the field $(\mathbb{R}, +, \cdot)$ or a subset of it. Then we shall deal with optimization problems; and we shall return to the discussion of finite automata from section 2.2. Finally, we shall present some examples of "non-standard" semirings, which occur in data flow analysis of programs and in two graph-theoretic problems.

6.1. *Inversion of matrices*

When the semiring is a field, equation (7) can be rewritten as $(I - A)X = I$, or $X = (I - A)^{-1}$ if the matrix $I - A$ is invertible. Then the elimination algorithm corresponds exactly to the Gauß-Jordan algorithm of linear algebra (without pivoting). The only difference is that we get the inverse of $I - A$ and not the inverse of A. This is reflected in the pivoting operation, where we set $a_{kk}^{(k)} := (a_{kk}^{(k-1)})^* = 1/(1 - a_{kk}^{(k-1)})$ and not $a_{kk}^{(k)} := 1/a_{kk}^{(k-1)}$.

This problem has originally nothing to do with path problems. We can just pose the equation (7) without reference to a particular graph or to sums of path weights. Nevertheless, the matrix $A^* = (I - A)^{-1}$ has some significance for path problems, as is exemplified in the following two subsections.

6.2. Partial differentiation

Many numerical problems, like finding the minimum of a function f over some domain, can be solved more efficiently if the algorithm has access to the derivative of f. When the function f can be written as a simple expression of one variable, computing the derivative is no problem, but when $f(z_1, \ldots, z_k)$ is a function of several variables, which is computed by a complicated program involving loops and conditional branches, the computation of all partial derivatives $\partial f/\partial z_1, \partial f/\partial z_2, \ldots, \partial f/\partial z_k$, seems to be a difficult task. Therefore, one used to resort to methods which do not require the derivatives, or they differentiated numerically, which presents new problems of numerical stability.

In this section, we show how the problem of computing the partial derivatives can be solved efficiently as a path problem in a graph, by applying the chain rule.

The graph on which we will work is the *computational graph* of the function f. f is given by a program like the following two-line program, which computes the real root $y = f(z_1, z_2, z_3)$ of the equation $z_1^2 y = z_2 z_3 \sqrt{y} + z_2^2$:

$$y := z2 * z3 + \text{SQRT}((z2 * z3)^2 + 4 * z1^2 * z2^2);$$

$$y := (y/(2 * z1^2))^2;$$

We can resolve this into a sequence of elementary operations, as follows:

1. $a := z2 * z3$;	5. $e := c * d$;	9. $y := a + i$;
2. $b := a^2$;	6. $g := 4 * e$;	10. $j := 2 * c$;
3. $c := z1^2$;	7. $h := b + g$;	11. $k := y/j$;
4. $d := z2^2$;	8. $i := \text{SQRT}(h)$;	12. $y := k^2$;

Imagine now that this sequence of elementary steps is executed. In the beginning, the graph consists only of k isolated vertices which correspond to the input variables. Each time a variable is assigned a new value, we add a new vertex to the graph, and arcs from this vertex to the one or two operands of this elementary computation (cf. figure 4). When one of the operands is a constant, we first generate a vertex corresponding to this constant. When a variable is assigned several values in succession, we generate different vertices for each assignment (y and $\bar{y}$ in the example).

In our case, the computational graph has a static structure, since it corresponds to a straight-line program. However, we can also handle programs with loops and conditional branches, since the computational graph is generated dynamically.

Now let us look at some vertex w with two outgoing arcs leading to vertices u and v. Then we can determine $\partial w/\partial z_i$ by the chain rule:

$$\frac{\partial w}{\partial z_i} = \frac{\partial w}{\partial u} \cdot \frac{\partial u}{\partial z_i} + \frac{\partial w}{\partial v} \cdot \frac{\partial v}{\partial z_i}$$

w is determined from u and v by some elementary operation, and hence $\partial w/\partial u$ and

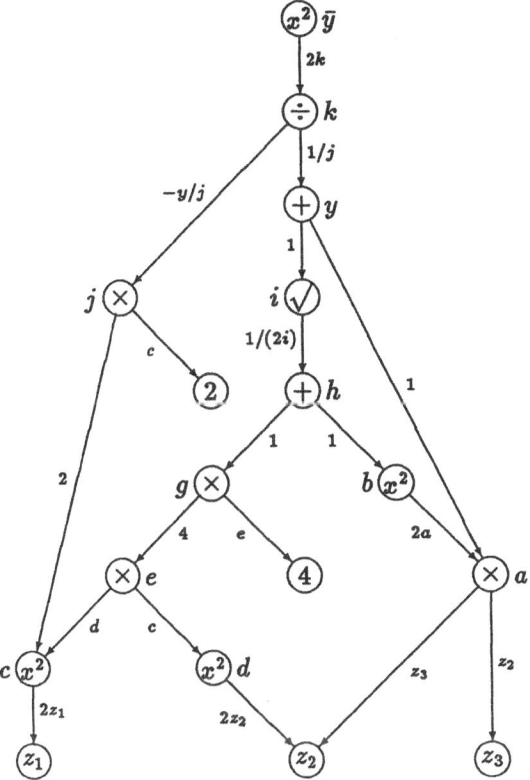

Figure 4. A computational graph

$\partial w / \partial u$ can be calculated directly from u, v, and w in a few elementary basic steps, taking constant time. For example, if $w = u/v$, then $\partial w / \partial u = 1/v$ and $\partial w / \partial v = -u/v^2 = -w/v$. When we associate the value $\partial w / \partial u$ with the arc (w, u) and the value $\partial w / \partial v$ with the arc (w, v), we can write the above equation as follows:

$$x_{wz_i} = a_{wu} x_{uz_i} + a_{wv} x_{vz_i}. \tag{17}$$

Here, a_{wu} and a_{wv} are numbers that can be calculated directly, and x_{wz_i} are the unknowns representing $\partial w / \partial z_i$. In figure 4, the arc weights are shown as small numbers.

We can see that the problem of computing the unknowns is just an instance of the path problem equation (3). In section 2, we have derived (3) starting from a path problem (4). Now arguing in the reverse direction, we obtain:

Theorem 5. $\partial f / \partial z_i$ is the sum of the weights of all paths from f to z_i in the computational graph, where the weight of a path is the product of its arc weights.

Since the computational graph is acyclic, we can compute $\partial u / \partial v$, for some fixed vertex u and all other vertices v, or for some fixed vertex v and all other vertices u, in time proportional to the number of arcs of the graph. Since each vertex has at most two outgoing arcs, this is proportional to the number of vertices of the graph, i.e., the number of steps of the algorithm for computing f. Thus, in time which is proportional to the time which the original program takes, we can

- compute $\partial f/\partial z_i$, for all input variables z_i, or
- compute $\partial v/\partial z_i$, for all intermediate variables and output variables v, and for some fixed z_i.

The second problem is solved by a bottom-up pass with a straightforward application of the chain rule (17). This case is interesting if we have a set of functions $f_1(z_1,\ldots,z_k), f_2(z_1,\ldots,z_k), \ldots, f_l(z_1,\ldots,z_k)$, with l output variables of the program.

The first problem is solved by a top-down pass through the tree, starting from f. A drawback of this method is that the computational tree must be stored, and hence storage requirement is also proportional to the time complexity of the original program for computing f alone.

Computation of the Jacobi matrix $(\partial f_i/\partial z_j)$ may take much longer than the original program for computing only the l values f_i, since kl values have to be computed. It corresponds roughly to solving the all-pairs path problem.

We can also iterate the procedure for computing derivatives and compute second-order derivatives. Again, note that the time for the computation of the whole Hessian matrix $(\partial^2 f/\partial z_i \partial z_j)$ is also longer than the computation of f, by more than a constant factor, since the Hessian has k^2 entries. However, one can compute the product of the Hessian matrix or the Jacobi matrix with a particular vector in time proportional to the original number of steps of the program.

Note that, in the algorithm, we also determine partial derivatives of f with respect to all intermediate variables. These values can be used to estimate the total rounding error which has been accumulated during the computation of f. For more information, the reader is referred to the survey of Iri and Kubota [8], or to Iri [7], Sawyer [14], or Baur and Strassen [3].

6.3. Markov chains—the number of paths

When the matrix A is the (ordinary) adjacency matrix of a graph, i.e., a_{ij} is 1 if the arc (i,j) exists and 0 otherwise, then the weight of every path is 1, if we use the ring of integers $(\mathbb{Z}, +, \cdot)$. Thus, x_{ij} represents the *number* of different paths from i to j. Of course this makes sense only when the graph is acyclic, because otherwise there will be infinitely many paths. On the other hand, from the considerations in section 5.1, we know that the power A^l contains the number of paths of length l between every pair of vertices.

A slight generalization of this is used in the theory of Markov chains. In a Markov chain, there is a finite set $\{1, 2, \ldots, n\}$ of states of a system, and the system changes between states in a random way in discrete time intervals. The probability that the system is in state j at some step t depends only on the state of the machine at step $t - 1$, and it is independent of t and of previous state transitions. Let a_{ij} be the probability that the system is in state j at step t if it was in state i at step $t - 1$. Then the probability that a system passes through a sequence of states $(v_0, v_1, \ldots, v_l)$ is the product $a_{v_0 v_1} a_{v_1, v_2} \cdots a_{v_{l-1} v_l}$. Thus the (i, j) entry of the matrix A^l is the probability that the system is in state j at step t if it was in state i at step $t - l$.

6.4. Optimal paths

6.4.1. Best paths

We have considered the shortest path problem as the first instance of a path problem. There are several other problems, where the set S of path weights is linearly ordered, and the weight of a best path is desired, i.e., $\oplus$ is the operation min or max. Besides the shortest path problem, we have the following examples:

- Maximum capacity paths. The solution uses the semiring $(\mathbb{R}_+ \cup \{\infty\}, \max, \min)$.
- Most reliable paths in networks with possible arc failures, where it is assumed that arc failures of different arcs are independent. Here we use the semiring $([0, 1], \max, \cdot)$. The entry a_{ij} of the initial data matrix represents the probability that the arc (i, j) is all right. We are looking for the path with the smallest failure probability.

For the case of maximum-capacity paths, the all-pairs problem can be solved more efficiently by constructing the maximum spanning tree.

The simplest kind of path problem arises when we only ask for the *existence* of a path. Here we take the simplest non-trivial semiring, the Boolean semiring with two elements $(\{0, 1\}, \max, \min)$. a_{ij} is 1 if and only if the arc (i, j) exists, and x_{ij} is 1 if j is reachable from i in the graph, i.e., the matrix X represents the transitive closure.

In all cases mentioned above and in the following subsections, the algorithms can be modified such that they will not only compute the weight of an optimal path, but produce the optimal path itself. To achieve this, the algorithms must store how the optimal path weight and each intermediate result was obtained. In some cases, this can only be done at the expense of an increased storage requirement. We will not discuss this in detail.

6.4.2. Multicriteria problems—lexicographic optimal paths

In many applications, paths are not selected according to one criterion, but according to several criteria. In the simplest case, we have a definite order of importance between different criteria. This leads to lexicographic optimization problems.

Imagine that a traveler plans a car trip from one city to another. For each street connecting two points i and j he knows the time t_{ij} to travel from i to j and the amount of sprit s_{ij} that his car needs for this distance. He wants to use as little fuel as possible, but if there are several paths which are equal in this respect, he wants to take the one which takes the shortest time.

Thus, he has a *lexicographic* preference relation $\preceq$ on the set of vectors (s, t):

$$(s_1, t_1) \preceq (s_2, t_2) \Leftrightarrow s_1 < s_2 \text{ or } (s_1 = s_2 \text{ and } t_1 \leq t_2).$$

This is a linear order of the vectors $(s, t) \in \mathbb{R}_+^2$. In the semiring, the operation $\oplus$ is the lexicographic minimum, whereas $\otimes$ is the ordinary elementwise vector addition:

$$(s_1, t_1) \oplus (s_2, t_2) = \begin{cases} (s_1, t_1), & \text{if } s_1 < s_2 \\ (s_2, t_2), & \text{if } s_2 < s_1 \\ (s_1, \min\{t_1, t_2\}), & \text{if } s_1 = s_2. \end{cases}$$

$$(s_1, t_1) \otimes (s_2, t_2) = (s_1 + s_2, t_1 + t_2)$$

One can even use a different $\otimes$-operation for the components. For example, imagine that the trip goes through the desert. If several journeys have the same time *and* the same fuel consumption, the traveler wants to select the safest trip among them, i.e., he wants his minimum safety reserve, below which his tank will never be emptied, to be as high as possible. This means that the sprit requirement between any two successive visits to filling stations on his journey should be as low as possible. Thus, we have a different semiring, where $\otimes$ is defined as

$$(s_1, t_1, s_1') \otimes (s_2, t_2, s_2') = (s_1 + s_2, t_1 + t_2, \max\{s_1', s_2'\}).$$

As before, $\oplus$ is the lexicographic minimum operation (of three components, this time).

Note however, that if the primary goal of our traveler is safety, whereas total fuel consumption and time have second and third priority, the corresponding operation $\otimes$, in which the third component would come in the first place, would not yield a semiring, because it violates the associative law. (The structure $(\mathbb{R}_+^3, \otimes, \preceq)$ would not be an ordered semigroup.)

If there is no clear preference between the objectives (of fuel over time, or vice versa), we can still eliminate from consideration a path which is worse than some other path in both respects. What remains is the set of *efficient* or *Pareto-optimal* or *minimal* paths; i.e., we are looking for all path weights (s, t), for which there is no other path with weight (s', t') such that $s' \leq s$ and $t' < t$, or $s' < s$ and $t' \leq t$.

This problem can also be formulated as a path problem, with *sets* of pairs (s, t) as elements of the semiring. The $\otimes$ operation for sets is the elementwise $\otimes$-product of the elements, and $\oplus$ is set union. However, after every operation, we can reduce the resulting sets by throwing away pairs (s, t) which are not efficient. Since these sets of efficient values can become very large, this approach is limited to small problems.

6.4.3. k-best paths

Another extension of the ordinary best path problem is the determination of the k best different paths between every pair of vertices. We can solve this by a semiring which operates on vectors of k elements. For simplicity we will assume that we want to compute the k shortest paths in the ordinary sense, i.e., we are working with the semiring $(\mathbb{R}_\infty, \min, +)$. However, the underlying semiring for the corresponding best path problem can be any semiring in which $\oplus$ is the min or the max operation (cf. the examples in the previous subsections). We are going to create a semiring $(S^k, \oplus, \otimes)$, whose elements are k-tuples of S. The vector $(a_1, a_2, \ldots, a_k)$ is meant to represent the lengths of the best, the second-best, ..., the k-best path in a certain set of paths. The operations for the semiring are defined as follows:

- $(a_1, a_2, \ldots, a_k) \oplus (b_1, b_2, \ldots, b_k)$ is the sequence of the k smallest values in the combination (union) of the two given sequences.
- $(a_1, a_2, \ldots, a_k) \otimes (b_1, b_2, \ldots, b_k)$ is the sequence of the k smallest values in the (multi-)set of k^2 elements $\{a_i + b_j | i = 1, \ldots, k; j = 1, \ldots, k\}$.

Note that a sequence $(a_1, a_2, \ldots, a_k)$ can contain repeated elements, which correspond to different paths with the same length. Moreover, addition in this semiring is not idempotent. However, the order of the elements in the sequence is immaterial, and thus we might just as well assume that they are sorted. Thus, by the way we treat the sequences in S^k, they are in fact multisets.

- $\mathbb{1} = (0, \infty, \ldots, \infty)$ and $\mathbb{0} = (\infty, \ldots, \infty)$. The initial entries of the matrix are $(a_{ij}, \infty, \ldots, \infty)$, where a_{ij} is the weight of the arc (i, j). (Initially, the path (i, j) is the only path from i to j that we know of. The second-, third-best, etc., paths do not exist.)
- We have to define the *-operation, i.e., the solution of

$$x = (0, \infty, \ldots, \infty) \oplus a \otimes x. \tag{18}$$

Let us assume that $a = (a_1, a_2, \ldots, a_k)$ with $a_1 \le a_2 \le \cdots \le a_k$. When we write x as $\mathbb{1} \oplus a \oplus a^2 \oplus \cdots$, we see the following:

- If $a_1 < 0$, there is no solution, except perhaps $(-\infty, -\infty, \ldots, -\infty)$.
- If $a_1 = 0$, we get $x = (0, 0, \ldots, 0)$ as the largest solution. However, any constant vector $x = (c, c, \ldots, c)$ with $c \le 0$ is also a solution of (18). (These are not the only solutions.)
- If $a_1 > 0$, there is a unique solution, which can be determined by looking at equation (18): $x = (x_1, x_2, \ldots, x_k)$ consists of the k smallest values in the (multi-) set

$$\{0\} \cup \{a_i + x_j | i = 1, \ldots, k; j = 1, \ldots, k\}. \tag{19}$$

We see that the smallest element in this set is $x_1 = 0$, since the elements of the right-hand set of the union are all positive. Let us assume that we have determined the l smallest elements of the set (19). x_{l+1}, the $(l + 1)$-smallest element of this set is the l-smallest element of the right-hand set. However, in order to determine the l-smallest element of $a \otimes x$, we need only know the l smallest elements of x, which we know already. Thus, we can successively determine $x_1, x_2, \ldots, x_k$.

Let us discuss the complexity of these operations. $\oplus$ can clearly be carried out in $O(k)$ time. The operation $\otimes$ can also be carried out in (theoretical) $O(k)$ time, using a sophisticated algorithm of Frederickson and Johnson [5]. The determination of a^* as described above, which occurs only n times during the elimination algorithm, can be carried out in $O(k \log k)$ steps, using priority queues.

Note that we do not get the k best *elementary* paths by this approach; i.e., the paths that we get can contain repeated vertices and arcs. The problem of finding the k best elementary paths is considerably more difficult, but also for this problem, algorithms which use the algebraic framework of path problems have been proposed.

The k-best path problem is an example of a path problem where non-elementary paths can have an influence on the solution. However, in case there are no negative cycles, the longest path that has to be taken into account has $kn - 1$ arcs. (For a longer path, one can construct at least k different shorter paths by successively eliminating elementary cycles from the path.)

Thus, the iterative algorithms of section 5 should converge after at most $kn - 1$ iterations, unless there are negative cycles.

6.5. Regular expressions

For our second example from the beginning, the determination of the language accepted by a finite automaton (cf. section 2.2), the elimination algorithm seems to be useless, since the *-operation will probably very soon lead to infinite sets. However, the algorithms provides us with a way to describe the language. In order to explain this, we need one more definition: A *regular language* is a set of words which is built starting from finite sets of words using only the operations $\cdot$ (concatenation), $\cup$, and *. For example, $e(fh*(f \cup \{g, h\}*))* \cup \{\varepsilon, ggg\}$ is a regular language. (Here, single words denote singleton sets.) Now the Gauß-Jordan elimination algorithm successively constructs a regular expression for each pair of states i and j, which describes the language x_{ij} leading from i to j: It starts from the finite sets a_{ij}, and as it proceeds, it uses only the operations $\cdot$, $\cup$, and *.

Since the language accepted by the automaton is just the union of several x_{ij}, we have proved the following theorem:

Theorem 6. *The language accepted by a finite automaton is regular.*   ∎

This is one half of Kleene's theorem about the equivalence of finite automata and regular expressions. The other direction, the construction of a finite automaton which accepts a given regular language, is even easier. Our proof by Gauß-Jordan elimination is in fact the standard proof of this result.

6.6. *Flow analysis of computer programs*

When a compiler wants to optimize the code for a computer program, for example by detecting common subexpressions or by moving invariant expressions out of loops, it needs to know whether the value of an expression remains unchanged between two uses of this expression. If this is the case, the expression need not be evaluated the second time.

In order to investigate this problem, for one particular expression $f(z_1, \ldots, z_k)$ which occurs in the program, we represent the program by its *flow graph*. The vertices correspond to basic blocks of the program, i.e., blocks of consecutive statements with one entry point at the beginning and one exit point at the end. The arcs indicate possible transfers of control between basic blocks. (This is similar to a flowchart.)

The execution of a basic block may have one of the following effects on the value of f:

- It may *generate* f, i.e., the value of f is computed in the block and is available on exit from this block.
- It may *kill* f, for example by assigning a new value to one of the input variables $z_1, \ldots, z_k$ of f.
- It may leave f *unchanged*.

We give an arc (i, j) the label G, K, or U, depending on whether the value of f is generated, killed, or left unchanged between the entry to block i and the entry to block j, (i.e., during the execution of block i). In addition, we need an element $\mathbb{O}$ for the arcs which are not present.

Now we can use the following semiring on the set $\{\mathbb{O}, G, U, K\}$.

$\oplus$	$\mathbb{O}$	G	U	K
$\mathbb{O}$	$\mathbb{O}$	G	U	K
G	G	G	U	K
U	U	U	U	K
K	K	K	K	K

$\otimes$	$\mathbb{O}$	G	U	K
$\mathbb{O}$	$\mathbb{O}$	$\mathbb{O}$	$\mathbb{O}$	$\mathbb{O}$
G	$\mathbb{O}$	G	G	K
U	$\mathbb{O}$	G	U	K
K	$\mathbb{O}$	G	K	K

a	a^*
$\mathbb{O}$	U
G	U
U	U
K	K

U is the $\mathbb{O}$-element of this semiring. All semiring axioms hold. Note that in this semiring the operation $\otimes$ is not commutative. The operation $\oplus$ is just the min operation for the order $K < U < G < \mathbb{O}$. This is typical of data flow problems, because when we unite two sets of possible paths from i to j, we can only keep the weaker information of the information that the two sets give about f.

To solve our original problem, let 1 be the start vertex of the program. We can eliminate an evaluation of the expression f in block j if and only if x_{1j} is G.

Further examples of data flow problems and references can be found in Tarjan [15].

o.7. *Some graph-theoretic problems*

The following examples are mentioned mainly as curiosities, in order to illustrate the broad range of applicability of the path problem formulation. For each of the problems, there are in fact linear-time algorithms to solve them directly.

The transitive closure of a graph, which also falls into this category, has already been mentioned briefly in section 6.4.1.

Path problem formulations have also been proposed for problems of enumerating elementary paths or cutsets of a graph. Such problems are exponential by their output size alone. There the solution procedures by elimination algorithms can be applied only to graphs of moderate size.

6.7.1. Testing whether a graph is bipartite

An undirected graph is bipartite if it contains no odd cycle (i.e., no cycle containing an odd number of edges). Since a cycle is just a special case of a path from a vertex to itself, we can formulate this as a path problem. Let the weight of a path be E or O according to whether its length is even or odd, and let the weight of a set of paths be $\emptyset$, E, O, or EO, according to whether the set is empty, contains only even paths, only odd paths, or paths of both types. Then we get the following semiring on the four-element set $S = \{\emptyset, E, O, EO\}$:

$\oplus$	$\emptyset$	E	O	EO
$\emptyset$	$\emptyset$	E	O	EO
E	E	E	EO	EO
O	O	EO	O	EO
EO	EO	EO	EO	EO

$\otimes$	$\emptyset$	E	O	EO
$\emptyset$	$\emptyset$	$\emptyset$	$\emptyset$	$\emptyset$
E	$\emptyset$	E	O	EO
O	$\emptyset$	O	E	EO
EO	$\emptyset$	EO	EO	EO

a	a^*
$\emptyset$	E
E	E
O	EO
EO	EO

We initialize the data matrix by setting $a_{ij} = O$ if the edge $\{i,j\}$ exists and $a_{ij} = \emptyset$ otherwise. Then, if any $x_{ii} = O$ or EO when the algorithm stops, the graph is not bipartite; otherwise it is. (Of course, as soon as the element EO appears somewhere in the matrix, we know already that the graph is not bipartite.)

We can also apply this algorithm to *directed* graphs and test for the existence of paths of given parity. Using a generalization of this idea, one can find shortest even paths or shortest odd paths, if one does not insist that the paths should be elementary. One could even find a shortest path whose number of arcs is, for example, congruent to 4 modulo 7, if one wishes to do so.

6.7.2. Finding the bridges and the cut vertices of a graph

A *bridge* in an undirected graph $G = (V, E)$ is an edge whose removal causes some connected component of G to break into two components. Similarly, a *cut vertex* or *articulation point* is a vertex whose removal causes some connected component of G to become disconnected.

For finding bridges, we use the semiring $(2^E \cup \{\textcircled{0}\}, \cap, \cup)$, which operates on the set of subsets of edges augmented by a zero element. $\cap$ and $\cup$ are the ordinary set intersection and set union operations, except that their interaction with $\textcircled{0}$ is specified by the semiring axioms. As unity element we have $\textcircled{1} = \emptyset$, and thus the *-operation presents no problem: $x = \textcircled{1} \oplus ax = \emptyset \cap (a \cup x)$ always has the same unique solution $x = \emptyset$, even when $a = \textcircled{0}$.

As the weight of an arc (i, j) we take simply the singleton set $\{e\}$, if $e \in E$ is the (undirected) edge corresponding to the (directed) arc (i, j), and we take $\textcircled{0}$ if no such edge exists. The weight of a path is then just the set of its edges. Using the formulation (4) of the algebraic path problem, we get:

$$x_{ij} = \bigcap_{\substack{p\,is\,a\,path \\ from\,i\,to\,j}} w(p).$$

In other words, x_{ij} is the set of edges which belong to *every* path from i to j. Such edges are clearly bridges, since their removal causes i to become disconnected from j. Conversely, every bridge must appear in some set x_{ij}.

Note that the semiring axioms would allow us to take the set E as the zero element instead of $\textcircled{0}$. But then we could not distinguish the case where all edges are bridges ($x_{ij} = E$) from the case when i and j are not connected ($x_{ij} = \textcircled{0}$).

The determination of cut vertices proceeds in essentially the same way. We use the semiring $(2^V \cup \{\textcircled{0}\}, \cap, \cup)$, and the weight of (i, j) is $\{i\}$, if $\{i, j\} \in E$, and $\textcircled{0}$ otherwise. All elements of the set x_{ij}, except for i, are cut vertices.

7. Conclusion—comparison of different approaches

The general path problem can be approached in several different ways. They are characterized by different formulations and by different assumptions about the underlying algebraic structure.

We have taken a purely algebraic approach: Solve the system of equations (3).

The usual approach is more direct. It involves the formulation of the problem as an infinite sum (4) of path weights and building up this sum by computing sums $a_{ij}^{(k)}$ of larger and larger path sets, using the equations (12). We have seen this approach in section 4.3. With suitable axiomatic assumptions for infinite sums this derivation of the solution can be made precise. In some semirings infinite sums do not always exist, although they can be defined for *some* sequences. These semirings include the important case of the real numbers $(\mathbb{R}, +, \cdot)$ with their rich structure of convergence. Such semirings can also be dealt with quite satisfactorily. This approach has for example been taken in Rote [13].

A variation of this method specifies the solution in the free semiring generated by the arc set. This is the semiring of multisets of paths with set union and concatenation as addition and multiplication operations. The solution for a specific semiring is then obtained by applying a homomorphism from the free semiring to the specific semiring. An approach like this is taken by Tarjan [15].

Lehmann [9] has taken the 2×2 block decomposition algorithm (14) as the recursive definition of A^* for matrices in terms of the operation a^* for scalars. He shows that the result is independent of how the matrix is decomposed into blocks. A similar approach is taken by Abdali and Saunders [1] who introduce the concept of eliminants to define A^*. Their definition corresponds to a particular way of computing A^* in terms of the a^* operation, very similar to our elimination algorithm.

A comparison of different approaches can be found in Mahr [12].

For some applications, like shortest paths, the formulation (4) involving sums of path weights is more natural, whereas the algebraic formulation (3) is more convenient for other applications such as the inversion of matrices. However, the relationship between the two formulations is not so close: The system (3) may have a solution although the infinite sum (4) makes no sense (consider the case of matrix inversion), or it may have several solutions (cf. the discussion of the shortest path example at the beginning of section 4). However, the desired solution of (3) can often be characterized as the smallest (or largest) solution. Theorems 3 and 4 of section 4.2 show that this desired solution can be obtained by defining a^* appropriately.

We hope that the broad range of applications from which we could draw our examples has convinced the reader of the importance and the general usefulness of path problems.

References

[1] S. K. Abdali and B. D. Saunders, Transitive closure and related semiring properties via eliminants, Theoret. Comput. Sci. **40** (1985), 257–274.

[2] A. V. Aho, J. E. Hopcroft, and J. D. Ullman, The Design and Analysis of Computer Algorithms, Addison-Wesley, Reading (Mass.) etc. 1974.

[3] W. Baur and V. Strassen, The complexity of partial derivatives, Theoret. Comput. Sci. **22** (1983), 317–330.

[4] B. A. Carré, Graphs and Networks, The Clarendon Press, Oxford University Press, Oxford 1979.

[5] G. N. Frederickson and D. B. Johnson, The complexity of selection and ranking in $X + Y$ and matrices with sorted columns, J. Comput. Syst. Sci. **24** (1982), 197–208.

[6] M. Gondran and M. Minoux, Graphes et algorithmes, Editions Eyrolles, Paris 1979; English translation: Graphs and Algorithms, Wiley, Chichester etc. 1984.

[7] M. Iri, Simultaneous computation of functions, partial derivatives, and estimates of rounding errors, Japan J. Appl. Math. **1** (1984), 223–252.

[8] M. Iri and K. Kubota, Methods of fast automatic differentiation and applications, Research Memorandum RMI 87-02, University of Tokyo, Faculty of Engineering, Hongo 7-3-1, Bunkyo-ku, Tokyo, July 1987.

[9] D. J. Lehmann, Algebraic structures for transitive closure, Theoret. Comput. Sci. **4** (1977), 59–66.

[10] R. J. Lipton, D. J. Rose, and R. E. Tarjan, Generalized nested dissection, SIAM J. Numer. Anal. **16** (1979), 346–358.

[11] R. J. Lipton and R. E. Tarjan, A separator theorem for planar graphs, SIAM J. Appl. Math. 36 (1979), 177–189.

[12] B. Mahr, Iteration and summability in semirings, in: R. E. Burkard, R. A. Cuninghame-Green, U. Zimmermann (eds.), Algebraic and Combinatorial Methods in Operations Research (Proc. Workshop on Algebraic Structures in Operations Research, Bad Honnef, Germany, April 1982), Annals of Discrete Mathematics 19, North-Holland, Amsterdam 1984, pp. 229–256.
[13] G. Rote, A systolic array for the algebraic path problem, Computing 34 (1985), 191–219.
[14] J. W. Sawyer, jr., Fast partial differentiation by computer with an application to categorical data analysis, The American Statistician 38 (1984), 300–308.
[15] R. E. Tarjan, A unified approach to path problems, J. Assoc. Comput. Mach. 28 (1981), 577–593.
[16] R. E. Tarjan, Fast algorithms for solving path problems, J. Assoc. Comput. Mach. 28 (1981), 594–614.
[17] U. Zimmermann, Linear and combinatorial optimization in order algebraic structures, Annals of Discrete Mathematics 10, North-Holland, Amsterdam 1981.

Günter Rote
Technische Universität Graz
Institut für Mathematik
Kopernikusgasse 24
A-8010 Graz, Austria

Computing Suppl. 7, 191–208 (1990)

Heuristics for Graph Coloring

D. de Werra, Lausanne

Abstract — Zusammenfassung

Heuristics for Graph Coloring. Some sequential coloring techniques are reviewed. A few general principles for designing heuristics are outlined and recent coloring techniques baaed on tabu search are discussed.

AMS Subject Classification: 05/90.

Key words: Graph coloring, chromatic scheduling, heuristics, combinatorial optimization, tabu search.

Heuristiken für Graphenfärbungen. In dieser Arbeit wird eine Übersicht über einige sequentielle Färbungstechniken gegeben. Ferner werden einige allgemeine Prinzipien zur Erstellung von Heuristiken aufgeführt und eine jüngst entwickelte Färbungstechnik auf der Basis von Tabu-Suchlisten diskutiert.

1. Introduction

As it happened for almost all graph-theoretical problems, interest in coloring was initially motivated by a kind of puzzle: is it possible to color the regions of a geographical map with at most four colors? This is by far not the most exciting application of graph coloring models. To-day many other situations are known where it is needed to color the nodes (or the edges) of a graph with a number of colors which is as small as possible.

We shall just mention a few: course scheduling [28, 31] school timetabling [24, 29], cluster analysis [13], group technology in production [4], tests of electronic circuits [9], VLSI (minimization of the number of layers in channels), transportation (optimization of a fleet of aircrafts), etc.

Besides these practical aspects, node coloring problems are also known to be NP-complete [10]; therefore many attempts have been made to attack these problems; as a by-product of this intensive research many heuristic procedures have been proposed for getting reasonable good colorings in general graphs.

In this paper we shall present a collection of heuristic methods (or shortly heuristics); we will try to give a sketch of classification as well as general observations about heuristics. The purpose of this review is not to compare heuristics from a computational point of view but to examine the basic ideas of the various techniques. Some experiences with the famous tabu search technique will be reported and extensions of colorings will be presented. All graph-theoretical concepts not defined here can be found in Berge [1].

2. Formulation and some basic ideas

A graph $G = (V, E)$ consists of a finite set V of nodes and a family E of edges (unordered pairs $[x, y]$ of distinct nodes). A *k-coloring* is a partition of the set V into k independent sets $S_1, S_2, \ldots, S_k$. Here a set S of nodes is *independent* if no two nodes in S are linked by an edge. The smallest k for which there exists a k-coloring of G is the *chromatic number* of G; it is denoted by $\chi(G)$.

As mentioned above finding whether an arbitrary graph has a k-coloring for a given k is generally an NP-complete problem. Moreover the determination of a $(2 - \varepsilon)\chi(G)$-coloring of an arbitrary graph for any $\varepsilon > 0$ is NP-complete [10]. On the other hand there is a simple method which gives in almost all cases a $(2 + \varepsilon)\chi(G)$-coloring [12].

If $\omega(G)$ is the maximum size of a clique (a set of nodes which are all linked pairwise) in G, then clearly $\chi(G) \geq \omega(G)$. But $\omega(G)$ can be a poor lower bound as can be seen from various classical constructions; in fact one can construct graphs G containing no triangles (hence $\omega(G) \leq 2$) and having an arbitrarily large value of $\chi(G)$. Other lower bounds can be found in Berge [1].

In order to derive upper bounds on the chromatic number, we simply have to find feasible colorings; this is precisely what heuristics will do.

For some heuristics, an analytical formula can be derived for obtaining an explicit form of the corresponding upper bound; for some others the derivation of such a formula is not as easy. We shall see methods of both types.

An idea which has been fruitful for developing heuristics in various contexts is the study of special cases which are solvable. More precisely when a problem type P for which no exact procedure is known has to be solved, we may consider a simpler problem type $\hat{P}$ (generally a special case of P); depending on the choice of a suitable $\hat{P}$ an exact procedure $A(\hat{P})$ can be found. Then an adaptation $\hat{A}$ of $A(\hat{P})$ may be developed for handling the general problem P. $\hat{A}$ will then be a heuristic method for P; we expect that its "quality" will be good if the special case $\hat{P}$ is sufficiently close to the initial problem P.

In our case, we may consider as a simplified problem $\hat{P}$ the coloring of the nodes of a special graph (for instance a bipartite graph or a special class of perfect graphs).

Then an exact coloring algorithm can be "generalized" in some sense in order to be able to deal with the case of arbitrary graphs. Several examples of this situation will be presented below.

Notice that on the other hand when a heuristic procedure is known, it may be useful to determine the simplified problem type $\hat{P}$ for which the procedure is an exact one.

For us, this amounts to finding the largest possible class of graphs G for which the procedure gives a coloring in a minimum number of colors (i.e. in $\chi(G)$ colors). The identification of such a class may give an idea of how far we are from the general case and also make explicit the situations where the heuristic will be efficient.

Along this line, *perfect* graphs will be appearing several times; these are defined as the graphs G for which the equality $\chi(G') = \omega(G')$ is satisfied in any induced subgraph G' of G.

A special type of coloring will be considered later; it is in some sense a concept corresponding to a schedule at the earliest dates. A k-coloring $(S_1, \ldots, S_k)$ is *canonical* if for every color $h \le k$ any node x in S_h belongs to some clique K with $K \cap S_i \ne \varnothing$ $(i - 1, \ldots, h)$. In other words we are using for each node a color (i.e. a positive integer) which is as small as possible: if x has received color h $(x \in S_h)$, it is because it belongs to some clique K which contains for each color $i \le h$ a node of color i.

Not all graphs have a canonical coloring; in fact we have the following:

Property 1.1 [26] *A graph G and all its subgraphs have a canonical coloring if and only if G is perfect.*

For an arbitrary graph we say that a k-coloring is *pseudo canonical* if for any color $h \le k$ the following holds: let $x_h \in S_h$, then the subgraph induced by $S_1 \cup \cdots \cup S_{h-1} \cup \{x\}$ has chromatic number h (see [27]).

For perfect graphs, the notions of pseudocanonical and canonical colorings coincide.

Finally let us define *strongly canonical k-colorings* as k-colorings such that for any clique K the following holds: if h is the smallest color occurring in K, there exists a clique $K' \supset K$ such that $K' \cap S_i \ne \varnothing$ for $i = 1, \ldots, h$.

Notice that if K is restricted to be a single node, then we simply get canonical colorings. Not all perfect graphs have strongly canonical colorings. For instance $\bar{C}_6$ (the complement of the cycle C_6 on 6 nodes) has no strongly canonical k-coloring; it can be shown that the existence of strongly canonical k-colorings characterizes the subclass of perfect graphs called strongly perfect graphs [26]. They are defined as graphs for which in any subgraph there exists an independent set which meets all (inclusionwise) maximal cliques.

3. Exact procedures

Although our purpose here is to deal mainly with heuristic procedures, we shall just mention a few exact methods for obtaining the chromatic number and an optimal coloring of a graph.

All these procedures work by implicit enumeration of all colorings of a given graph. An efficient enumeration scheme has been described by J. Randall Brown [3] who has been among the first ones to use the term "chromatic scheduling".

These enumeration procedures use systematically lower bounds of the chromatic number of specific subgraphs of the graph G to be colored. Such bounds can be obtained by finding a clique in G; such a clique is easily determined by some heuristic coloring algorithm (see [2]).

In order to increase the efficiency of the enumeration, several tricks have been proposed: for instance when a color i is introduced in the coloring process, all nodes x with degree smaller than i can be repeatedly removed from the graph.

Furthermore some "look-ahead" features are also worth being implemented; as an example when a node x requires a new color, one chooses the color which is already forbidden for the largest possible number of nodes which are uncolored yet [2].

A generalized implicit enumeration algorithm based on the version of J. Randall Brown [3] has been devised by Kubale et al [20]; it was used for implementations of several exact algorithms. Computational experiments for random graphs having up to 60 nodes and edge density up to 0.9 are reported in [20].

These experiments include the method of Brélaz [3], its corrected version given in [25] and the procedure of Korman [19].

In the next sections we shall examine some types of heuristic methods adapted to graph coloring.

4. Sequential colorings

We shall first consider a general type of coloring procedure which can be specialized in many ways. It proceeds as follows

a) determine an order O: $v_1 < v_2 < \cdots v_n$ of the nodes of G
b) color node v_1 with color 1; generally if $v_1, \ldots, v_{i-1}$ have been colored, give node v_i the smallest color which has not been used on any node v_j ($j < i$) linked to v_i.

This general procedure is called a *sequential coloring procedure* based on order O (or shortly $SC(O)$). Needless to say that the coloring obtained with an SC will depend on the order chosen in a). The use of such a procedure is justified by the following:

Proposition 4.1: *For any graph G, there exists an order O of the nodes for which $SC(O)$ produces a coloring in $\chi(G)$ colors.*

This can be seen easily by taking any coloring $S_1, \ldots S_k$ of G in $\chi(G) = k$ colors and ordering the nodes in such a way that whenever $i < j$ the nodes in S_i come in the order before the nodes in S_j.

More generally we shall say that a (heuristic) procedure $H(p)$ characterized by a family p of parameters is *acceptable* for solving a problem P if the set of solutions $S(H)$ obtained by varying the parameters in all possible ways contains some optimal solution of P.

In the case of $SC(O)$ we have $p = O$; for solving the coloring problem P, the method $SC(O)$ is acceptable from Proposition 4.1.

We may now ask what are the graphs G for which with any order O the procedure $SC(O)$ will give a coloring in $\chi(G)$ colors. We notice that no nice characterization can be found since for any graph G with n nodes we can hang a clique K_n at some node of G and for the resulting graph $\hat{G}$ any order O will produce with $SC(O)$ a coloring with $\chi(\hat{G}) = n$ colors.

Howeever, if we require the above property to hold for G and for all its induced subgraphs we have the following (a P_4 is a chordless path on four nodes):

Proposition 4.2: *The following statements are equivalent for an arbitrary graph G:*

a) for all induced subgraphs G' of G (including G itself) $SC(O)$ based on any order O gives a coloring in $\chi(G')$ colors
b) G has no induced P_4

Proof. A) Clearly if G' is a P_4 with edges $[a,b]$, $[b,c]$, $[c,d]$, with $O = a < d < b < c$ we get a coloring with $3 > 2 = \chi(G')$ colors.

B) Conversely assume that G has no induced P_4, we use induction on $k = \chi(G)$ to show that a) holds. In fact, we show that $SC(O)$ gives a strongly canonical coloring; this is clearly true for $k \leq 2$. So assume that we have a graph G with $\chi(G) = k \geq 3$. Consider an order O and a coloring $S_1, S_2, \ldots, S_h$ given by $SC(O)$ with $h \geq k$. If the coloring obtained is not strongly canonical, there exists some node $x_r \in S_r$ $(r \leq h)$ for which there is no clique $K \ni x_r$ such that $K \cap S_i \neq \emptyset$ $(i = 1, \ldots r)$. Let $i(r)$ be the smallest color for which there is a clique $K \ni x_r$ with $K \cap S_i \neq \emptyset$ $(i = i(r), i(r) + 1,$ $\ldots, r)$. We may choose the node x_r by taking the smallest r for which $i(r) > 1$. Let $x_r, x_{r-1}, \ldots, x_{i(r)}$ be the nodes forming K; by the minimality of r and by the induction hypothesis, $x_{r-1}, x_{r-2}, \ldots, x_{i(r)}$ belong to some clique K' with $K' \cap S_{i(r)-1} \neq \emptyset$. Let $x_{i(r)-1} = K' \cap S_{i(r)-1}$. Clearly $x_{i(r)-1}$ is not linked to x_r (by the minimality of $i(r)$). Since we used an $SC(O)$ procedure x_r must have a neighbor $x'_{i(r)-1}$ in $S_{i(r)-1}$. Then there is at least one node x_i $(i(r) \leq i \leq r - 1)$ in K to which $x'_{i(r)-1}$ is not linked (by the minimality of $i(r)$). So $x_{i(r)-1}, x_i, x_r, x'_{i(r)-1}$ induce a P_4; this is a contradiction. So the coloring obtained by $SC(O)$ is strongly canonical (and hence it uses $h = \chi(G)$ colors). $\square$

A graph is *complete k-partite* if there exists a partition $A_1, \ldots, A_k$ of the node set such that $x \in A_i$, $y \in A_j$ are linked if and only if $i \neq j$.

As a consequence of Proposition 4.2 we have:

Corollary 4.2: *For a complete k-partite G SC(O) gives a coloring in $\chi(G)$ colors with any order O.*

Remark 4.3 The proof of proposition 4.2 gives in fact a stronger result of Chvàtal [5] on perfectly orderable graphs; an order $O: v_1 < v_2 < \cdots v_n$ of the nodes of G is *perfect* if for all induced subgraphs G' the procedure $SC(O')$ based on the order O' induced by O on G' gives a coloring in $\chi(G')$ colors. Chvàtal has shown that an order of the nodes of G is perfect if and only if there is no induced P_4 with edges $[a,b]$, $[b,c]$, $[c,d]$ with $a < b$, $d < c$ [5]. Since in the proof of proposition 4.2 we have $x_{i(r)-1} < x_i$ and $x'_{i(r)-1} < x_r$, we have in fact shown that $SC(O)$ based on a perfect order gives a strongly canonical coloring. Graphs with a perfect order will be called *perfectly orderable*; they form a subclass of strongly perfect graphs. Some classes of perfectly orderable graphs are characterized in [6]. □

For any order $O: v_1 < v_2 < \cdots < v_n$ of the nodes of G we can get an upper bound on the number of colors used by $SC(O)$. Let $d_G(v)$ be the degree of node v in G. If G_i is the subgraph of G induced by nodes $v_1, v_2, \ldots, v_i$, one gets with $SC(O)$ a coloring in at most

$$B(O) = 1 + \max_{1 \leq i \leq n} (\min\{i - 1, d_{G_i}(v_i)\})$$

$$= 1 + \max_{1 \leq i \leq n} d_{G_i}(v_i) \tag{4.1}$$

From this we deduce an upper bound for $\chi(G)$:

$$WP(O) = 1 + \max_{1 \leq i \leq n} (\min\{i - 1, d_G(v_i)\}) \tag{4.2}$$

It is easy to see that $WP(O)$ is minimized when an order $O_{wp}: v_1 < \cdots < v_n$ such that $d_G(v_1) \geq \cdots \geq d_G(v_n)$ is chosen (see Welsh and Powell [28]).

By analogy in order to minimize $B(O)$ in (4.1) one should try to find an order $v_1 < \cdots < v_n$ such that

$$d_{G_1}(v_1) \geq d_{G_2}(v_2) \geq \cdots \geq d_{G_n}(v_n) \tag{4.3}$$

Here however the values $d_{G_i}(v_i)$ are not known beforehand and in fact there may exist no order $v_1 < \cdots < v_n$ satisfying (4.3).

An order O_{SL} (smallest last) can be constructed as follows (see Matula [22]).

1) Let v_n be a node with minimum degree in $G = G_n$
2) for $i = n - 1, n - 2, \ldots, 1$ let v_i be a node with minimum degree in G_i (the graph generated by all yet unnumbered nodes when $v_n, v_{n-1}, \ldots, v_{i+1}$ have been chosen).

So from (4.1) we have

$$B(O_{SL}) = 1 + \max_{1 \leq i \leq n} (\min_{v \in G_i} d_{G_i}(v)) \tag{4.4}$$

Let us now define the following function which is independent of the order O in G.

$$\Lambda(G) = 1 + \max_H (\min_{v \in H} d_H(v)) \tag{4.5}$$

where the maximum is taken over all induced subgraphs H of G; clearly $\Lambda(G) \geq B(O_{SL})$ since in (4.4) only subgraphs $G_1, G_2, \ldots, G_n$ are considered.

On the other hand, let F be a subgraph of G for which the maximum is obtained in (4.5) and let i be the last node of F in the order O_{SL}; then

$$d_{G_i}(v_i) \geq d_F(v_i) \geq \min_{v \in F} d_F(v) \text{ so } B(O_{SL}) \geq \Lambda(G)$$

Hence we have $B(O_{SL}) = \Lambda(G)$ and $SC(O_{SL})$ gives a coloring in at most $\Lambda(G)$ colors.

Matula and Beck have shown that $SC(O_{SL})$ can be implemented in $O(m + n)$ time (where m is the number of edges) [23] (see also [27a]).

It was observed that $SC(O_{SL})$ gives a coloring in at most 2 colors if G is a forest, at most 5 colors if G is planar and at most 3 colors if G is outerplanar (i.e. G can be embedded in the plane in such a way that all nodes are on the outer face) [23]. This means that the algorithm $SC(O_{SL})$ is an exact procedue for forests; it gives a coloring in at most $\chi(G) + 1$ colors for planar graphs. This brings us back to the discussion in section 2 : $SC(O_{SL})$ is exact for the special case $\hat{P} = $ "coloring of forest" of the general problem $P = $ "coloring of an arbitrary graph".

In the next section we shall review some heuristics which are exact for a more general problem than the above $\hat{P}$.

5. Some sequential coloring techniques

In order to develop heuristic procedures which are hopefully efficient for general graphs, we may try to extend some SC methods which are exact for a simplified problem type $\hat{P}$. It is reasonable to concentrate on the bipartite graph coloring problem $\hat{P}$ since for such graphs finding the chromatic number is easy.

A simple labeling technique for checking whether a graph is bipartite will give us directly a bicoloring:

1) label an arbitrary node x with $+$; x is then labelled, all remaining nodes are unlabeled
2) apply as long as possible the folowing rule: if a node is labelled with $+$ (resp. $-$) then label its neighbors with $-$ (resp. $+$).

   If all nodes receive a unique label, the graph is bipartite; otherwise it contains an odd cycle.

The nodes will be labelled one after the other and if the graph is bipartite, there will be some order O in which the nodes are labelled. The procedure can be extended to an $SC(O)$ procedure for an arbitrary graph as follows: at each step i of the coloring procedure we define for each node x the *degree of saturation* $ds_i(x)$ as the number of different colors already used for the neighbors of node x when nodes $v_1, v_2, \ldots,$ v_{i-1} have been colored. Initially (i.e. when no node is colored yet we take $ds_0(x) = 0$).

Consider now an order $O: v_1 < v_2 < \cdots < v_n$ and apply the $SC(O)$ procedure. We will get a coloring of G with at most

$$DS(O) = 1 + \max_{1 \leq i \leq n} ds_i(v_i) \qquad (5.1)$$

colors. Clearly $DS(O) \leq B(O)$. Furthermore $DS(O)$ depends on the order O. For getting a value of $DS(O)$ which is small, a reasonable procedure consists in choosing at each step i of the $SC(O)$ procedure a node v_i such that

$$ds_i(v_i) = \max_{x \text{ uncolored}} ds_i(x)$$

This is precisely the DSATUR algorithm described in Brélaz [2]. It differs from the previous $SC(O)$ techniques by the fact that the order is constructed dynamically during the coloring process.

Proposition 5.1 [2] *The $SC(O)$ procedure DSATUR is exact for bipartite graphs.*

Proof. If G is a nontrivial connected bipartite graph, DSATUR will first give color 1 to an arbitrary node v_1. Then as long as there are uncolored nodes a node v_i with $ds_i(v_i) = 1$ can be given color 1 or 2. We have for each i $ds_i(v_i) \leq 1$, otherwise G would contain an odd cycle. So we obtain a bicoloring of G. $\square$

Remark 5.2 Brélaz [2] suggests to start from a node v_1 of maximum degree when using DSATUR for an arbitrary graph. Comparisons of DSATUR with other $SC(O)$ procedures are reported in [2, 20, 25]. Refinements such as recursive DSATUR procedures are discussed in [18]. $\square$

The above described method DSATUR is a special case of a more general $SC(O)$ procedure. Essentially what happens in a connected graph G when DSATUR is used is the following: for each i, the subgraph G_i induced by nodes $v_1, v_2, \ldots, v_i$ is connected. We shall say that for an arbitrary graph G the order $O: v_1 < v_2 < \cdots < v_n$ is *connected* if for each i, the set of colored nodes $\{v_1, \ldots, v_i\}$ induces a connected subgraph (possibly empty) in each connected component of G.

We may now ask what are the graphs G such that for any connected order O and for any induced subgraph G', the $SC(O)$ procedure gives a coloring in $\chi(G')$ colors.

For instance we notice that for any odd cycle C_{2k+1} the procedure will give a 3-coloring, i.e. a $\chi(C_{2k+1})$-coloring. The procedure will also give an optimum coloring for any induced subgraph of C_{2k+1}. Hence there are nonperfect graphs which belong to this class. On the other hand the graphs of Fig. 1 (Tent and Fish) are perfect, but with the given connected orders O the $SC(O)$ procedure does not give an optimum coloring.

Before considering graphs where any connected order will give a $\chi(G)$-coloring, we make a simple observation which will give the following consequence of Proposition 4.1.

Proposition 5.3 *For any bipartite graph G there is a connected order O such that the $SC(O)$ procedure gives a coloring in $\chi(G)$ colors.*

G. Tinhofer has constructed a graph with 18 nodes for which any $SC(O)$ based on a connected order does not give an optimum coloring.

This shows in fact that if O is restricted to be a connected order of G, the heuristic $SC(O)$ procedure is generally not acceptable for the graph coloring problem.

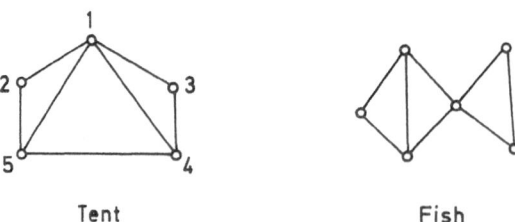

Tent   Fish

Figure 1

It follows however from Proposition 4.2 that if G has no induced P_4 then any $SC(O)$ based on a connected order O will produce an optimum coloring.

Characterizing graphs for which any connected order gives a $\chi(G)$-coloring seems more difficult. We can however characterize a subclass of these. For this purpose we need a few definitions.

A graph G will be called (strongly) *SCORE-perfect* if the $SC(O)$ based on a Connected *ORdEr* gives for any i a (strongly) canonical coloring of G_i.

We recall the definition of a special class of perfect graphs; the *parity graphs* are graphs where every odd cycle of length at least five has at least two crossing chords. We can then state

Proposition 5.4 [17]: For a graph G the following statements are equivalent:
a) G is SCORE-perfect
b) G is strongly SCORE-perfect
c) G is a parity graph without any induced Fish

This result is an extension of proposition 5.1 since bipartite graphs satisfy c). We shall not give here the proof of proposition 5.3, it can be found in [17].

Remark 5.5 One should observe that for the various heuristic procedures described here, bounds on the number of colors used have been derived. These represent the worst cases that can happen. There are relations between these bounds; but if a bound of a method M is better than a bound of method N, it does not mean that M will in general use less colors than N.   □

A variation of the $SC(O_{WP})$ technique is described by Leighton [21]. Given a graph $G = (V, E)$ let v_1 be a node with maximum degree and assign color 1 to v_1. The order $v_1 < \cdots < v_n$ is constructed dynamically (as in the DSATUR algorithm).

Assume i nodes (i.e. $v_1, \ldots, v_i$) have been given color 1. Let N_o (resp N_1) be the set of uncolored nodes not adjacent to any (resp. adjacent to at least one) colored node. In other words N_o contains nodes which are possible candidates for color 1 and N_1 nodes which cannot get color 1. Let $d(v, N_1)$ be the number of edges between node v in N_o and nodes in N_1.

Node v_{i+1} is chosen in N_o in such a way that

$$d(v_{i+1}, N_1) = \max_{v \in N_o} d(v, N_1)$$

Ties are broken by choosing a node with minimum degree in the subgraph generated by N_o.

This is repeated until $N_o = \varnothing$. Then the process is iterated with color 2 on the graph generated by the uncolored nodes and so on.

The above procedure is called RLF (Recursive Largest First); it can be implemented in $O(n^3)$ time for general graphs. Experiments are described in [21] and in [27a].

The RLF procedure is a sequential coloring algorithm based on a dynamic order; it does not use a connected order. It is nevertheless exact for bipartite graphs. The reasons for this is that RLF tries to choose for v_{l+1} (next node to get color 1) a node which has many neighbors in common with nodes having already received color 1. This idea will be exploited in the algorithm to be described next.

6. Additional methods

There are many different algorithms which can be devised as extensions of exact procedures for bipartite graphs. Some may not be in a straightforward way based on the idea of sequential colorings. We first review the technique of Dutton and Brigham [7] which runs as follows:

As long as there are nonadjacent nodes in G, repeat the following steps:

a) compute for each pair of nonadjacent nodes v_i, v_j the number c_{ij} of common neighbors
b) determine the pair v_r, v_s for which c_{rs} is maximum
c) merge v_r and v_s

So the procedure finds at each step a pair of nodes v_r, v_s which will have the same color in the coloring which is constructed. The algorithm stops when the graph is reduced to a clique K_p. Then a p-coloring is obtained for the initial graph by assigning each node v_i of G the color of the node of K_p which represents it.

The above procedure is exact for bipartite graphs as can be seen easily: two nonadjacent nodes x, y which are in different sets of a connected bipartite graph $G = (X, Y, E)$ have no common neighbor; hence these will not be chosen for merging. So after each merging operation the graph will still be bipartite and it will be reduced to K_2 at the end.

A variation of the merging algorithm has been suggested by Hertz [16]; it consists in using the same node v_r as long as possible (i.e. until it is linked to all remaining nodes) in the merging operation. The COSINE algorithm obtained in this way has the following property:

Proposition 6.1 [16]: The COSINE algorithm gives a $\chi(G)$-coloring for any Meyniel graph G.

We recall the defintion of Meyniel graphs: these are characterized by the existence of at least two chords in each odd cycle (of length at least five).

Remark 6.1: Hertz shows that a slightly more general class of graphs can be colored in $\chi(G)$ colors by COSINE. These are the graphs containing one node adjacent to all edges which are the unique chord of some odd cycle of length at least five [16]. □

A coloring procedure based on the similarity of neighborhoods of nodes was suggested by Wood [31]. It runs as follows: a pair of nodes i, j with maximum similarity c_{ij} is given color 1. Then the pairs are considered in order of nonincreasing similarities. (Observe that the pairs are generally not disjoint). For pair v_i, v_j we have the three cases (assume k colors have been used).

Table 1. The BIPCOL technique

Input: a graph G *Output*: a coloring of G

Initialization: $k = 1$; $C = \emptyset$ (C = set of colored nodes)

While there are some uncolored nodes in G, *do*
begin
construct a bipartite induced graph $\hat{B}$ in G-C;
color the nodes of $\hat{B}$ with colors k and $k + 1$
(or k only if $\hat{B}$ has no edges);
$C = C \cup \{\text{nodes of } \hat{B}\}$;
replace k by $k + 2$

end

---

a) if both v_i and v_j are colored, go to the next pair
b) if v_i has color $g \leq k$ and v_j is uncolored then
  1) if $d_G(v_j) < k$, then v_j can be colored without problem; ignore it.
  2) if v_j can be colored with g, give v_j color g.
c) if neither v_i nor v_j is colored then
  1. if $d_G(v_i) < k$, $d_G(v_j) < k$, ignore them
  2. find the smallest color g such that both v_i and v_j can get color g (introduce a new color if needed)

This procedure has been applied to construct examination schedules (with about 500 exams) [31].

Let us now briefly sketch another procedure which is by nature an exact method for bipartite graphs. It runs in the following way (see Table 1): in the set of uncolored nodes we construct a bipartite graph by applying the simple labeling technique given in section 5. We repeat the labeling procedure until no more (uncolored) node can be labelled. We have then a bipartite induced graph $\hat{B}$; we color it with colors 1 and 2 and we repeat the whole process in the graph generated by the uncolored nodes; the bipartite graph obtained is colored with colors 3 and 4. This is repeated until all nodes are colored.

This method (which we may call BIPCOL is a sequential coloring procedure; the order O is determined dynamically.

We shall now describe another technique which is again an exact procedure for bipartite graphs; it is closely related to the previous procedure where a bipartite graph B is constructed at each step (see Table 2). The procedure is called CANABIS (Coloring Algorithm for Networks Acting on Bipartite Induced Subgraphs). The difference with BIPCOL lies in the fact that we choose in each connected component B of the bipartite subgraph $\hat{B}$ one of the two node sets and we color its nodes with color k; the other node set of B is uncolored. $m(A, B)$ is the number of edges between node sets A and B.

In each connected component $B = (V, W, E_B)$ of $\hat{B}$ with node sets V, W and edge set E_B we color with color k the set among V, W which has the largest cardinality. The other set is then considered as uncolored. These simple procedures are faster then

Table 2. The CANABIS Technique

Input: a graph G *Output*: a coloring of G

Initialization: $k = 1$; $C = \emptyset$ ($C =$ set of colored nodes)

while there are some uncolored nodes in G, *do*
begin
construct a bipartite induced graph $\hat{B}$ in G-C;
for each connected component $B = (V, W, E_B)$ of $\hat{B}$
do if $m(W, C) > m(V, C)$ *then* $C := C \cup W$
           color the nodes in W with color k

    else C: $C \cup V$
           color the nodes in V with color k;

replace k by $k + 1$

end

Table 3

	DSATUR	BIPCOL	CANABIS	
Normal random graphs	18.9	19.6	20.0	edge density 0, 5
euclidian graphs	31.9	30.25	33.25	nodes in (1×1)-square; linked if distance ≤ 0.5

Average number of colors
(100 samples with 100 nodes)

DSATUR and the colorings produced are almost as good. Table 3 shows a few computational results on random graphs with 100 nodes (euclidean graphs are obtained by generating random nodes in a square of size 1 and linking two nodes by an edge if their distance is at most 0.5; the normal random graphs are obtained by introducing each possible edge with probability 0.5).

7. Tabu Search

In an entirely different direction we may view the coloring problem as an instance of minimization of a certain function.

Let us first describe briefly the general form of the optimization techniques that we will use later. Suppose we have a (finite) set X of feasible solutions; we are given a function $f: X \to Z^+$ and we have to find some solution s in X for which $f(s)$ is minimum.

Now X generally has some structure; we shall assume that for each feasible solution s a neighborhood $N(s)$ can be defined. This amounts to representing the elements s of X as the nodes of a graph and if $s' \in N(s)$ we introduce an arc (s, s'). Notice that we may have $s' \in N(s)$, but $s \notin N(s')$.

Many minimization procedures work as follows: they start from an initial solution s; then as long as a better solution s' (i.e, a solution with $f(s') < f(s)$) can be found in $N(s)$, one moves to s', i.e, s is replaced by s' and one repeats the step. In general such a technique will reach a local minimum of f and will be trapped there.

In order to avoid such troubles, some refinements have been proposed; among other techniques the famous simulated annealing technique has been constructed by exploiting analogy with some physical systems. When a solutions s' in $N(s)$ is found, the move to s' is accepted if $f(s') < f(s)$ and if $f(s') > f(s)$ the move is accepted with a probability $p(\Delta, t) = \exp(-\Delta/t)$ where $\Delta = f(s') - f(s)$ and t is a parameter corresponding to temperature; t is decreased as the iterations are performed. This amounts to reducing the probability of accepting a solution s' which is worse than s.

The general procedure is described in Table 4. References on the technique can be found in [4] where an application to graph coloring is described. Although this technique has appeared as extremely appealing for getting almost optimal solutions in various types of large size combinatorial optimization problems, it is no longer used as much as earlier; the main reason is that there is a procedure which keeps some of the basic ideas of simulated annealing but which is much more simple and, as numerous experiences have shown, much more efficient.

It is the Tabu Search procedure; the basic ideas of the technique are developed in Glover [11].

Table 4. The Simulated Annealing procedure

```
Set t to a suitably high value;
choose a feasible solution s in X;
compute f(s);
change := true;
while   change do
begin

    change := false
    repeat rep times
    begin
    choose an s'εN(s);
    Δ = f(s') − f(s);
    p(Δ) = exp(−Δ/t);
    generate random variable x (uniform on [0, 1]);
    if x < p(Δ) then go to accept
        else go to exit;

    accept: s := s';
        f(s) = f(s) + Δ;
        if Δ ≠ 0 then change := true;

    exit:

      end;

    t := t*a   (0 < a < 1)

end
```

Before going into the details of the procedure, let us mention briefly how coloring problems were transformed into a minimization of a function $f: X \to Z$ in order to apply simulated annealing.

Suppose we are interested in finding a k-coloring of a graph G for a given k. So for this purpose a feasible solution s is simply a partition $s = (S_1, S_2, \ldots, S_k)$ of the node set of G into k subsets. Such a partition will be a coloring if each S_i is an independent set of nodes, i.e. if each set $E(S_i)$ of edges of G with both endpoints in S_i is empty. It is therefore natural to consider as a function f to be minimized the function given by $f(s) = \sum_{i=1}^{k} |E(S_i)|$. This is precisely what was used in [4] for simulated annealing and in [15] for tabu search. We shall have a k-coloring if and only if $f(s) = 0$. Now the neighborhood $N(s)$ of a solution $s = (S_1, S_2, \ldots S_k)$ contains every solution s' obtained from s by moving some node x in some S_i (where x is adjacent to at least one other node in S_i) to some other subset S_j of the partition.

For tabu search (TS) we used the same formulation as above. The TS procedure starts again from an initial solution. Whenever we are at some solution s, we generate a sample of rep (rep is a parameter) solutions s_i in $N(s)$. We move to the best s_i, say s' (even if $f(s') > f(s)$). As in simulated annealing we will so have a chance to escape from local minima. But such a procedure may now cycle. Therefore in order to prevent the process from cycling, we give a value k and we do not allow the algorithm to go back to a solution which has been visited in the last k steps. This could be done by introducing a so-called *tabu list* T of size $|T| = k$ which would contain the last k solutions visited. But since keeping track of these solutions would be extremely space consuming, it is simpler to store a move $[s' \to s]$ from a solution s' to a solution s. More precisely when we move from s to $s' \in N(s)$, we just take a node x in G with color i and we give color j to x; so what we will do with the list T is simply to forbid node x to get back to color i during the next k iterations. The list T will then contain k forbidden "moves"; each one consists of a node x and a forbidden color i.

Usually the length k of T is kept constant. So, whenever it is needed, the oldest element of T is dropped.

By introducing a tabu list, we reduce the risk of cycling (in fact we eliminate the possibility of going through cycles of length at most k).

The existence of tabu moves may sometimes be an obstacle in the search for a global minimum.

A move $[s' \to s]$ may be a tabu move and so be forbidden. However applying this move to the current solution s'' may not bring us back to a solution visited in the last k iterations (simply because between the step where we visited s and moved to s' and the present step many other moves were made; this caused many local changes in the current solution).

So we may wish to allow a tabu move to be made in some circumstances; recognizing whether it would bring us back to an already visited solution is difficult. But we can simply say that we accept a tabu move if the value $f(s)$ of the resulting solution s is

Table 5. The tabu search procedure

Choose a feasible solution s in X;
compute $f(s)$;
take an arbitrary tabu list T;
nbiter := 0;

while $f(s) > 0$ *and* nbiter < nbmax *do begin*

generate rep solutions $s_i \in N(s)$
    with $[s \to s_i] \notin T$ or $f(s_i) \le A(f(s))$;

(as soon as an s_i with $f(s_i) < f(s)$
is found, stop the generation)

Let s' be the best s_i generated
update tabu list T, update $A(f(s))$
    (remove oldest tabu move and
    introduce move $s' \to s$)

$s := s'$;
nbiter := nbiter + 1

endwhile

---

small enough. More precisely, if we suppose that f takes integral values, we may define an aspiration $A(z)$ for each integral z; initially we set $A(z) = z - 1$ for all z. Then a tabu move leading from s to s' will be accepted if $f(s') < A(f(s))$; after this acceptance we update the aspiration function by setting $A(f(s)) = f(s') - 1$. So the next time we will be at some solution giving a value $z = f(s)$, a tabu move will be accepted only if it gives a larger improvement than the last time we left s by accepting a tabu move. The idea behind this computation is again to avoid cycling.

The whole process is then repeated (see Table 5) until we get $f(s) = 0$ (i.e, we have a k-coloring or no improvement of the best value of f has occurred during a given number of steps or simply until a fixed number of iterations have been performed).

The TS method was applied to coloring of random graphs having up to 1000 nodes; computational results are reported in [15]. It is worth mentioning that with some refinements, TS was the only procedure which could give k-colorings for large graphs with a value of k very close to the estimated chromatic number (values of these estimations are given in [18].

Experiments have also shown that TS gives colorings which are as good as the ones produced by any known heuristic procedure in the same computational time.

No general convergence properties are known at the moment; further research should be carried out for understanding better the apparent efficiency of TS. This is the more needed because for many of the difficult combinatorial optimization problems, TS seems to beat simulated annealing by far ...

Remark 7.1 Among the many heuristics which have been described here, only a few are not exact procedures for bipartite graphs. It is not known whether TS will always give a $\chi(G)$-coloring of a bipartite G. It is easy to find examples where $f = |E(S_1)| + |E(S_2)|$ has some local minima for the neighborhood structure described above. The

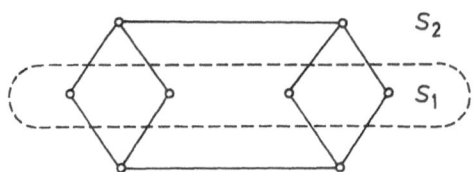

A local minimum of $f = |E(S_1)| + |E(S_2)|$

Figure 2

graph in fig. 2 is bipartite and the partition $s = (S_1, S_2)$ shown gives a value $f(s) = 2$; it is a local minimum. ☐

The efficiency of TS for graph coloring was increased by combining it with some other techniques (see [15]); improvements will depend on the efficiency of procedures for constructing large independent sets. TS has been adapted to this case and computational results are reported in [8].

8. More on sequential colorings

For application purposes many other types of colorings have been considered by various authors (see [14] for some examples). Among these variations we mention the interval colorings. We shall just show how the $SC(O)$ techniques can be adapted to deal with this extension.

Let $G = (X, E, c)$ be a finite graph in which each node i is associated with a positive integer c_i. An *interval k-coloring* is an assignment of a set $S(i)$ of c_i consecutive colors (chosen in $\{1, 2, \ldots, k\}$) to each node i such that $S(i) \cap S(j) = \varnothing$ for any two adjacent nodes i, j [30].

Such colorings may be needed in scheduling a collection of jobs i with processing times c_i in such a way that incompatible jobs (i.e. jobs represented by nodes which are all linked together) are not processed simultaneously and that no preemption occurs.

If we take an order $O: 1 < 2 < \cdots < n$, then by applying an obvious adaptation of $SC(O)$ we get a k-coloring with $k \leq IB(O)$ where

$$IB(O) = \max_{1 \leq i \leq n} \left(\min \left\{ \sum_{k=1}^{i} c_k, c_i + (c_i - 1)d_{G_i}(i) + \sum (c_j : j \in N_{G_i}(i)) \right\} \right)$$

Here $N_{G_i}(i)$ is the set of neighbors of node i in G_i. If $\chi_{\text{int}}(G)$ is the smallest k for which $G = (X, E, c)$ has an interval k-coloring, then $IB(O)$ is an upper bound on $\chi_{\text{int}}(G)$.

By analogy with the $SC(O_{SL})$ procedure, we may construct an order O_{SL} by starting from the end as follows:

assume we have already numbered nodes v_n, v_{n-1}, ..., v_{i+1} then we call G_i the graph generated by the remaining nodes. We take for v_i a node p for which

$$\Delta(G_i, p) = (d_{G_i}(p) + 1)(c_p - 1) + \sum (c_j : j \in N_{G_i}(p))$$

is minimum. Then one gets another upper bound for $\chi_{int}(G)$ [30]:

$$IM(O_{SL}) = 1 + \max_{G' \subseteq G} \min_{i \in G'} \left[(d_G(i) + 1)(c_i - 1) + \sum_{j \in N_{G'}(i)} c_j \right]$$

As in the case of classical k-colorings (where $c_i = 1$ for each node i), $IB(O_{SL})$ is independent of the order O_{SL}.

It is worth mentioning that in the general case of interval colorings the bounds $IB(O)$ and $IM(O_{SL})$ are unrelated; examples can be constructed where one of those is better than the other.

A lower bound for $\chi_{int}(G)$ can be obtained by considering the cliques K in G and computing

$$\omega_{int}(G) = \min_K \left(\sum_{i \in K} c_i \right)$$

It would be interesting to characterize graphs G for which $\chi_{int}(G) = \omega_{int}(G)$ for any choice of values $c_i \geq 0$.

Observe that if $c_i \in \{0, 1\}$ for each i, the above graphs are precisely the perfect graphs.

References

[1] C. Berge, Graphes, Gauthier-Villars, Paris, 1983.
[2] D. Brélaz, New Methods to Color the Vertices of a Graph, Communications of the Association for Computing Machinery 22 (1979) 251–256.
[3] J. Randall Brown, Chromatic scheduling and the chromatic number problem, Management Science 19 (1972) 456–463.
[4] M. Chams, A. Hertz, D. de Werra, Some experiments with simulated annealing for coloring graphs, European J. of Operational Research 32 (1987) 260–266.
[5] V. Chvàtal, Perfectly ordered graphs in: Topics on Perfect Graphs (C. Berge, V. Chvàtal, eds) Annals of Discrete Mathematics 21 (1984) 253–277.
[6] V. Chvàtal, C. T. Hoang, N. V. R. Mahadev, D. de Werra, Four classes of perfectly orderable graphs, J. of Graph Theory 11 (1987) 481–495.
[7] R. D. Dutton, R. C. Brigham, A new graph coloring algorithm, Computer J. 24 (1981) 85–86.
[8] C. Friden, A. Hertz, D. de Werra, Stabulus: a technique for finding stable sets in large graphs with tabu search, Computing 42 (1989) 35–44.
[9] M. R. Garey, D. S. Johnson, H. G. So, An Application of Graph Coloring to Printed Circuit Testing, IEEE Transactions on circuits and systems 23 (1976) 591–598.
[10] M. R. Garey, D. S. Johnson, Computers and Intractability: a Guide to the Theory of NP-Completeness, W. H. Freemann, San Francisco, 1978.
[11] F. Glover, Tabu Search, CAAI Report 88–3, University of Colorado, Boulder 1988.
[12] G. R. Grimmett, C.J.H. Mc Diarmid, On Coloring Random Graphs, Mathematical Proceedings of the Cambridge Philosophical Society 77 (1975) 313–324.
[13] P. Hansen, M. Delattre, Complete-Link Cluster Analysis by Graph Coloring, J. of the American Statistical Association 73 (1978) 397–403.
[14] A. Hertz, D. de Werra, eds., Graph Coloring and Variations, Annals of Discrete Mathematics 39 (North Holland, Amsterdam, 1989).
[15] A. Hertz, D. de Werra, Using Tabu Search Techniques for Graph Coloring, Computing 39 (1987) 345–351.

[16] A. Hertz, A fast algorithm for coloring Meyniel graphs, to appear in J. of Combinatorial Theory

[17] A. Hertz, D. de Werra, Connected sequential colorings, in [14], pp. 51–59.

[18] A. Johri, D. W. Matula, Probabilistic bounds and heuristic algorithms for coloring large random graphs, Southern Methodist University, Dallas, Texas, 1975.

[19] S. Korman, The Graph-Coloring Problem, in : N. Christofides et al, ed. Combinatorial Optimization, (J. Wiley, New-York, 1979) pp. 211–235.

[20] M. Kubale, B. Jackowski, A Generalized Implicit Enumeration Algorithm for Graph Coloring, Communications of the Association for Computing Machinery 28 (1985) 412–418.

[21] F. T. Leighton, A graph Coloring Algorithm for Large Scheduling Problems, J. of research of the National Bureau of Standards 84 (1979) 489–503.

[22] D. W. Matula, G. Marble, J. D. Isaacson, Graph Coloring algorithms, in: R. C. Read, ed. Graph Theory and Computing (Academic Press, New-York, 1972) pp. 108–122.

[23] D. W. Matula, L. L. Beck, Smallest-Last Ordering and Clustering and Graph Coloring Algorithms, J. of the Association for Computing Machinery 30 (1983) 417–427.

[24] N. Mehta, The application of a graph coloring method to an examination scheduling problem, Interfaces 11 (1981) 57–64.

[25] J. Peemöller, A correction to Brélaz's modification of Brown's coloring algorithm, Communications of the Association for Computing Machinery 26 (1983) 595–597.

[26] M. Preissmann, D. de Werra, A note on strong perfectness of graphs, Mathematical Programming 31 (1985) 321–326.

[27] B. Roy, Nombre chromatique et plus longs chemins d'un graphe. Revue française d'informatique et de Recherche Opérationnelle 5 (1967) 129–132.

[27a] M. M. Syslo, N. Deo, J. S. Kowalik, Discrete Optimization Algorithms with Pascal Programs (Prentice-Hall, Englewood Cliffs, N. J. 1983).

[28] D. J. A. Welsh, M. B. Powell, An upper bound on the chromatic number of a graph and its application to timetabling problems, Computer J. 10 (1967) 85–87.

[29] D. de Werra, An introduction to timetabling, European Journal of Operational Research 19 (1985) 151–162.

[30] D. de Werra, A. Hertz, Consecutive colorings of graphs, Zeitschrift für Operations Research 32 (1988) 1–8.

[31] D. C. Wood, A technique for coloring a graph applicable to large scale timetabling problems, Computer J. 12 (1969) 317–319.

A. Gyarfas, J. Lehel, Outline and FF colorings of graphs, J. of Graph Theory 12 (1988) 217–227.

D. de Werra
Ecole Polytechnique Fédérale de Lausanne
Département de Mathématiques
Chaire de Recherche Opérationnelle
CH-1015 Lausanne

Computing, Supp. 7, 209–233 (1990)

Computing
© by Springer-Verlag 1990

Probabilistic Analysis of Graph Algorithms

A. M. Frieze, Pittsburgh, Pa.

Abstract — Zusammenfassung

Probabilistic Analysis of Graph Algorithms. We review some of the known results on the average case performance of graph algorithms. The analysis assumes that the problem instances are randomly selected from some reasonable distribution of problems. We consider two types of problem. The first sort is polynomially solvable in the worst case but there are algorithms with better average case performance. In particular we consider the all-pairs shortest path problem, the minimum spanning tree problem, the assignment problem and the cardinality matching problem in sparse graphs. Our second category of problems consists of problems which seem hard in the worst-case but still have algorithms with good average case performance. In particular we consider three NP-Complete problems; the Hamilton cycle problem, the graph bisection problem and graph colouring. In addition we consider the graph isomorphism problem whose exact complexity is still undetermined.

AMS Subject Classifications: 68Q25, 05C80.

Key words: Graph Algorithms, Probabilistic Analysis.

Probabilistische Analyse von Graphenalgorithmen. Die Arbeit bietet einen Überblick über Ergebnisse zur durchschnittlichen Leistungsfähigkeit von Graphenalogrithmen, wobei stets vorausgesetzt wird, daß die Problembeispiele zufällig gemäß einer 'vernünftigen' Wahrscheinlichkeitsverteilung gewählt werden. Wir betrachten zwei Problemtypen. Der erste Typ ist polynomial lösbar im schlechtesten Fall, jedoch existieren nicht-polynomiale Lösungsalgorithmen mit besserer durchschnittlicher Leistungs-fähigkeit. Insbesondere betrachten wir das Problem der kürzesten Wege zwischen allen Knotenpaaren, das Problem der Minimalbäume, das Zuordnungsproblem und das ungewichtete Matchingproblem in dünn belegten Graphen. Der zweite Problemtyp besteht aus Problemen, die im schlechtesten Fall äußerst schwer zu lösen erscheinen, wofür es aber weiterhin Lösungsalgorithmen mit guter durch-schnittlicher Leistungsfähigkeit gibt. Insbesondere betrachten wir drei NP-vollständige Probleme: das Problem der Hamiltonkreise, das Bisektionsproblem für Graphen und das Färbungsproblem. Darüber hinaus wird das Graphenisomorphie-problem behandelt, dessen exakte Komplexität noch unbestimmt ist.

1. Introduction

Graph theory is an important source of computational problems and as such has played a significant part in the development of a theory of algorithms and their analysis. We find here as elsewhere that the analysis of the execution times of algorithms has concentrated in the main on that of their worst-case. There is nevertheless a sizeable literature on the average case performance of algorithms.

The analysis assumes that the problem instances are randomly selected from some reasonable distribution of problems and an attempt is made to estimate the expected running time of algorithms for these problems. The analytical difficulties are compounded by the fact that algorithms condition their data quickly. Consequently,

the statistical independence which is required by most common forms of probabilistic analysis is hard to come by. Probabilistic algorithm analysis has therefore
necessitated the development of its own, often indirect, techniques.

In this paper we will try to review some cases where probabilistic analysis has
something positive to say about the performance of algorithms. We will look in
some detail at eight problems. The first four: the all pairs shortest path problem,
the assignment problem, the matching problem in general graphs and the minimum
spanning tree problem are all solvable in polynomial time in the worst-case. Nevertheless we will find that algorithms can be constructed whose average performance
on natural distributions is significantly better than the worst-case of any known
algorithm. The next three: the Hamilton cycle problem, the graph colouring problem and the graph bisection problem are all known to be NP-hard. We will, in
spite of this, be able to describe polynomial time algorithms which have a high
probability of finding solutions to these problems. Our final example will be that
of graph isomorphism whose exact complexity is at present unknown. Here we will
find that a simple algorithm works with high probability. Thus, in these examples
and many others, the average case is a long way from the worst-case.

In the next section we introduce some notation and state some basic results needed
from probability theory. The next eight sections cover the problems we have
mentioned above. Following this we will mention some results with a different
flavour.

2. Notation and Basic Probabilistic Inequalities

We first define what we mean by a random graph. Let $V_n = \{1, 2, \ldots, n\}$ and suppose
$1 \leq m = m(n) \leq N = \binom{n}{2}$. The random graph $G_{n,m}$ has vertex set V_n and its edge
set $E_{n,m}$ is a randomly chosen subset of m edges. Thus if G is a graph with vertex
set V_n and m edges then $Pr(G_{n,m} = G) = \binom{N}{m}^{-1}$.

There is a closely related model $G_{n,p}$ where $0 \leq p = p(n) \leq 1$. This has vertex set V_n
and edge set $E_{n,p}$ where each of the N possible edges is independently included
with probability p. Hence if G is a graph with vertex set V_n and m edges then
$Pr(G_{n,p} = G) = p^m(1 - p)^{N-m}$. Observe that if $p = \frac{1}{2}$ then $Pr(G_{n,1/2} = G) = 2^{-N}$ and
so each graph with vertex set V_n is equally likely.

These models have been studied extensively since the pioneering work or Erdös
and Renyi [ER1] − [ER4]. The book of Bollobás [Bo1] gives a systematic and
extensive account of this subject. A gentler introduction is provided by Palmer [P].

When $m \approx Np$ $\left(\text{i.e. } \lim_{n \to \infty} \left| \frac{m}{Np} - 1 \right| = 0\right)$ the graphs $G_{n,m}$ and $G_{n,p}$ have similar
properties. Indeed for any graph property $\mathcal{A}$ we have

$$Pr(G_{n,p} \in \mathcal{A}) = \sum_{m=0}^{N} Pr(G_{n,p} \in \mathcal{A} \mid \mid E_{n,p}\mid = m) Pr(\mid E_{n,p}\mid = m)$$

$$= \sum_{m=0}^{N} Pr(G_{n,m} \in \mathcal{A}) Pr(\mid E_{n,p}\mid = m), \qquad (1.1)$$

since $G_{n,p}$, given $\mid E_{n,p}\mid = m$, is precisely $G_{n,m}$. Now $\mid E_{n,p}\mid$ is distributed as the binomial random variable $B(N, p)$. So for example, if $m = \lceil Np \rceil$

$$Pr(G_{n,p} \in \mathcal{A}) \geq Pr(G_{n,m} \in \mathcal{A}) \binom{N}{m} p^m (1 - p)^{N-m}$$

$$\geq Pr(G_{n,m} \in \mathcal{A})(2\pi p(1 - p)N)^{-1/2} \qquad (1.2)$$

on using Stirling's inequalities for factorials. (1.2) can often be used to show that $Pr(G_{n,m} \in \mathcal{A})$ is small when $Pr(G_{n,p} \in \mathcal{A})$ is small.

We are mainly concerned with asymptotic results in this paper and in all cases we will be concerned with what happens as $n \to \infty$. So let $\mathcal{E}_n$ be some event (dependent on n). We say that $\mathcal{E}_n$ occurs *with high probability* (**whp**) if

$$\lim_{n \to \infty} Pr(\mathcal{E}_n) = 1.$$

Finally we will note the following bounds on the tails of the binomial

$$Pr(\mid B(n, p) - np\mid \geq \varepsilon np) \leq 2e^{-\varepsilon^2 np/3}. \qquad (1.3)$$

(See e.g. [Bo1]).

Thus if $np \to \infty$ and we take $\varepsilon = (np)^{-1/4}$ then we see that $B(n, p) \approx np$ **whp**. By using this in (1.1) we can see that $G_{n,p}$ and $G_{n,\lceil Np \rceil}$ are "similar". We will refer to (1.3) as the Chernoff bound.

3. Shortest Path Problem

In this section we consider the problem of finding a shortest path between all pairs of nodes in a digraph $D = (V, A)$ with non-negative arc lengths $\ell(u)$ for $u \in A$. For notational convenience we assume that D is the complete digraph on V_n. The arc lengths are random and satisfy an "endpoint independence" condition. More precisely the lengths of arcs with different start vertices are independent and if for a given $v \in V_n$ we have $\ell(vw_1) \leq \ell(vw_2) \leq \cdots \leq \ell(vw_{n-1})$ (ties broken randomly) then $w_1, w_2, \ldots, w_{n-1}$ is a random permutation of $V_n - \{v\}$.

We present here an algorithm of Moffat and Takaoka [MT1] which solves the problem in $0(n^2 \log n)$ time. This is to be contrasted with $0(n^3)$ for the best worst-case performance. (See for example Lawler [La] or Papadimitrion and Steiglitz [PS]). The algorithm in [MT1] proceeds as follows:

A: sort the arcs incident with each $v \in V_n$ into increasing order to create list $L(v)$.
B: for each $s \in V_n$ find a shortest path from s to every other vertex.

Since A requires $0(n^2 \log n)$ time we need only prove an $0(n \log n)$ expected time bound for each single source problem in B. The algorithm used is based on one originally attributable to Dantzig [D] and improved and analysed in the average-case by Spira [Sp]. We first describe this version and then given the contribution of Moffat and Tokakoa.

The algorithm works with a set S. Initially $S = \{s\}$, finally $S = V_n$ and at any stage $v \in S$ means that a shortest path of length $D(v)$ has been found from s to v. If $v \notin S$ then $D(v)$ is an estimate of the shortest path length.

Suppose now that for $v \in S$, $w \notin S$ we let $\lambda(vw) = D(v) + \ell(vw)$ and $D(w) = \min\{\lambda(vw): v \in S\}$. It is easy to show that if $D(x) = \min\{D(w): w \notin S\}$ then $D(x)$ is the length of a shortest path from s to x.

In the algorithm that follows we keep a priority queue AQ of items $(xy, \lambda(xy))$, one for each $x \in S$, ordered by increasing value of λ.

Algorithm SHORTPATH(s)

begin
    Initialise AQ with $(st, \lambda(st))$ where st is the first arc on $L(s)$; $D(s) := 0$;
    $S := \{s\}$;
L1: **while** $S \neq V_n$ **do**
    begin
L2:   remove the first item $(xy, \lambda(xy))$ of AQ; add the item $(xy', \lambda(xy'))$
        to AQ where xy' succeeds xy on $L(x)$;
L3:   **if** $y \notin S$ **then do**
        begin
           $S := S \cup \{y\}$; $D(y) := D(x) + \ell(xy)$
L4:   add $(yz, \lambda(yz))$ to AQ where yz is the first arc on $L(y)$ with head not in
           S
        end
    end
end

The above algorithm spends too much time at L3 with $y \in S$. Building on an idea of Fredman (Fd) (rediscovered independently later by Frieze and Grimmett [FG]) Moffat and Takaoka [MT1] "clean up" AQ at line L1 when $|S|$ reaches $n - \dfrac{n}{2^k}$ for $k = 1, 2, \ldots, L = \lfloor \lg \lg n \rfloor$. We shall use lg to denote $\log_2$ and reserve log for $\log_e$.

Procedure CLEANUP

begin
    $E := \phi$;
    for each $xy \in AQ$ **do**
    begin
    if $y \notin S$ **then** $E := E \cup \{xy\}$

C1:   **else** $E := E \cup \{xy'\}$ where xy' is the first arc after xy on $L(x)$
        with head not in S.
     end
C2: rebuild AQ out of the arcs in E.
end

Analysis

Let Stage k run from $|S| = n - \dfrac{n}{2^{k-1}}$ to $|S| = n - \dfrac{n}{2^k}$ for $1 \le k \le L = \lfloor \log\log n \rfloor$ and let Stage $L + 1$ denote the final part of the algorithm.

$1 \le k \le L$

Let T_k denote $V_n - S$ at the start of Stage k. The probability that $y \notin S$ at $L2$ of SHORTPATH is always at least $\dfrac{1}{2}$ since $y \in T_k$, $|V_n - S| \ge \dfrac{1}{2}|T_k|$ throughout Stage k and y is equally likely to be any member of T_k.

Since $\dfrac{n}{2^k}$ vertices are added to S in Stage k we expect to execute $L3$ and hence $L2$ at most $\dfrac{n}{2^{k-1}}$ times. Since $L2$ requires $0(\log n)$ time we have

$$E(\text{time spent at } L2 \text{ in Stage } k) = 0\left(\frac{n}{2^{k-1}} \log n\right) \qquad 1 \le k \le L. \qquad (3.1)$$

To choose z in $L4$ we expect to examine at most 2^k entries in the list of arcs leaving y. This is beacuse $|V_n - S| \ge \dfrac{n}{2^k}$ throughout Stage k and the next vertex of y's list is equally likely to be any vertex not encountered so far on this list. Hence

$$E(\text{time spent at } L4 \text{ in Stage } k) = 0\left(\frac{n}{2^k} 2^k\right) \qquad 1 \le k \le L \qquad (3.2)$$

Now consider CLEANUP. At line C1 we expect to examine at most 2^k arcs before y' is found (same argument as for $L4$) and so

$$E(\text{time spent at C1 in Stage } k) = 0(n2^k) \qquad 1 \le k \le L \qquad (3.3)$$

It takes $0(n)$ time to rebuild AQ at C2 and so from (3.1), (3.2), (3.3) we obtain

$$E(\text{time spent in first } L \text{ stages})$$

$$= 0\left(\sum_{k=1}^{L} \frac{n}{2^{k-1}} \log n + \sum_{k=1}^{L} n + \sum_{k=1}^{L} n2^k + \sum_{k=1}^{L} n \right)$$

$$= 0(n \log n).$$

Let us now consider Stage $L + 1$.

First consider $L2$. Vertex $y \in T_{L+1}$ and is equally likely to be any member of T_{L+1} that has not yet been examined on x's list. Suppose $|S| = n - s$ at some point. Then we expect to repeat $L2$ at most $\dfrac{|T_{L+1}|}{s} \approx \dfrac{n}{s\lg n}$ times before finding $y \notin S$. Hence

$$E(\text{time spent at } L2 \text{ in Stage } L + 1)$$

$$= 0\left(\frac{n}{\log n} \sum_{s=1}^{n/\lg n} \frac{1}{s}\log n\right)$$

$$= 0(n\log n)$$

(the final $\log n$ factor is the time to delete the first element of AQ). Now consider $L4$ and suppose again that $|S| = n - s$. This time we expect to examine at most $\dfrac{n}{s}$ edges before finding z. Hence

$$E(\text{time spent at } L4 \text{ in Stage } L + 1)$$

$$= 0\left(n \sum_{s=1}^{n/\lg n} \frac{1}{s}\right)$$

$$= 0(n\log n)$$

We have thus shown that algorithm SHORTPATH runs in $0(n^2 \log n)$ expected time. In [MT2] Moffat and Takaoka gave another $0(n^2 \log n)$ expected time algorithm for the same problem. It is not known whether $o(n^2 \log n)$ expected time is achievable for this problem.

4. Assignment Problem

In this section we discuss the result of Karp [Ka1] that the $m \times n$ assignment problem $(m \le n)$ can be solved in $0(mn\log n)$ expected time. The analysis can be applied when the matrix of costs $\|c(i,j)\|$ is such that (i) the costs in different rows are independent and (ii) for each i, if $c(i,j_1) \le c(i,j_2) \le \cdots \le c(i,j_n)$ then $j_1, j_2, \ldots, j_n$ is a random permutation of $\{1, 2, \ldots, n\}$. (This is the endpoint independence condition of § 3). The proposed algorithm starts with an empty matching and then uses shortest augmenting paths to increase it to size m. The idea of Edmonds and Karp [EK] and Tomizawa [To] is used to ensure that the shortest path problems that need to be solved have non-negative arc lengths.

Let G be the bipartite graph with vertex set $V = X \cup Y$ where $X = \{x_1, x_2, \ldots, x_m\}$, $Y = \{y_1, y_2, \ldots, y_n\}$ and the cost of edge $x_i y_j$ is $c(i, j)$. We are looking for a minimum cost matching that covers X. If M is any matching of G let $D(M)$ be the digraph with vertex set V and arcs

$$x_i y_j \quad \text{whenever } \textit{edge } x_i y_j \notin M \quad \text{forward arc}$$

$$y_j x_i \quad \text{whenever } \textit{edge } x_i y_j \in M \quad \text{backward arc}.$$

Let $A = A(M)$ (resp. $B = B(M)$) denote the vertices of X (resp. Y) not covered by M.

The following algorithm can be implemented to solve the assignment problem in $0(m^2 n)$ *worst-case* time (for a proof see [EK] or [To]).

Algorithm ASSIGN

begin

$M := \phi$

for $v \in V$ **do** $\alpha(v) = 0$ {α is the potential function used to keep arc lengths ≥ 0}

while $|M| < m$ **do**

begin

A.   Find a shortest path P from A to B in $D(M)$ where the arc-lengths are given by

$$\bar{\ell}(x_i y_j) = c(i,j) + \alpha(x_i) - \alpha(y_j) \qquad x_i y_j \notin M$$

$$\bar{\ell}(y_j x_i) = -c(i,j) + \alpha(y_j) - \alpha(x_i) \qquad x_i y_j \in M$$

{update M}

Use the alternating path P to alternately add and delete edges to and from M in the normal way.

for $v \in V$ **do** $\alpha(v) := \alpha(v) + \gamma(v)$

where $\gamma(v)$ is the minimum of $\bar{\ell}(P)$ and the length of a shortest $\bar{\ell}$ path from s to v

end

end

To find the shortest paths in A we use a modification of algorithm SHORTPATH of § 3

Changes to SHORTPATH

We create adjacency lists $L(x_i)$, $x_i \in X$, sorted by increasing $c(i, \cdot)$. (For $y \in Y$ either $L(y) = \phi (y \in B)$ or $L(y)$ consists of the unique vertex of X matched with y by M). We only have time to do the sorting once for each $x \in X$ but on the other hand, at Statement A we need them sorted according to $\bar{\ell}$ and not c. Karp's solution to this problem is rather nice. Define

$$\ell^*(x_i y_j) = c(i,j) + \alpha(x_i) - \alpha^* \qquad x_i y_j \notin M$$

where $\alpha^* = \max\{\alpha(v): v \in V\}$.

Observe that

$$\ell^*(x_i y_j) \leq \bar{\ell}(x_i y_j) \qquad x_i y_j \notin M \tag{4.1}$$

and

$$\ell^*(x_i y_j) = \bar{\ell}(x_i y_j) \qquad y_j \in B(M). \tag{4.2}$$

When an item $(uv, D(u) + \bar{\ell}(uv))$ is added to AQ in L2 or L4 we also add a special item $(uv, D(u) + \ell^*(uv))$ unless $v \in B(M)$ or uv is a backward arc of $D(M)$. Also, if the item removed from AQ is special, then it is ignored and the next item of AQ is removed. The point is that we are not necessarily examining the arcs leaving a vertex $x \in X$ in increasing $\bar{\ell}$ order. We want to be sure that the "real" items get to the front of AQ in the order they would in the unmodified SHORTPATH algorithm. Thus we want to be sure that when an item $(xy, D(x) + \bar{\ell}(xy))$ gets to the front it has a lower value than all competing arcs. But this follows from the fact that if this item precedes $(uv, D(u) + \ell^*(uv))$ then

$$D(x) + \bar{\ell}(xy) \leq D(u) + \ell^*(uv) \leq D(u) + \bar{\ell}(uw)$$

for all $w \notin S$.

We can also make the simplification that yz in $L4$ is now to be the first item on $L(y)$. Finally, we will of course start each execution of SHORTPATH with $S = A$ and AQ made from the first items of $L(a)$, $a \in A$ and terminate when $S \cap B \neq \phi$.

Analysis

We say that xy is a *virgin* edge if it has not been selected in $L2$ in any execution of SHORTPATH. The key observation is that if the selection xy in $L2$ is a virgin edge then

$$Pr(y \in B) \geq \frac{|B|}{|Y|}. \tag{4.3}$$

This is because the virgin edges with start node x come to the head of AQ in their (original random) order on $L(x)$ and none of the non-virgin edges with start node x have an end node in B. For when $y \in B$ an augmentation is triggered which means that y gets covered by the new M.

Let Stage k denote the k'th execution of SHORTPATH and σ_k denote the number of virgin edge selections at $L2$ in Stage k. Then by (4.3) we have

$$E(\sigma_k) \leq \frac{n}{n - k + 1}.$$

If an edge ceases to be virgin in Stage k then it can be selected at most $2(m - k + 1)$ times altogether. Hence the expected number of executions of $L2$ overall is bounded above by

$$2n \sum_{k=1}^{m} \frac{m - k + 1}{n - k + 1} \leq 2mn \qquad (\text{since } m \leq n)$$

Each such selection requires $0(\log n)$ time. The cost of $L2$ selections and initial sorting dominates the execution time and Karp's result follows.

5. Matchings in Sparse Random Graphs

Karp and Sipser [KSp] analysed a simple heuristic algorithm for finding a large matching in $G_{n,p}$, $p = \frac{c}{n}$ for a constant $c > 0$. The algorithm runs in $0(n)$ time and produces a near optimal matching **whp**. This is to be compared with the asymptotically most efficient $0(n^{1.5})$ algorithm of Micali and Vazirani [MV] which is much more complex. The analysis is difficult and we will only be able to outline what is going on. (Even so, our treatment is technically at variance in some places with what is said in [KSp]).

First the algorithm: here $\delta(G)$ is the minimum degree of graph G.

Algorithm MATCH

(i) Remove isolated vertices—if G is now empty, stop.
(ii) if $\delta(G) = 1$ choose a random degree 1 vertex v and let vw be its incident edge. Otherwise $(\delta(G) \geq 2)$ let vw be a random edge of G.
(iii) add edge vw to the output matching M and then remove vertices, v, w from G. Goto (i).

Phase 1 of the algorithm lasts until the first time that $\delta \geq 2$ in Step (ii) and Phase 2 constitutes the remainder of the algorithm.

Let a vertex be *lost* by the algorithm if it is deleted in Step (i) and so is not covered by M. Let $L_i(n, c)$ denote the number of vertices lost in Phase i, $i = 1, 2$. Let $R(n, c)$ denote the number of vertices remaining after Phase 1.

Karp and Sipser prove the following:

Theorem 5.1
For every $\varepsilon > 0$

(a) $\lim\limits_{n \to \infty} Pr\left(\left| \dfrac{L_1(n, c)}{n} - \alpha(c) \right| > \varepsilon \right) = 0$
for some $\alpha(c) > 0$.

(b) $\lim\limits_{n \to \infty} Pr(L_2(n, c) \geq \varepsilon n) = 0$

(c) $Pr\left(\left| \dfrac{R(n, c)}{n} - p(c) \right| > \varepsilon \right) = 0$
for some $p(c) \geq 0$.

Also $p(c) = 0$ iff $c \leq e = 2.71828\ldots$

$\square$

Now any maximum matching must leave at least $L_1(n, c)$ vertices isolated and so (b) above shows that M is usually of almost optimum size. The final property that $p(c) = 0$ iff $c \leq e$ (the *e-phenomenom*) is remarkable.

Analysis

Let $\mathscr{G}(n_1, n_2, m)$ denote the set of graphs (i) with vertex set $V \subseteq V_n$, (ii) with n_1 vertices of degree 1, (iii) n_2 vertices of degree ≥ 2 and (iv) m edges. Suppose that after first removing the isolated vertices of $G = G_{n, p}$ we have a graph in $\mathscr{G}(n_1, n_2, m)$. It is easy to see that each graph in $\mathscr{G}(n_1, n_2, m)$ is equally likely. (Each such graph arises from a unique $G_{n, p}$ with m edges). More importantly, if we stop the algorithm at the end of any Step (i) and observe the values of n_1, n_2, m then the graph we have is still equally likely to be any of $\mathscr{G}(n_1, n_2, m)$. This is proved inductively by showing that each $G \in \mathscr{G}(n_1, n_2, m)$ can arise from the same number of graphs in $\mathscr{G}(n_1', n_2', m')$ via a single execution of Steps (i)–(iii).

Knowing this, we examine the Markov chain with state space $\left\{ (n_1, n_2, m): n_1 + n_2 \le n, n_1 + 2n_2 \le m \le \binom{n}{2}, n_1, n_2 \ge 0 \right\}$ with transition probabilities defined by the algorithm. Using this we can, for example, examine the length of Phase 1 by seeing how long it is before n_1 becomes zero.

Consider Phase 1. Let $n_1(t)$, $n_2(t)$, $m(t)$ denote the values of n_1, n_2, m at the start of the tth iteration of the algorithm. If in Step (ii) w has degree k and k_i neighbours of degree i, $i = 1, 2$ then we have

$$n_1(t) - n_1(t + 1) = k_1 - k_2 + \delta_{k,1} \qquad \text{[Kronecker delta]}$$

$$n_2(t) - n_2(t + 1) = k_2 + 1 - \delta_{k,1} \qquad (5.1)$$

$$m(t) - m(t + 1) = k + 1$$

Now consider a period of time $t \in [\tau n, (\tau + \delta\tau)n]$. If $\delta\tau$ is small one imagines that **whp** the values $(n_1(t), n_2(t), m(t))$ will be close to some values $ny_1(\tau), ny_2(\tau), ny_3(\tau)$ where y_1, y_2, y_3 are functions of τ only. It would also be reasonable to assume that **whp**.

$$n_1(\tau n + \delta\tau n) - n_1(\tau n) \approx n(y_1(\tau + \delta\tau) - y_1(\tau)) \approx n\delta\tau y_1'(\tau)$$

$$\approx \sum_{t=\tau n}^{(\tau+\delta\tau)n} (k_1 - k_2 + \delta_{k,1}) \approx n\delta\tau E(k_1 - k_2 + \delta_{k,1}).$$

In summary we expect that **whp** the Markov chain (n_1, n_2, m) closely follows a path $n(y_1(\tau), y_2(\tau), y_3(\tau))$ for $0 \le \tau \le T = \inf(\tau: y_1(\tau) = 0)$, where $y(\tau)$ satisfies

$$y_i'(\tau) = u_i(\tau), \qquad i = 1, 2, 3, \qquad 0 \le \tau \le T \qquad (5.2)$$

and the u_i are the expected values of the RHS of (5.1) at $t = n\tau$. Furthermore

$$y(0) = (ce^{-c}, 1 - e^{-c}, -ce^{-c}, c/2) \qquad (5.3)$$

since the degree of a given vertex in $G_{n, c/n}$ is asymptotically Poisson with mean c.

The formal justification for (5.2) can be obtained by applying a theorem of Kurtz [Kz].

The next question is how to compute the u_i. Consider a graph G chosen randomly from $\mathscr{G}(n_1, n_2, m)$. Suppose we know that **whp** G has approximately v_i vertices of degree i, for $i = 1, 2, \ldots$ (thus $v_1 = n_1$ and $\Sigma v_i = 2m$). The study of random graphs with a fixed degree sequence is most easily handled by the configuration model of Bollobás (see [Bo1]: if vertex i is of degree d_i then it gives rise to a set W_i of cardinality d_i. $W = \bigcup W_i$. A configuration F is a random partition of W into 2-element sets. From F we obtain a multigraph $\mu(F)$ by mapping $\{\alpha, \beta\} \in F$ to uv where $\alpha \in W_u$, $\beta \in W_v$. Conditional on $\mu(F)$ being simple each such graph with the given degree sequence is equally likely. Also if $|W|/(n_1 + n_2) = 0(1)$ then the probability of being simple is bounded away from zero by a constant and so we can study random F in place of random $G \in \mathcal{G}(n_1, n_2, m)$.

Returning to the evaluation of u_1, u_2, u_3 in (5.2), we take any i such that $d_i = 1$ and pair the unique $x \in W_i$ with a random element in $W - W_i$. This yields

$$Pr(k = 1) = \frac{v_1}{2m}; \qquad Pr(k = t \geq 2) = \frac{v_t t}{2m}. \qquad (5.4)$$

By similar reasoning we obtain

$$E(k_1|k) = 1 + (k - 1)\frac{v_1}{2m} \Rightarrow E(k_1) = 1 + \frac{v_1}{m}(E(k) - 1) \qquad (5.5)$$

$$E(k_2|k) = (k - 1)\frac{2v_2}{2m} \Rightarrow E(k_2) = \frac{v_2}{m}(E(k) - 1) \qquad (5.6)$$

$$E(k) = \frac{\mu_1}{2m} + \sum_{t=2}^{\infty} \frac{v_t t^2}{2m}. \qquad (5.7)$$

Thus we can compute u_1, u_2, u_3 once we have a handle on $v_1, v_2, \ldots$. Now it is well known that in a random graph with n vertices and average degree d constant that the degree of vertex 1, say, is asymptotically Poisson with mean d. We should not be surprised that if we condition on minimum degree at least $d_0 \leq d$ then the degree of vertex 1 is asymptotically *truncated* Poisson with parameter θ, i.e.

$$Pr(\text{the degree of vertex 1 is } t \geq d_0) \approx \pi_{\theta, d_0}(t) = \frac{e^{-\theta}\theta^t}{t!} \bigg/ \left(e^{-\theta} \sum_{k=d_0}^{\infty} \frac{\theta^k}{k!} \right). \quad (5.8)$$

θ must be chosen so that the average degree is still d (to get the number of edges correct) i.e.

$$\mu_{\theta, d_0} = \sum_{t=d_0}^{\infty} t\pi_{\theta, d_0}(t) = d. \qquad (5.9)$$

(The proof of (5.8), (5.9) is rather long).

Now in our case we can show that, ignoring vertices of degree 1, the degree sequence of what remains is precisely that of a graph with $2m-n_1$ edges, n_2 vertices and minimum degree at least 2. So we now have enough to compute the u_i for (2). Unfortunately, these equations have not been solved explicitly, but at least Part (*a*) of Theorem 5.1 follows.

The analysis of Phase 2 is more complicated. There we define *clean* states to be those with $y_1 = 0$ and consider transitions from clean state to clean state so that each such transition corresponds to a sequence of iterations of MATCH in which all but the first iteration deletes vertices of degree 1. It is possible to establish differential equations as in (2), (3) which describe the process with high probability. We will not try to establish them here but instead aim to give the barest justification of part (*b*) of the Theorem.

This will be quite easy if we accept that **whp** a random graph in $\mathcal{G}(0, n_2, m)$, $m \leq cn$ satisfies

no two (*small*) cycles of length $\leq \sqrt{\log n}$ are within distance $\sqrt{\log n}$ of each other.
$$(5.10a)$$

The number of vertices within distance $\sqrt{\log n}$ of a small cycle is $o(n)$. (5.10b)

It then follows that **whp** the number of lost vertices in a transition from a clean state
to a clean state is $0\left(\dfrac{\#\ \text{matching edges found}}{\sqrt{\log n}}\right)$. We leave the justification of this
last remark to the reader and note that it implies part (*b*) of the theorem. We will
not attempt to justify the *e*-phenomenon.

6. Minimum Spanning Forests

In this section we consider the problem of finding a minimum weight spanning forest
of a graph. Our model of randomness is $G_{n,m}$ with edge weights which when ordered
define a random permutation of the edge-set. Remember that it is the edge weight
order that defines the minimum weight forest.

Karp and Tarjan [KT] gave an $0(m + n)$ expected time algorithm for this problem
based on an algorithm of Cheriton and Tarjan [CT]. This should be compared with
the best deterministic algorithm which runs in $0(m\omega(m, n))$ time ([FT] and [GGS]).
Here ω is a very slowly growing function of m and n which nevertheless tends to
infinity with n. McDiarmid [M2] gave an alternative treatment of a key lemma.
The algorithm of [KT] is in two stages:

Stage 1

Step 1a: construct a queue Q of n trees each consisting of a single vertex.

Step 1b: if the queue has at most $\sqrt{n}$ trees go to Stage 2, otherwise delete the first
tree T from Q.

Step 1c: 4. let vw be the unexamined (by Step 1c) edge of least weight with one
endpoint, v say, in T. (If there are no such edges, go to Step 1b). If $w \in T$
then delete vw and restart Step 1c. Otherwise add vw to the minimum
forest F_0 and go to Step 1d.

Step 1d: Let tree T' be the tree containing w. If T' is *small* ($\leq \sqrt{n}$ vertices) then
delete it from Q. Merge T, T' into a single tree T''. If T'' is small then add
it to the rear of Q and go to Step 1b.

At the end of Stage 1 there are at most $2\sqrt{n}$ subtrees. In $0(m + n)$ time we can
contract each such tree to a single vertex and reduce the problem to that of finding
a minimum forest on $\leq 2\sqrt{n}$ vertices. This requires $0((\sqrt{n})^2) = 0(n)$ time. The
validity of the algorithm follows, for example, from Lemma 5.2 of Aho, Hopcroft
and Ullman [AHU]. The most interesting question from the view of probabilistic
analysis is answered by

Lemma 6.1

$$Pr(w \in T \text{ in Step 1c}) \leq \frac{1}{2}.$$

Proof
Suppose Q contains trees $T_1, T_2, \ldots, T_k$. A vertex $v \in T_i$ is *virgin* if it has never belong to a tree T in Step 1c. It is simple to show by induction that each tree T_i in Q contains exactly one virgin vertex v_i. Now if $v, w \in T \neq T_i$ and $vw < vv_i$ then interchanging their order in the permutation of edges will not affect the course of the algorithm to this point. On the other hand $k > \sqrt{n} \geq |T|$ and the result follows with a little work. $\square$

The remainder of the analysis is mainly nonprobabilistic. The sets of vertices of the trees of Q are treated as the sets in the UNION-FIND problem in [AHU]. Each set is represented by a tree so that given edge xy say, where $x \in T$, $y \in T'$ it takes $0(\text{height}(T'))$ to find out that $y \in T'$ and $0(1)$ time to merge T, T' if $T \neq T'$. When merging, if height $(T) \geq$ height (T') we make the root of T' a child of the root of T and vice-versa. The sets of unexamined edges incident with trees in Q are represented as priority queues. Karp and Tarjan used binomial queues (Vuillemin [V]), but the analysis will be easier, if we use *bottom up skew heaps* from Sleator andTarjan [ST]. Then if there are k unexamined edges indident with T then it takes $0(\log k)$ (amortized) time to remove the one of minimum weight and $0(1)$ time to merge two queues.

The final concept is that of level. Initially imagine a marker placed at the back of Q. All trees (single vertices) are level zero. The marker continually moves to the front and then is placed at the back. If the marker has reached the front ℓ times then we say the trees behind it in Q are at level $\ell + 1$ and those in front are at level ℓ. The following are easy to justify inductively:

(a) A tree of level ℓ contains at least 2^ℓ vertices.

(b) Trees of same level are disjoint $\left(\rightarrow \text{ at most } \dfrac{n}{2^\ell} \text{ trees of level } \ell \right)$.

(c) height $(T) \leq$ level (T) for $T \in Q$.

Let us now bound the (expected amortized) running time of Phase 1

(i) Total time of find the tree containing w in Step 1c and merge trees in Step 1d.

$$0\left(\sum_{\ell=0}^{\infty} \frac{n}{2^\ell}(\ell + 1) \right) = 0(n)$$

(ii) Total time to find vw in Step 1c

$$0\left(\sum_{\ell=0}^{\infty} \sum_{\text{level}(T)=\ell} \log e(T) \right)$$

(where $e(T) = |\{$unexamined edges incident with T when it reaches front of $Q\}|)$

$$= 0\left(\sum_{\ell=0}^{\infty} \frac{n}{2^\ell} \log\left(\frac{m2^\ell}{n} \right) \right) \text{ by concavity of log}$$

$$= 0(m).$$

(iii) Total time to merge priority queues in Step 1d
= 0(n).

One can show that it takes 0(m) time to initialize the data structures and that amortized time (with a suitable potential function) is within 0(m) of actual time. This completes the analysis of the algorithm.

7. Hamilton Cycles

Komlós and Szemerédi [KSz] prove the following

Theorem 7.1

Let $m = \dfrac{1}{2} n \log n + \dfrac{1}{2} n \log \log n + \dfrac{1}{2} c_n n.$ *Then*

$$\lim_{n \to \infty} Pr(G_{n,m} \text{ is Hamiltonian}) =$$

$$\lim_{n \to \infty} Pr(\delta(G_{n,m}) \geq 2) = \begin{cases} 0 & c \to -\infty \\ e^{-e^{-c}} & c_n \to c \\ 1 & c_n \to +\infty \end{cases}$$

☐

The aim of this section is to prove the result of Bollobás, Fenner and Frieze [BFF] that there exists an $0(n^{3+o(1)})$ time algorithm HAM satisfying

$$\lim_{n \to \infty} Pr(\text{HAM finds a Hamilton cycle in } G_{n,m})$$

$$= \lim_{n \to \infty} Pr(G_{n,m} \text{ is Hamiltonian}). \qquad (7.1)$$

The most interesting case is where $c_n \to c$. For this we can reformulate (7.1) as

$$\lim_{n \to \infty} Pr(\text{HAM finds a Hamilton cycle} \mid \delta(G_{n,m}) \geq 2) = 1. \qquad (7.2)$$

The following idea has been used extensively: given a path $P = (v_1, v_2, \ldots, v_k)$ plus an edge $e = v_k v_i$ where $1 \leq i \leq k - 2$, we can create another path of length $k - 1$ by deleting edge $v_i v_{i+1}$ and adding e. Thus let

$$\text{ROTATE}(P, e) = (v_1, v_2, \ldots, v_i, v_k, v_{k-1}, \ldots, v_{i+1}).$$

HAM proceeds in a sequence of stages. At the beginning of the k'th stage we have a path P_k of length k, with endpoints w_0 and w_1. Stage k ends when we have constructed a path of length $k + 1$ or created a Hamilton cycle. We try to extend P_k from either w_0 or w_1. If we fail, but $w_0 w_1 \in E(G_{n,m})$ then, assuming $G_{n,m}$ is connected, as it is **whp**, we can find a longer path than P_k. Failing this, we can create a set of paths of length k by constructing all possible paths of the form ROTATE(P_k, e). These paths at *rotation depth* 1 from P_k are then tested for extension or closure. If none of these yield a path of length $k + 1$ or form a Hamilton cycle then we create all possible paths at rotation depth 2 and so on. The algorithm only gives up trying to find a path of length $k + 1$ or close a Hamilton path (and fails) when it has created all paths at rotation depth $2T$ where $T = \lceil \log n/(\log\log n - \log\log\log n)\rceil$.

Now it can be shown that **whp** the number of distinct pairs of endpoints of the paths created grows by a factor of at least $\log n/1000$ as we create each set of paths at a given rotation depth. Thus if HAM fails at any stage there will be a set of $\alpha n^2 (\alpha > 0$ constant) pairs of vertices Z (the distinct pairs of endpoints of the paths created) which depends on the execution of the algorithm, such that if $(v, w) \in Z$ then $vw \notin E(G_{n,m})$.

The final part of the proof is rather unintuitive. It is based on a counting argument of Fenner and Frieze [FF]. In order to get the main idea across we will omit to mention certain technical conditions which hold **whp** and are required for the proof.

Suppose now that HAM fails on $G_{n,m}$ during Stage k. Now P_k is derived from P_0 ($=$ vertex 1) by a sequence of at most $2nT$ rotations and extensions. Let $W = W(G_{n,m})$ denote the set of at most $2nT + n$ edges which are involved in these operations.

Now consider the deletion of $\omega = \lceil \log n\rceil$ random edges X from $G_{n,m}$ and the following events which are all made conditional on $\delta(G_{n,m}) \geq 2$.

$\mathscr{E}_0 = \{$HAM fails to find a Hamilton cycle$\}$

$\mathscr{E}_1 = \{$HAM fails on $G_{n,m} - X$ in the same stage as on $G_{n,m}\}$.

Observe first that

$$Pr(\mathscr{E}_1|\mathscr{E}_0) \geq \left(1 - \frac{3nT}{m}\right)^{\omega} \tag{7.3}$$

Since if X avoids $W(G_{n,m})$ then $\mathscr{E}_1$ will occur. But on the other hand, for any fixed graph H with $m - \omega$ edges

$$Pr(\mathscr{E}_1|G_{n,m} - X = H) \leq (1 - \alpha)^{\omega} \tag{7.4}$$

This is because given H, X is a random ω-subset of $\overline{E(H)}$ and in order that $\mathscr{E}_1$ occur, X must avoid $Z(H)$ which will be of size αn^2. (7.3) and (7.4) together show $Pr(\mathscr{E}_0) = o(1)$ which yields (7.2).

Modifications of these ideas have been used to find Hamilton cycles in sparse random graphs [F1], random directed graphs [F2] and to solve travelling salesman problems [F3].

8. Graph Colouring

In this section we discuss an algorithm which tries to 3-colour graphs. If a graph is chosen uniformly at random from the set of 3-colourable graphs with vertex set V_n then it succeeds **whp**. Our discussion is based on work of Dyer and Frieze [DF] and Turner [Tu].

Before getting into this discussion it is as well to briefly state what is known about colouring random graphs in general, say for $G_{n,.5}$ where each graph with vertex set V_n is equally likely. Now it has been shown by Bollobás [Bo2] that

$$\chi(G_{n,.5}) \approx \frac{n}{2\log_2 n} \text{ whp}$$

(See also Luczak [L] for the sparse graph case). In spite of a great deal of effort the best polynomial time algorithms tend to use roughly twice as many colours as are really needed (see e.g. Grimmett and McDiarmid [GM], Bollobás and Erdős [BE], Shamir and Upfal [SU]).

Having explained this sorry situation we can turn to 3-colourable graphs. (Actually, the proposed methods extend to k-colourable graphs, k fixed—see [DF] or [Tu] for details). It is not obvious how to deal with the probability space $\mathcal{G}(n: \chi = 3) = $ the set of 3-colourable graphs with vertex set V_n. We must deal with it indirectly.

First consider a simple way of constructing a random 3-colourable graph. Suppose B_1, B_2, B_3 is a random partition of V_n into sets of size $\approx \frac{n}{3}$. For each $e \in (B_1 \times B_2) \cup (B_1 \times B_3) \cup (B_2 \times B_3)$ independently put in the coresponding edge with probability $p \approx \frac{1}{2}$. $\left(\text{We need to allow } p \text{ close to } \frac{1}{2} \text{ as well as } = \frac{1}{2}\right)$. Call the resulting random graph G_1. Clearly G_1 is 3-colourable. Can we 3-colour G_1 whp without knowing B_1, B_2, B_3? The answer is yes. In the following algorithm X_1, X_2, X_3 will (hopefully) denote a 3-colouring of G_1. We use the notation $d_S(v)$ to denote the number of neighbours of a vertex v in a set S.

Algorithm COLOUR

begin

    for $i := 1$ **to** 2 **do**
    begin
        $X_i := \phi; \; Y_i := V_n - \bigcup_{j<i} X_i$

        repeat
            choose $v \in Y_i$ such that $d_{Y_i}(v)$ is minimal;
            $X_i := X_i \cup \{v\}; \; Y_i := Y_i - \{v\} - \Gamma(v)$
        until $Y_i = \phi$
    end

    if $X_3 = V_n - \bigcup_{j=1}^{2} X_i$ is independent **then** X_1, X_2, X_3 is a 3-colouring
        else COLOUR has failed.

end

(Replace 2 by $k - 1$ and 3 by k to get an algorithm for colouring k-colourable graphs.)

Lemma 8.1
$Pr(\text{COLOUR fails}) = o(1)$.

Proof
It is only necessary to show that the first repetition of the for-loop in COLOUR terminates with $X_1 = B_1$ or B_2 or B_3. If this is the case then we are effectively re-applying the algorithm having replaced 3 by 2.

Without loss of generality assume that the first $v \in Y_1$ is in B_1. Suppose inductively that $r \geq 1$ vertices have been selected in X_1 and suppose $X_1 \subseteq B_1$. Note that $Y_1 = V_n - \Gamma(X_1) - X_1$. If $r \leq 3$ then we can show using the Chernoff bound that for any r-subset X of B_1, $|B_j - \Gamma(X)| \approx \dfrac{n}{3 \times 2^r}$, $j = 2, 3$, **whp** and if $v \in B_i$ then it has degree $\approx \dfrac{n}{3 \times 2^{r+1}}$ in $B_j - \Gamma(X)$, $j \neq 1$, Similarly $v \in B_i$, $i \neq 1$ has degree $\approx \dfrac{n}{6}$ in B_1.

Hence under these assumptions

$$d_{Y_1(v)} \approx 2\frac{n}{3 \times 2^{r+1}}, \qquad v \in B_1$$

$$d_{Y_1(v)} \approx \frac{n}{6} + \frac{n}{3 \times 2^{r+1}}, \qquad v \notin B_1$$

and so the next choice is also in B_1. For $r > 3$ we use the fact that **whp** $|(B_2 \cup B_3) \cap Y_1| \lesssim \dfrac{n}{12}$ while $v \notin B_1$ retains $\approx \dfrac{n}{6}$ neighbours in B_1. □

Now we have not yet proved that COLOUR works with high probability on graphs chosen uniformly at random from $\mathcal{G}(n: \chi = 3)$ and we do not have the space here to give all the details of how to "translate" the result of Lemma 8.1 to obtain this result. On the other hand it is easy to show that **whp** G_1 is uniquely 3-colourable, a fact which is of interest in its own right and vital to the "translation".

Lemma 8.2
$Pr(G_1$ is not uniquely 3-colourable$) = o(1)$.

Proof(Outline)
Consider a vertex $v \in B_1$. **Whp** it has $\approx \dfrac{n}{6}$ neighbours $N_i \subseteq B_i$, $i = 1, 2$. **Whp** $N_1 \cup N_2$ induces a connected, and hence uniquely 2-colourable, bipartite graph. But then **whp** each $w \in B_1 - \{v\}$ is adjacent to a vertex in both N_1 and N_2 and so B_1 is determined as one colour class. Finally, **whp** $B_2 \cup B_3$ induces a connected bipartite graph which then determines B_2, B_3 uniquely. □

We can now discuss the "translation" of Lemma 8.1 Let G_2 be the random graph in which we randomly choose $m \approx \dfrac{n^2}{6}$ edges from $(B_1 \times B_2) \cup (B_1 \times B_3) \cup (B_2 \times B_3)$ where B_1, B_2, B_3 are as for G_1. If we let $p = \dfrac{3m}{n^2} \approx \dfrac{1}{2}$ then we have

$$Pr(\text{COLOUR fails on } G_1)$$
$$\geq Pr(\text{COLOUR fails on } G_1 \mid |E(G_1)| = m)Pr(|E(G_1)| = m)$$
$$= Pr(\text{COLOUR fails on } G_2)Pr(|E(G_1)| = m).$$

Now if the calculations are made explicit in Lemma 8.1 then we can prove that, say,

$$Pr(\text{COLOUR fails on } G_1) \leq e^{-\sqrt{n}}$$

and it is easy to see that

$$Pr(|E(G_1)| = m) = \Omega(1/n) \tag{8.1}$$

for p, m close enough to $\dfrac{1}{2}$, $\dfrac{n^2}{6}$ respectively. Hence, with these caveats, one sees immediately that

$$Pr(\text{COLOUR fails on } G_2) = o(1).$$

Similarly

$$Pr(G_2 \text{ is not uniquely 3-colourable}) = o(1). \tag{8.2}$$

Now some rather tedious calculations show that almost all graphs in $\mathscr{G}(n: \chi = 3)$ have $\approx \dfrac{n^2}{6}$ edges and have 3 colour classes of size $\approx \dfrac{n}{3}$ only, (the approximations here are good enough for (8.2) to hold). Thus we really only have to show that Lemma 8.1 can be translated to G_3 chosen uniformly from $\mathscr{G}' = $ the set of 3-colourable graphs with $m \approx \dfrac{n^2}{6}$ edges and a set of colour classes of size $n_1, n_2, n_3 \approx \dfrac{n}{3}$.

Now while G_3 is chosen uniformly from $\mathscr{G}'$, G_2 is chosen from $\mathscr{G}'$ with probability proportional to the number, $v(G_2)$ of different (unordered) 3-colourings G_2 (with colour-classes of the appropriate size).

Now for any $\mathscr{A} \subseteq \mathscr{G}'$

$$Pr(G_2 \in \mathscr{A}) = \sum_{G \in \mathscr{A}} \frac{v(G)}{v(\mathscr{G}')} \geq \frac{|\mathscr{A}|}{|\mathscr{G}'|} \cdot \frac{|\mathscr{G}'|}{v(\mathscr{G}')}, \quad v(\mathscr{G}') = \sum_{G \in \mathscr{G}} v(G)$$

$$\geq (1 - o(1)) Pr(G_3 \in \mathscr{A})$$

since the result of Lemma 8.2 can be expressed as

$$\sum_{v(G) \geq 2} \frac{v(G)}{v(\mathscr{G}')} = o(1).$$

Thus

$$Pr(G_3 \in \mathscr{A}) \leq (1 + o(1)) Pr(G_2 \in \mathscr{A}).$$

We obtain the result we want by taking $\mathscr{A} = \{G \in \mathscr{G}': \text{COLOUR fails on } G\}$.

The failure probability of COLOUR is not quite small enough so that one has a polynomial expected time algorithm if one handles exceptional cases by enumeration. In [DF] we construct another algorithm COLOUR1 to handle the exceptional cases of COLOUR. It has polynomial running time and failure probability $0(e^{-\Omega(n \log n)})$. Thus if both COLOUR and COLOUR1 fail we can then resort to enumeration of all possible 3-colourings and we will have a polynomial expected time algorithm for colouring 3-colourable graphs.

9. Graph Isomorphism

In this section we give the earliest and simplest result concerning the graph isomorphism problem for random graphs. It is due to Babai, Erdös and Selkow [BES]. Suppose we are given graphs $G_i = (V_n, E_i)$, $i = 1, 2$, where G_1 is the random graph $G_{n,.5}$ and G_2 is any graph. Can we quickly tell whether or not $G_1 \cong G_2$ i.e. whether there exists a bijection $f: V_n \to V_n$ such that $vw \in E_1$ iff $f(v)f(w) \in E_2$. The answer in [BES] is that **whp** we can check this in $0(n^2)$ time.

The method is based on the fact that **whp** G_1 has the properties (9.1) and (9.2) below. Let the vertices of G_1 be relabelled so that $d(i) \geq d(i + 1)$, $i = 1, 2, ..., n - 1$. Let $r = \lceil 3 \log_2 n \rceil$. Then **whp**

$$d(i) > d(i + 1) \qquad \text{for} \qquad 1 \leq i < r. \tag{9.1}$$

Next, for $j > r$ let $X_j = \{i: 1 \leq i < r \quad \text{and} \quad ij \in E_1\}$. then **whp**

$$X_j \neq X_k \qquad \text{for} \qquad j, k > r. \tag{9.2}$$

Thus we can relabel the vertices $r + 1, ..., n$ so that

$$X_i \text{ is lexicographically larger than } X_{i+1}, \qquad i = r + 1, ..., n - 1. \tag{9.3}$$

Given (9.1) and (9.2) it is easy to check if G_1 and G_2 are isomorphic.

1. Compute the degree sequence of G_2. If the largest r degrees do not coincide with those of G_1 then G_1 and G_2 are not isomorphic.
   If they are then by (9.1) we can identify $f(1), ..., f(r)$ in any possible isomorphism. By relabelling vertices of G_2 we can assume $f(i) = i$ for $1 \leq i \leq r$.
2. For each vertex $v > r$ of G_2 compute $Y_v = \{i: 1 \leq i \leq r \text{ and } iv \in E(G_2)\}$. Sort these n-r sets into lexicographic order. If these exists $i > r$ such that $Y_i \neq X_i$ then by (9.2) and (9.3) G_1 and G_2 are not isomorphic. Otherwise the only possible isomorphism is now $f(i) = i$.
3. Finally, check if $f(i) = i$ is an isomorphism i.e. check if now $G_1 = G_2$.

The proof of (9.1) requires a lot of calculation. Babai, Erdös and Selkow proved considerably more than this. They showed that

$$d(i) - d(i + 1) \geq n^{.03} \qquad \text{for} \qquad 1 \leq i < n^{.15}. \tag{9.4}$$

For an even stronger result see Theorem III.15 of Bollobás [Bo1]. Given (9.4) it is quite easy to prove (9.2). If (9.4) holds and $X_i = X_j$ for some $i, j > r$ then i and j have the same set of neighbours among the r largest vertices in $H_{ij} = G_1 - \{i,j\}$. Denote the this event by $\mathcal{E}_{ij}$. Now since the graph H_{ij} is independent of i, j the probability of $\mathcal{E}_{ij}$ is $\left(\frac{1}{2}\right)^r$. Hence

$$Pr(\exists i, j > r: X_i = X_j) \leq \sum_{i=1}^{n-1} \sum_{j=1}^{n} Pr(\mathcal{E}_{ij}) + Pr((9.4) \text{ fails}) \leq n^2 \cdot \left(\frac{1}{2}\right)^r + o(1) = o(1).$$

The result of [BES] has been strengthened by Karp [Ka2], Lipton [Li] and Babai and Kucera [BK]. In particular Babai and Kucera handle exceptional graphs in such a way that graph isomorphism can be tested in *linear expected time* on $G_{n,.5}$. For regular graphs, the above algorithm(s) would be particularly ineffective. However, Kucera [Ku] has recently devised an algorithm for regular graphs which runs in linear expected time, i.e. $O(nd)$, assuming the degree d does not grow with n.

10. Graph Bisection

Here we are given a graph $G = (V, E)$ with n vertices, n even, and the problem is to find the partition of V into two equal sized subsets S_1, S_2 so that the number of S_1 : S_2 edges is minimised. The minimum such number of edges is called the *bisection width* of G. The problem is useful in VLSI design problems (see Bhatt and Leighton [BL]), but is NP-hard (Garey, Johnson and Stockmeyer [GJS]).

If we take the graph $G_{n,m}$ as a model of random input then we find that all relevant cuts have $\approx \frac{m}{2}$ edges **whp** provided m is sufficiently large e.g. $m = \Omega(n \log n)$. Finding the bisection width in these circumstances is still open.

Positive results can be obtained if we consider sampling uniformly from $\mathscr{G}(n, m, b)$, the set of graphs with vertex set V_n, m edges and bisection width b. Basically, the idea is to have b significantly smaller than $\frac{m}{2}$ and then **whp** there will be a unique cut of size b which will be easy to find. Bui, Chaudhuri, Leighton and Sipser [BCLS] considered this approach for regular graphs, Dyer and Frieze [DF] considered $\mathscr{G}(n, m, b)$ with $m = \Omega(n^2)$ and Boppana [Bp] considered the case $m = \Omega(n \log n)$. We will outline Boppana's approach here.

First of all we remark that it is not easy to work directly with $\mathscr{G}(n, m, b)$. Instead one chooses a random partition of V_n into S_1, S_2 of equal size, and then add edges between S_1 and S_2 with probability $q = 4b/n^2$ and within each S_i with probability $p = 4(m - b) / \binom{\frac{1}{2}n}{2}$. Results are proved for this "independent" model and then translated to $\mathscr{G}(n, m, b)$-see § 8 on colouring.

For $S \subseteq V_n$ we define $x = x(S) \in \mathbb{R}^n$ by $x_i = 1$, $i \in S$ and $= -1$ otherwise. Given $d \in \mathbb{R}^n$ we let $B = B(d) = A + D$ where A is the adjacency matrix of G and $D = \text{diag}(d)$. Also let sum $(B) = 2|E| + \sum_{i=1}^{n} d_i = $ the sum of the entries of B. Next let

$$f(G, D, x) = \sum_{(i,j) \in E} \frac{1 - x_i x_j}{2} - \frac{1}{4} \sum_{i=1}^{n} d_i(x_i^2 - 1) = \frac{1}{4}(\text{sum}(B) - x^T B x).$$

The significance of this function is that

$$f(G, d, x(S)) = |(S : V_n - S)| \qquad \text{for} \qquad S \subseteq V_n. \tag{10.1}$$

Observe that $\|x(S)\| = \sqrt{n}$ (Euclidean norm) and $e^T x(S) = 0$ when $|S| = \frac{1}{2} n$. ($e^T = (1, 1, \ldots, 1)$). So from (10.1) it is natural to consider

$$g(G,d) = \min_{\substack{e^T x = 0 \\ \|x\| = \sqrt{n}}} f(G,d,x) = \min_{\|x\| = \sqrt{n}} \frac{1}{4}(\text{sum}(B) - x^T \hat{B} x)$$

where $\hat{B} = \hat{B}(d) = \left(I - \frac{1}{n}ee^T\right)B$. (Observe that the matrix $I - \frac{1}{n}ee^T$ projects $\mathbb{R}^n$ onto $\{x \in \mathbb{R}^n: e^T x = 0\}$.)

Bopanna's idea is that one can find d for which the x minimising $f(G,d,x)$ is $x(S)$ for a minimum bisection S.

Note that

$$g(G,d) = \frac{1}{4}(\text{sum}(B) - n\lambda(\hat{B})) \qquad (10.2)$$

where $\lambda(\hat{B})$ is the largest eigenvalue of $\hat{B}$.

Now $g(G,d)$ being the infimum of a collection of linear functions is concave in d and so

$$h(G) = \max_{d \in \mathbb{R}^n} g(G,d)$$

can be computed in polynomial time. (Grotschel, Lovasz and Schrijver (GLS)).

Since $g(G,d) \le f(G,d,x(S))$ for $S \subseteq V_n$ we see that

$$h(G) \le \text{bisection width of } G.$$

The nice probabilistic result of Boppana is that if G is sampled uniformly from $\mathcal{G}(n,m,b)$ and

$$0 \le b \le \frac{1}{2}m - 5\sqrt{mn\log n} \qquad (10.3)$$

then **whp** the bisection width of G is $b = h(G) = g(G,d^*)$ and the eigenvector corresponding to $\lambda(\hat{B}(d^*))$ yields the minimum bisection. The proof of this result is as follows. First of all it is straightforward to show given (10.3), that in the independent model there is **whp** a unique minimum bisection of size b. Next let S_1, S_2 be a minimum bisection. For $i \in S_1$ let $d_i^* = d(i,S_2) - d(i,S_1)$ and for $i \in S_2$ let $d_i^* = d(i,S_1) - d(i,S_2)$ where $d(v,S) = |\{w \in S: vw \in E\}|$. Now sum $(B(d^*)) = 4b$ and so by (10.2) $g(G,d^*) = b$ iff $\lambda(\hat{B}(d^*)) = 0$. Observe also that $\hat{B}(d^*)x(S_1) = 0$ and so Boppana's result is reduced to showing that **whp** $\hat{B}(d^*)$ has a unique eigenvalue of zero and every other eigenvalue is negative.

Now we have $E(B) = M - \frac{1}{2}(p-q)nI$ where $M = E(A)$. Also $\hat{M}x(S_1) = 0$, $x(S_1)^T \hat{M} = 0$ and so if $\xi^T(\hat{B} - \hat{M})\xi \le 0$ always then $\xi^T \hat{B}\xi \le 0$ always, which is what we need. This follows a *fortiori* if $B - M$ has non-positive eigenvalues or equivalues or equivalently if $B - E(B)$ has eigenvalues bounded above by $\frac{1}{2}(p-q)n$. Now

$$\lambda(B - E(B)) \le \lambda(A - E(A)) + \lambda(D - E(D)).$$

The eigenvalues of $D - E(D)$ are precisely its diagonal entries and using Chernoff's bound we find that $\lambda(D - E(D)) \le 5\sqrt{pn\log n}$.

Extimating $\lambda(A - E(A))$ is more difficult, but the eigenvalues of random matrices have been intensively studied. By modifying a result due to Furédi and Komlós [FuK] Boppana shows that $\lambda(A - E(A)) \le 3\sqrt{pn}$ **whp** and so $\lambda(B - E(B)) \le 6\sqrt{pn\log n}$ **whp**. The reader can now check that if (10.3) holds then $6\sqrt{pn\log n} < \frac{1}{2}(p - q)n$ as required.

The probability that Boppana's algorithm fails to work is not sufficiently small that exceptional cases can be dealt with more crudely and still yield a polynomial expected time algorithm which handles all graphs. For $m = \Omega(n^2)$ however, there is a polynomial expected time algorithm, [DF].

11. Other Aspects

We have concentrated here on positive results that arise in probabististic analysis. This field also has its share of negative results. We mention three: Chvatal [C] showed that a certain class of approaches to finding the largest independent set in a graph took exponential time **whp**; McDiarmid [M1] proved a similar result for graph colouring as did Ahn, Cooper, Cornuéjols and Frieze [ACCF] for finding a small dominating set.

There is an increasing interest in finding fast parallel algorithms. There are a few results here of interest to us: Frieze and Rudolph [FR] gave an 0(loglogn) expected time parallel algorithm for the shortest path problem of § 3; Frieze [F4] gave an $0((\log\log n)^2)$ expected time parallel algorithm for the Hamilton cycle problem in $G_{n,p}$, p constant; Frieze and Kucera [FrK] give a polylog expected time algorithm for colouring graphs; Coppersmith, Raghavan and Tompa [CRT] give polylog expected time algorithms for graph colouring, finding maximal independent sets and finding Hamilton cycles: Calkin and Frieze [CF] deals with maximal independent sets.

Finally we mention an area of particular interest to probabilists. Given a weighted optimisation problem, determine the properties of the (random) optimal value. We first mention two similar results: consider the $n \times n$ assignment problem in which the costs $c(i,j)$ are independent uniform $[0,1]$ random variables. Let W_n denote the minimal value of an assignment. Walkup [W] showed that, rather surprisingly, $E(W_n) < 3$ for all n. Karp [K] improved this to $E(W_n) < 2$ (see also Dyer, Frieze and McDiarmid [DFM]). A lower bound of $1 + e^{-1}$ was proved by Lazarus [Lz].

Consider next the minimum spanning tree problem where the edge weights are also independent uniform $[0,1]$ random variables. Let L_n denote the minimum length of a spanning tree. We showed [F5] that

$$\lim_{n \to \infty} E(L_n) = \zeta(3) = \sum_{k=1}^{\infty} k^{-3},$$

(see also Steele [St1] and Frieze and McDiarmid [FM]).

Probabilists have found Euclidean problems even more interesting. For example, suppose $X_1, X_2, \ldots, X_n$ are independently chosen uniformly at random in the unit square $[0, 1]^2$. In a very important paper Beardwood, Halton and Hammersley [BHH] showed that if T_n is the minimum length of a travelling salesman tour through these points then there exists a constant $\beta > 0$ such that

$$Pr\left(\lim_{n \to \infty} \frac{T_n}{\sqrt{n}} = \beta \right) = 1.$$

Steele [St2] has generalised this result considerably and the paper by Karp [Ka4] was very influential in generating interest in the probabilistic analysis of algorithms.

For a bibliography on topics related to this paper see Karp, Lenstra, McDiarmid and Rinnooy Kan [KLMR].

Acknowledgement

I would like to thank Colin McDiarmid for a thorough reading of this paper.

References

[ACCF] S. Ahn, C. Cooper, G. Cornuejols and A. M. Frieze, 'Probabilistic analysis of a relaxation for the k-median problem', Mathematics of Operations Research *13* (1988) 1–31.

[AHU] A. Aho, J. E. Hopcroft and J. D. Ullman, 'The design and analysis of computer algorithms', Addison-Wesley, Reading MA, 1974.

[BES] L. Babai, P. Erdos and S. M. Selkow, 'Random graph isomorphisms', SIAM Journal on Computing *9* (1980) 628–635.

[BHH] J. Beardwood, J. H. halton and J. M. Hammersley, 'The shortest path through many points', Proceedings of the Cambridge Philosophical Society *55* (1959) 299–327.

[Bo1] B. Bollobas, Random graphs, Academic Press, 1985.

[Bo2] B. Bollobas, 'The chromatic number of random graphs', Combinatorica *8* (1988) 49–55.

[BE] B. Bollobas and P. Erdos, 'Cliques in random graphs', Mathematical Proceedings of the Cambridge Philosophical Society *80* (1976) 419–427.

[BFF] B. Bollobas, T. I. Fenner and A. M. Frieze, 'An algorithm for finding Hamilton paths and cycles in random graphs', Combinatorica *7* (1987) 327–342.

[Bp] R. Boppana, 'Eigenvalues and graph bisection: an average case analysis', Proceedings of the 28th Annual IEEE Symposium on the Foundations of Computer Science (1987) 280–285.

[BK] L. Babai and L. Kucera, 'Canonical labelling of graphs in linear average time', Proceedings of the 20th Annual IEEE Symposium on the Foundations of Computer Science (1979) 39–46.

[BL] S. Bhatt and T. Leighton, 'A framework for solving VLSI graph problems', Journal of Computer and System Sciences *28* (1984) 300–343.

[BCLS] T. Bui, S. Chaudhuri, T. Leighton and M. Sipser, 'Graph bisection algorithms with good average case behaviour', Combinatorica 6.

[CF] N. Calkin and A. M. Frieze, 'Probabilistic analysis of a parallel algorithm for finding a maximal independent set', to appear.

[C] V. Chvatal, 'Determining the stability number of a graph', SIAM Journal on Computing *6* (1977) 643–662.

[CT] D. Cheriton and R. E. Tarjan, 'Finding minimum spanning trees', SIAM Journal on Computing *5* (1976) 724–742.

[CRT] D. Coppersmith, P. Raghavan and M. Tompa, 'Parallel graph algorithms that are efficient on average', Proceedings of the 28th Annual IEEE Symposium on the Foundations of Computer Science (1987) 260–270.

[D]   G. B. Dantzig, 'On the shortest route through a network', Management Science 6 (1960) 187–190.

[DF]   M. E. Dyer and A. M. Frieze, 'The solution of some random NP-hard problems in polynomial expected time' to appear in Journal of Algorithms.

[DFM]   M. E. Dyer, A. M. Frieze and C. J. H. McDiarmid, 'On linear programs with random costs', Mathematical Programming 35 (1986) 3–16.

[EK]   J. Edmonds and R. M. Karp, 'Theoretical improvements in algorithmic efficiency for network flow problems', Journal of the ACM 19 (1972) 248–264.

[ER1]   P. Erdos and A. Renyi, 'On random graphs 1', Publ. Math. Debrecen 6 (1959) 290–297.

[ER2]   P. Erdos and A. Renyi, 'The evolution of a random graphs', Publ. Math. Inst. Hungar. Acad. Sci. 5 (1960) 17–61.

[ER3]   P. Erdos and A. Renyi, 'On the strength of connectedness of a random graph', Acta Math. Acad. Sci. Hungar. 12 (1961) 261–267.

[ER4]   P. Erdos and A. Renyi, 'On the existence of a factor of degree one of a connected random graph', Acta Math. Inst. Acad. Sci. Hungar. 17 (1966) 359–368.

[FF]   T. I. Fenner and A. M. Frieze, 'On the existence of Hamilton cycles in a class of random graphs', Discrete Mathematics 45 (1983) 301–305.

[Fd]   M. L. Fredman, 'On the decision tree complexity of the shortest path problem', Proceedings of the 16th Annual IEEE Symposium on the Foundations of Computer Science ((1985) 101–105.

[FT]   M. L. Fredman and R. E. Tarjan, 'Fibonnacci heaps and their uses in network optimisation problems', Proceedings of the 25th Annual IEEE Symposium on the Foundations of Computer Science (1984) 338–346.

[F1]   A. M. Frieze, 'Finding Hamilton cycles in sparse random graphs', Journal of Combinatorial Theory B44 (1988) 230–250.

[F2]   A. M. Frieze, 'An algorithm for finding Hamilton cycles in random digraphs', Journal of Algorithms 9 (1988) 181–204.

[F3]   A. M. Frieze, 'On the exact solution of random travelling salesman problems with medium sized integer costs', SIAM Journal on Computing 16 (1987) 1052–1072.

[F4]   A. M. Frieze, 'Parallel algorithms for finding Hamilton cycles in random graphs', Information Proceesing Letters 25 (1987) 111–117.

[F5]   A. M. Frieze, 'On the value of a random minimum spanning tree problem', Discrete Applied Mathematics 10 (1985) 47–56.

[FG]   A. M. Frieze and G. R. Grimmett, 'The shortest path problem for graphs with random arc-lengths', Discrete Applied Mathematics 10 (1985) 57–77.

[FrK]   A. M. Frieze and L. Kucera, 'Parallel colouring of random graphs', to appear in Annals of Discrete Mathematics.

[FM]   A. M. Frieze and C. J. H. McDiarmid, 'On random minimum length spanning trees', to appear in Combinatorica.

[FR]   A. M. Frieze and L. Rudolph, 'A parallel algorithm for all-pairs shortest paths in a random graph', Proceedings of the 22nd Allerton Conference on Communication, Control and Computing (1985) 663–670.

[FuK]   Z. Furedi and M. Komlos, 'The eigenvalues of random symmetric matrices' Combinatorica 1 (1981) 233–241.

[GGS]   H. N. Gabow, Z. Galil and T. Spencer, 'Efficient implementation of graph algrithms using contraction', Proceedings of the 25th Annual IEEE Symposium on the Foundations of Computer Science (1984) 347–357.

[GJS]   M. R. Garey, D. S. Johnson and L. Stockmeyer, 'Some simplified NP-complete graph problems', Theoretical Computer Science 1 (1976) 237–267.

[GM]   G. R. Grimmett and C. J. H. McDiarmid, 'On colouring random graphs' Mathematical Proceedings of the Cambridge Philosophical Society 77 (1975) 313–324.

[Ka1]   R. M. Karp, 'An algorithm to solve the $m \times n$ assignment problem in expected time $O(mn \log n)$', Networks 10 (1980) 143–152.

[Ka2]   R. M. Karp, 'Probabilistic analysis of a canonical numbering algorithm for graphs', Proceedings of Symposia in Pure Mathematics, Volume 34, 1979, American Mathematical Society, Providence, RI, 365–378.

[Ka3]   R. M. Karp, 'An upper bound on the expected cost of an optimal assignment', Discrete Algorithms and Complexity: Proceedings of the Japan—US Joint Seminar (D. Johnson et al, eds.), 1–4, Academic Press, New York, 1987.

[Ka4]   R. M. Karp, 'Probabilistic analysis of partitioning algorithms for the traveling salesman problem in the plane', Mathematics of Operations Research 2 (1977) 209–224.

[KLMR]   M. Karp, J. K. Lenstra, C. J. H. McDiarmid and A. H. G. Rinnooy Kan', Probabilistic analysis of combinatorial algorithms: an annotated bibligoraphy', in Comminatorial Optimisation: Annotated Bibliographies, (M. O'Heigeartaigh, J. K. Lenstra and A. H. G. Rinnooy Kan, eds.), John Wiley and Sons, New York, 1984.

[KSp]   R. M. Karp and M. Sipser, 'Maximum matchings in sparse random graphs', Proceedings of the 22nd Annual IEEE Symposium on Foundations of Computer Science (1981) 364–375.

[KT]   R. M. Karp and R. E. Tarjan, 'Linear expected time algorithms for connectivity problems', Journal of Algorithms 1 (1980) 374–393.

[KSz]   M. Komlos and E. Szemeredi, 'Limit distributions for the existence of Hamilton cycles in a random graph', Discrete Mathematics 43 (1983) 55–63.

[Ku]   L. Kucera, 'Canonical labeling of regular graphs in linear expected time' Proceedings of the 28th Annual IEEE Symposium on the Foundations of Computer Science (1987) 271–279.

[Kz]   T. G. Kurtz, 'Solutions of ordinary differential equations as limits of pure jump Markov processes', Journal of Applied Probability 7 (1970) 49–58.

[La]   E. L. Lawler, Combinatorial optimization: networks and matroids, Holt, Rinehart and Winston, New York 1976.

[Lz]   A. J. Lazarus, 'The assignment problem with uniform (0, 1) cost matrix', B. A. Thesis, Department of Mathematics, Princeton University, Princton, N. J.

[Li]   R. J. Lipton, 'The beacon set approach to graph isomorphism', Yale University, pre-print.

[Lu]   T. Luczak, 'The chromatic number of random graphs', Combinatorica to appear.

[M1]   C. J. H. McDiarmid, 'Determining the chromatic number of a graph', SIAM Journal on Computing 8 (1979) 1–14.

[M2]   C. J. H. McDiarmid, 'On some conditioning results in the analysis of algorithms', Discrete Applied Mathematics 10 (1985) 197–201.

[MV]   S. Micali and V. V. Vazirani, 'An $O(\sqrt{|V|} \cdot |E|)$ algorithm for finding maximum matching in general graphs', Proceedings of the 21st Annual IEEE Symposium on Foundations of Computer Science (1980) 17–27.

[MT1]   A. Moffat and T. Takaoka, 'An all pairs shortest path algorithm with expected running time $O(n^2 \log n)$', Proceedings of the 26th Annual IEEE Symposium on the Foundations of Computer Science (1985) 101–105.

[MT2]   A. Moffat and T. Takaoka, 'An all pairs shortest path algorithm with expected time $O(n^2 \log n)$', SIAM Journal on Computing 16 (1987) 1023–1031.

[P]   E. M. Palmer, Graphical evolution, Wiley-Interscience, 1985.

[PS]   C. H. Papadimitriou and K. Steiglitz, Combinatorial optimization: algorithms and complexity, Prentice-Hall, 1982.

[SU]   E. Shamir and E. Upfal, 'Sequential and distributed graph colouring algorithms with performance analysis in random graph spaces', Journal of Algorithms 5 (1984) 488–501.

[Sp]   P. M. Spira, 'A new algorithm for finding all shortest paths in a graph of positive arcs in average time $O(n \log n)$' SIAM Journal on Computing 2 (1973) 28–32.

[St2]   J. M. Steele, 'On Frieze's $\zeta(3)$ limit for lengths of minimal spanning trees', Discrete Applied Mathematics.

[St2]   J. M. Steele, 'Subadditive Euclidean functionals and non-linear growth in geometric probability', Annals of Probability 9 (1981) 365–376.

[ST]   D. D. Sleator and R. E. Tarjan, 'Self-adjusting heaps', SIAM Journal on Computing 15 (1986) 52–69.

[To]   N. Tomizawa, 'On some techniques useful for solution of transportation problems', Networks 1 (1972) 173–194.

[Tu]   J. S. Turner, 'Almost all k-colorable graphs are easy to color', Journal of Algorithms 9 (1988) 63–82.

[V]   J. Vuillemin, 'A data structure for manipulating priority queses', Communications of the ACM 21 (1978) 309–315.

[W]   D. Walkup, 'On the expected value of a random assignment problem', SIAM Journal on Computing 8 (1979) 440–442.

Department of Mathematics
Carnegie Mellon University
Pittsburgh, PA 15213
U.S.A.

Computing, Supp. 7, 235–255 (1990)

Generating Graphs Uniformly at Random

G. Tinhofer, München

Abstract — Zusammenfassung

Generating Graphs Uniformly at Random. This paper deals with the problem of sampling from a uniform distribution on various classes of graphs of given size. We consider algorithms and restarting procedures for uniform generation of several kinds of trees, arbitrary unlabelled graphs and various kinds of labelled graphs. Most of the material discussed in this paper has been developed during the last decade by several authors. In section 4.3 some recent results on the generation of outerplanar graphs and maximal planar graphs are presented.

AMS Subject Classification: 05C.

Key words: uniform sampling, unlabelled graphs, labelled graphs, counting problems.

Erzeugung von gleichverteilten Zufallsgraphen. Diese Arbeit gibt einen Überblick über bekannte Verfahren zur Erzeugung von Zufallsgraphen gemäß einer Gleichverteilung über einer gegebenen Klasse von Graphen mit fester Knotenzahl. Wir betrachten Algorithmen und 'Restarting Procedures' für die Erzeugung verschiedener Arten von Bäumen, für allgemeine unlabelled Graphen und für diverse Arten von labelled Graphen. Der Großteil des behandelten Materials wurde von verschiedenen Autoren im letzten Jahrzehnt entwickelt. In Abschnitt 4.3 wird über jungste Ergebnisse bei der Erzeugung von outerplanaren und von maximal planaren Graphen berichtet.

1. Introduction

One of the most important aspects in research fields where mathematics is applied (physics, chemistry, computer science, operations research, social science, biology, and so on) is to construct a model of a concrete situation in order to understand it better and, possibly, to influence it. As for structural relations, in numerous situations, graphs have turned out to provide the most appropriate tool for setting up the mathematical model. This is certainly one reason why graph theory has expanded so rapidly during the last decades.

Having a mathematical model for a real world structure at hand it seems natural (and has been successfully performed) to look for variations of the model (by changing structural parameters) which possibly are abstract pictures of other still unknown real world structures. In this way it is sometimes possible to discover new structures by inspecting what models are available for them. A different aspect is the optimality of some structure with respect to various criteria. Again, often optimality can be studied best by inspecting all suitable models bearing the same typical features.

For the just mentioned reasons, in graph theory and its applications, there have been early attempts for listing and cataloguing graphs of a given size and type. Catalogues of graphs can be of great use for a variety of purposes. They provide stores of graphs which can be sampled whenever there is a need for a graph theorist to play with a set of graphs, or they can be exhaustively studied to settle some question by turning up a counterexample or to come up with plausible conjectures. Results on cataloguing graphs can be found for example in [5], [12], [17], [24] and [25].

However, when the size of the graphs which are to be considered is large then exhaustive listing is impossible. In such situations, to study typical graphical configurations or to evaluate the performance of an algorithm on such configurations, it is desirable to be able to generate these configurations uniformly at random (u.a.r. for short). For labelled graphs on n vertices, for instance, this is very simple to do, just choose the edges independently with probability $1/2$ for each. However, the situation changes drastically when unlabelled graphs are wanted or when a certain graph property has to be forced (such as regularity, hamiltonicity, chordality, etc.). In such cases the problem of random generation quickly becomes non-trivial.

Generating problems are strongly connected to counting problems. This will become clear in the following sections. Counting problems have a long and distinguished history. A standard text for graph counting problems is [10]. The study of counting problems as a class from a computational point of view, however, was initiated by Valiant in the late 1970's [32, 33]. A parallel approach to generation problems was proposed more recently in [13]. As one of the results algorithms for generating graphs of certain types approximately u.a.r. have been established [27, 28].

This paper presents a short review on the problem of generating graphs u.a.r. Most of the material has been developed during the last decade and can be found in the original papers only, with the exception of sections 2.1 and 2.2 which are included for completeness and the content of which can also be found in [22].

In what follows we use the standard notations found in [3] or [9]. Throughout the paper, unless otherwise stated, any graph $G = (V, E)$ under consideration is understood to be undirected, loopless and without parallel edges. V is the vertex set, E the edge set. The vertex number $n = |V|$ is also called the size of G, $m = |E|$ is the edge number.

We distinguish between labelled and unlabelled graphs. For simplicity, a labelled graph of size n is one whose vertex set is $V = \{1, 2, \ldots, n\}$, and two labelled graphs $G = (V, E)$ and $G' = (V, E')$ are considered the same iff $E = E'$. G and G' are called *isomorphic* iff there is a permutation $p\colon V \to V$ such that $\langle i, j \rangle \in E$ iff $\langle p(i), p(j) \rangle \in E'$, for all $i, j \in V$. An unlabelled graph G is an *isomorphism class* of labelled graphs and may be represented by any element of this class.

2. Trees

A *tree* is a connected graph without cycles. Because of their importance as combinatorial models in chemistry, social science, computer science, operations research and many other areas of applied mathematics and because of their relative structural simplicity trees are one of the most extensively studied classes of graphs. Therefore, generating trees is an important computational problem.

Trees have many equivalent characterizations. One of them involves the notion of an endpoint. An *endpoint* of a graph G is a vertex of degree 1. It is well known that every tree with more than one vertices has at least two endpoints. A 'recursive' characterization of trees is: A graph G of size $n > 1$ having the endpoint v is a tree iff after removing v and the unique edge incident with v the remaining graph is a tree of size $n - 1$.

A *rooted tree* is a tree in which one vertex, the *root*, has been distinguished from the others. Two rooted trees are considered isomorphic iff there is a $1 - 1$ adjacency preserving correspondence between them which maps the root of one onto the root of the other. A tree for which no root is distinguished is sometimes called a *free tree*. These definitions are used for both labelled and unlabelled trees.

In this section we deal with the generation of labelled trees, rooted unlabelled trees and free trees.

2.1. Labelled trees

There are exactly n^{n-2} labelled trees with n vertices. This result, due to Cayley, is one of the most celebrated counting results in graph theory. There are several proofs for it, one is due to Prüfer [23] who used a particular code for representing labelled trees, the so-called *Prüfer-code*.

The Prüfer-code of the smallest non-trivial tree, an edge connecting two vertices 1 and 2, is the empty string. For $n \geq 3$, the Prüfer-code is a string of $n - 2$ integers from V. It is found by the following algorithm.

Algorithm PRCODE(T)

(0) **Define** L to be the empty list;
(1) **Find** the endpoint v of T with the smallest label and its unique neighbour u;
(2) **Put** u at the end of the List L, **remove** v and the edge $\langle u, v \rangle$ from T; **If** T has still more than 2 vertices **then** goto (1);
(3) **Output** the list L;

As an example take the labelled tree in Fig. 1. Its Prüfer-code is $(1, 1, 3, 2)$. Prüfer's construction goes both ways. Using the recursive characterization of trees mentioned above it is immediately seen how the procedure PRCODE can be inversed to find the labelled tree from its code string. Thus the set of labelled trees on $V = \{1, 2, \ldots, n\}$ is in $1 - 1$ correspondence with the set of $n - 2$-tuples of natural

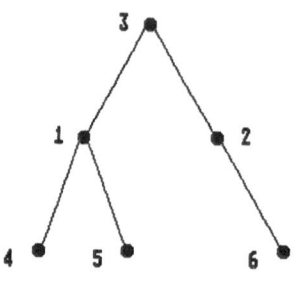

Figure 1

numbers not exceeding n. This is one way to prove Cayley's counting result. From this discussion we see that the following algorithm will generate labelled trees of size n u.a.r.

Algorithm LABTREE(n)

(1) **For** $1 \le i \le n - 2$ **do**
    select a number a_i from V u.a.r.;
(2) **Construct** the tree corresponding with the Prüfer-code $(a_1, \ldots, a_{n-2})$ and **output** it;

Both steps 1 and 2 of this algorithm can be implemented to run in time $0(n)$. Hence, Algorithm LABTREE is optimal with respect to time-complexity. The Prüfer-code may be used also to generate labelled trees with prescribed degree vector $(d(i)|i \in V)$ or labelled trees with prescribed number of endpoints u.a.r. For more details see [22].

2.2. Rooted Unlabelled Trees

The situation for rooted unlabelled trees (*ru*-trees for short) is substantially more complicated than for (free) labelled trees. The basic observation here is that any *ru*-tree of size n may be constructed in the following way: Take j copies of an *ru*-tree T'' of size $d(jd < n)$ and an *ru*-tree T' of size $n - jd$ and join the root of T' to the roots of each of the copies of T'' (see Fig. 2). This operation is indicated by $T := T' + j \times T''$.

One consequence of this observation is the following recurrence formula for t_n, the number of *ru*-trees of size n, namely

$$t_n = \sum_{1 \le m < n} \sum_{d|m} \frac{d \cdot t_{n-m} \cdot t_d}{n-1}, \qquad (2.2.1)$$

where $d|m$ means that d is a divisor of m, or equivalently

$$\sum_{d,j \ge 1} \frac{d \cdot t_{n-jd} \cdot t_d}{(n-1)t_n} = 1 \qquad (2.2.2)$$

$(t_1 = 1, t_k = 0 \text{ for } k \le 0)$.

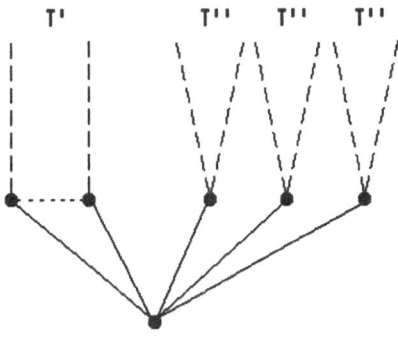

Figure 2

In [22] the following recursive algorithm for generating ru-trees u.a.r. is presented. The algorithm is based on formula (2.2.2) where the left side is interpreted as a sum of probabilities. The unique rooted unlabelled trees on 1 or 2 vertices are denoted by T_1 or T_2, respectively.

Algorithm RUTREE(n)

(1) **Choose** a pair $(j, d), j \geq 1, d \geq 1$, with probability
$$\text{Prob}(j, d) = \frac{d \cdot t_{n-jd} \cdot t_d}{(n-1)t_n};$$

(2) **If** $n - dj > 2$ **then** $T' :=$ RUTREE $(n - dj)$ **else** $T' := T_{n-jd}$;
 If $d > 2$ **then** $T'' :=$ RUTREE(d) **else** $T'' := T_d$;

(3) **Output** $T := T' + j \times T''$;

Algorithm RUTREE requires preprocessing for calculating the numbers $t_j, 1 \leq j \leq n$. On the base of (2.2.1) this can be done in time $0(n^2)$.

2.3. Free Unlabelled Trees

In this section a tree is a free unlabelled tree. The presentation of the generating process is based on [36].

Let $w_1, \ldots, w_k$ be the neighbours of a vertex v in a tree T. For each w_j there is a uniquely defined subtree T_j of T induced by the set of vertices which can be reached from v via the edge $\langle v, w_j \rangle$. The maximum size of a T_j is called the *weight* of v. A vertex v in T with minimum weight is called a *centroid* of T. It is well known that the number of centroids of a tree T may be 1 or 2. If T has two centroids then they are linked by an edge. Removing this edge leaves two ru-trees of equal size, the roots being the corresponding centroids.

The latter remark shows how one can reduce the problem of generating bicentroidal trees to the problem of generating ru-trees *u.a.r.* As before, let t_n be the number of ru-trees of size n. The following algorithm produces bicentroidal trees u.a.r. Clearly, the algorithm works for even $n > 0$ only.

Algorithm BICENTREE(n)

With probability $(t_{n/2} + 1)^{-1}$ **do** step (1) **else do** step (2)
(1) $T' := \text{RUTREE}(n/2)$; $T := T' + 1 \times T'$;
(2) $T' := \text{RUTREE}(n/2)$; $T'' := \text{RUTREE}(n/2)$; $T := T' + 1 \times T''$;
(3) **Output** T;

A vertex v is the unique centroid of a tree T iff its weight is at most $(n - 1)/2$, or with other words, iff none of the subtrees T_j corresponding to the neighbours w_j has size larger than $(n - 1)/2$. Thus there is a bijection between the set of trees with one centroid and the set of *ru*-trees for which all subtrees defined by the neighbours of the root are of size $(n - 1)/2$ at most.

A collection of disjoint *ru*-trees is called an *ru-forest*. Let $a(m, q)$ denote the number of *ru*-forests of size m whose trees are of size at most q. These numbers can be calculated using the formula

$$a(m, q) = \sum_{j \geq 1} \sum_{1 \leq d \leq q} \frac{d}{m} \cdot t_d \cdot a(m - jd, q)$$

$$(m \geq 1, a(0, q) = 1, a(k, q) = 0 \quad \text{for} \quad k < 0). \tag{2.3.1}$$

According to this recurrence equation an *ru*-forest of the type considered can be selected u.a.r. using the following algorithm.

Algorithm RUFOREST(m, q)

If $m = 0$ **then** exit with the empty forest **else do**
(1) **Choose** a pair of integers (j, d) with probability
$$\text{Prob}(j, d) = \frac{d \cdot a(m - jd, q)}{m \cdot a(m, q)} \cdot t_d, \quad 1 \leq j, \quad 1 \leq d \leq q;$$
(2) $F := \text{RUFOREST}(m - jd, q)$; $T := \text{RUTREE}(d)$;
(3) **Exit** with j copies of T adjoint to F;

Choosing a tree u.a.r. is now done by the following algorithm which combines the above two procedures into a single one.

Algorithm TREE(n)

(1) **If** n is odd **then do** $p := 0$ **elso do** $p := \left(\dfrac{1 + t_{n/2}}{2}\right) \cdot t_n^{-1}$;
(2) **With** probability p **do** $T := \text{BICENTREE}(n)$ and **output** T
    else do
    $$F := \text{RUFOREST}\left(n - 1, \frac{n - 2}{2}\right); T := \langle \text{a new vertex } v, \text{joint to all roots of } F \rangle;$$
    Output T;

The algorithms RUTREE and RUFOREST are based on the recurrence relations (2.2.1) and (2.3.1), respectively, which in a sense solve the associate counting problems. It was pointed out in [22] that any similar recurrency relation for the numbers of combinatorial configurations of a given type and size possibly leads to an efficient uniform generation algorithm. We will outline the idea behind in some detail in section 4.3.

3. Unlabelled Graphs

3.1. The Method of Dixon and Wilf

In their pioneering paper [6] Dixon and Wilf presented a method for generating unlabelled graphs of size n u.a.r. The method is applicable also to subclasses of graphs of a given type provided the corresponding counting problem can be solved. It is based on basic group-theoretical concepts, we outline it briefly in this section.

Let Φ be a finite group acting on a finite non-empty set Γ. The equivalence classes of Γ according to the relation $\alpha \sim \beta (\alpha, \beta \in \Gamma)$ iff $\alpha = \varphi\beta$ for some $\varphi \in \Phi$ are usually called *orbits* of Γ under Φ. For each $\varphi \in \Phi$ define $\text{Fix}(\varphi) = \{\alpha \in \Gamma | \varphi\alpha = \alpha\}$ (the set of elements fixed by the action of φ). Two group elements φ and φ' are called *conjugate* iff $\varphi' = \psi\varphi\psi^{-1}$ for some $\psi \in \Phi$.

The starting point in [6] is an algorithm which, given Γ and Φ, generates an orbit O of Γ under Φ u.a.r. The algorithm is based on the observation that if the elements in the sets $\text{Fix}(\varphi)$, $\varphi \in \Phi$ are listed then the combined list contains exactly $|\Phi|$ representatives of each orbit O. Furthermore, if φ and φ' are conjugate then for any orbit O we have $|\text{Fix}(\varphi) \cap O| = |\text{Fix}(\varphi') \cap O|$, and therefore, it is sufficient to list the elements in the sets $\text{Fix}(\varphi_i)$, $1 \le i \le r$, where the φ_i's are arbitrary elements in C_i, and $C_1, \ldots, C_r$ are the conjugacy classes of Φ. Put $c_i = |C_i|$ and let t be the number of orbits. Then, using the well-known 'Frobenius-Burnside lemma' which states that $t \cdot |\Phi| = \sum_{\varphi \in \Phi} |\text{Fix}(\varphi)|$, one easily proves that the following algorithm selects an orbit O of Γ under Φ u.a.r.

Algorithm RANDORB(Γ, Φ)

(1) **Choose** a number i from the set $\{1, 2, \ldots, r\}$ with probability
    $\text{Prob}(i) = c_i |\text{Fix}(\varphi_i)|/t|\Phi|$, $1 \le i \le r$;
(2) **If** i was chosen in step (1) **then choose** u.a.r. an element $\gamma \in \text{Fix}(\varphi_i)$;
(3) **Return** the orbit O which contains γ;

Now, specialize this algorithm to generate unlabelled graphs. Let n, the number of vertices, be fixed and let Γ be the set of all labelled graphs on n vertices. According to our convention, this is the set of all graphs with vertex set $V = \{1, 2, \ldots, n\}$. Let S be the full symmetric group acting on V. For each permutation $\pi \in S$ there is a corresponding bijective mapping $\varphi(\pi): \Gamma \to \Gamma$ defined by

$$\varphi(\pi)(G) = (V, \pi E),$$

$$\pi E = \{\langle \pi i, \pi j \rangle | \langle i, j \rangle \in E\}, \qquad \text{where } G = (V, E) \in \Gamma.$$

Let Φ be the group of all these mappings. Φ acts on Γ, and the set of orbits of Γ under Φ is, by definition, the set of unlabelled graphs of size n. Therefore, we may use Algorithm RANDORB to generate unlabelled graphs of given size. To do so one first has to solve some counting problems.

Let P be the set of all unordered pairs $\langle i,j \rangle$, $1 \leq i,j \leq n$. For each $\pi \in S$ there is a permutation π^* of P defined by $\pi^* \langle i,j \rangle = \langle \pi i, \pi j \rangle$. Write φ for the corresponding group element $\varphi(\pi) \in \Phi$. Fix(φ) consists of all those graphs for which π is an automorphism, i.e. maps their edge sets onto themselves. Thus $G = (V,E) \in$ Fix(φ) iff for each cycle q of π^* either all pairs in q are edges of G or none of them are. This implies $|\text{Fix}(\varphi)| = 2^{c(\pi)}$, where $c(\pi)$ is the number of cycles of π^*.

Let us use the notation $k = (k_1, \ldots, k_n)$ to denote the partition of n with k_i parts of size i, $1 \leq i \leq n$. Write $[k]$ to denote the corresponding conjugacy class of S consisting of all permutations that have exaktly k_i cycles of length i, $1 \leq i \leq n$. The number of elements in $[k]$ is known to be $n!/d(k)$ where

$$d(k) = \prod_{1 \leq i \leq n} (i^{k_i} k_i!).$$

Furthermore, in [26] it has been proved that for $\pi \in [k]$

$$c(\pi) = \frac{1}{2} \cdot \left\{ \sum_{1 \leq i,j \leq n} k_i k_j \cdot \gcd(i,j) - \sum_{1 \leq i \leq n} k_{2i+1} \right\}.$$

Now we are prepared to adapt Algorithm RANDORB for generating unlabelled graphs of size n. The number of orbits in our particular case is g_n, the number of unlabelled graphs of size n. We get the following algorithm.

Algorithm RANDGRAPH(n) (Dixon and Wilf)

(1) **Choose** a partition $k = (k_1, \ldots, k_n)$ of n with probability
    Prob$(k) = 2^{c(\pi)}/(g_n \cdot d(k))$ where π is any permutation in $[k]$;
(2) Let $\varphi = \varphi(\pi)$ and **choose** u.a.r. a graph $G = (V,E) \in$ Fix(φ);
(3) **Output** the unlabelled graph G^* underlying G;

The method outlined here is not restricted to our particular specification of Γ as the set of all unlabelled graphs on n vertices. We could deal also with smaller sets of unlabelled graphs, for instance with connected graphs, with graphs of a prescribed edge number, with regular graphs of some degree r, and so on. However, the corresponding counting problem (we need to know the numbers $|\text{Fix}(\varphi)|$ and the number of orbits t) often becomes an invincible obstacle in such cases, which means that the resulting algorithm allows no implementation that runs in polynomial average time.

The crucial point for an implementation of Algorithm RANDGRAPH is step (1). In general, suppose you have to generate the elements of a finite set K according to a given probability distribution Prob(k), $k \in K$. For this aim we may use the following well-known standard procedure which is based on a linear ordering of K, i.e. on an arrangement $(k^1, k^2, \ldots, k^{|K|})$ of the elements of K.

Algorithm RANDELEM(K)

(1) **Choose** a random number $x \in (0,1)$ and **initialize** $i := 1$ and $s_1 :=$ Prob(k^1);

(2) **While** $x > s_i$ **do** $i := i + 1$ and $s_i := s_{i-1} + \text{Prob}(k^i)$;
(3) **Output** k^i;

Let I be the random variable whose values are the actual values of i when the algorithm terminates, i.e. I is the number of times minus 1 RANDELEM performs step (2). We have

$$\text{Exp}(I) = \sum_{1 \leq i \leq |K|} i \cdot \text{Prob}(k^i).$$

Obviously, this term depends on the ordering of K and reaches its minimum for an ordering $(k^1, k^2, \ldots, k^{|K|})$ satisfying $\text{Prob}(k^1) \geq \text{Prob}(k^2) \geq \cdots \geq \text{Prob}(k^{|K|})$. In our case K is the set of all partitions $k = (k_1, \ldots, k_n)$ of n, and $|K| = p(n)$, the number of these partitions. Write I_n instead of I to indicate the dependence on n. Dixon and Wilf propose to use an ordering $(k^1, \ldots, k^{p(n)})$ of K satisfying $k_1^i \geq k_1^{i+1}$ for all $l = 1$, 2, $\ldots$, $p(n) - 1$. They show that under this condition $\text{Exp}(I_n) \leq 3$. Furthermore $\lim_{n \to \infty} \text{Exp}(I_n) = 1$. This latter result is substantially based on the well-known fact that for 'almost all' graphs G the automorphism group $\text{Aut}(G)$ is trivial.

Now for implementing step (1) of RANDGRAPH there is still the question how to calculate g_n, the number of unlabelled graphs of size n. While the numbers $p(n)$ are rapidly computable due to a simple recurrence relation, it is not known if there is a polynomial-time algorithm for the calculation of g_n. The best method known so far requires time proportional to $e^{\sqrt{n}}$.

Step (2) of the algorithm does not present difficulties. Once given $(k_1, \ldots, k_n)$ we may find a permutation $\pi \in [k]$ by writing down the symbols $1, 2, \ldots, n$ and then inserting brackets so as to obtain successively k_1 cycles of length 1, k_2 cycles of length 2, and so on. For example, the representative of the conjugacy class $[3, 0, 1, 1, 0, 0, 0, 0, 0, 0]$ is $(1)(2)(3)(4, 5, 6)(7, 8, 9, 10)$. Next we have to find the cycles of π^* and to assign edges or non-edges to them with equal probability $1/2$. This procedure will finally determine the output graph G^*.

Summarizing we may state the following theorem.

Theorem. *Assume that the universal number g_n has been precomputed. Then **there is** an implementation of Algorithm RANDGRAPH which runs in expected time $O(n^2)$.*

The method of Dixon and Wilf is in fact very fast. One may use this result for cataloguing graphs by generating them u.a.r. In case one aims a complete catalogue listing and checking for isomorphism is unavoidable, however, with small vertex numbers this does not present difficulties. In [17] Kerber et al. report on a successful effort for cataloguing all $g_{10} = 12005169$ unlabelled graphs of size 10 by generating them u.a.r. They used the method of Dixon and Wilf applied successively to sets of graphs of size 10 having prescribed edge number. The corresponding counting problem is solved by Polya's method. This approach, which is in contrast with the recursive methods used by Cameron et al. in [5] who have produced the same list of graphs earlier, however, does not seem to be efficient enough for listing graphs of larger size.

3.2. *Restarting Procedures—Wormald's Method*

All known algorithmic methods for selecting u.a.r. an element from a set S use the numerical value of the cardinality $|S|$ of S—at least implicitely (for example via a valid recurrence formula). It is often the case, however, that sufficient knowledge about $|S|$ is not available or is such that evaluation of $|S|$ can not be done efficiently (as in the case of the parameter g_n in section 3.1). This is certainly true if 'algorithmic' is understood in the very classical sense. The situation changes when we use a so-called *restarting procedure*, i.e. a procedure which accepts a result and outputs it only with some specified probability and which in the case of non-acceptance restarts the whole probabilistic process again. Let us consider such restarting procedures in more detail.

Let S be the set under consideration from which we want to select elements u.a.r. Assume that

$$S = \bigcup_{\alpha \in A} S_\alpha$$

is a partition of S into non-empty and mutually disjoint sets S_α where A is some set of indices. Let $P(\alpha)$, $\alpha \in A$, be a probability function on A, and for $\alpha \in A$ let $P_\alpha(s)$ be a probability function on $S_{\dot{\alpha}}$. Let a_α be an arbitrary number satisfying

$$a_\alpha \leq P(\alpha) \cdot \text{Min}\{P_\alpha(s) | s \in S_\alpha\}.$$

The following procedure is a restarting procedure selecting elements from S.

Algorithm RRANDELEM(S)

(1) **Choose** an element α in A with probability $P(\alpha)$;
(2) **If** α was chosen in step (1) **then choose** $s \in S_\alpha$ with probability $P_\alpha(s)$;
(3) **Accept** s with probability $P_{acc}(s) = a_\alpha/P(\alpha) \cdot P_\alpha(s)$ **else goto** (1);

Obviously, this is not an algorithm in the classical sense since there may be indefinitely many steps before normal termination occurs. Let E_s be the event 'RRANDELEM(S) selects s' and let T be the random variable which counts the number of starts before termination. We have for $s \in S_\alpha$

$$\text{Prob}(E_s) = a_\alpha \cdot \left(\sum_{\alpha \in A} a_\alpha \cdot |S_\alpha| \right)^{-1}.$$

Hence, RRANDELEM(S) selects elements in S u.a.r. iff a_α is independent of α, say $a_\alpha = a$ for all $\alpha \in A$. In this case

$$\text{Exp}(T) = (a \cdot |S|)^{-1}.$$

According to this and $a \leq \text{Min}\{P(\alpha) \cdot P_\alpha(s) | \alpha \in A, s \in S_\alpha\}$ the best result is obtained when $P(\alpha) = |S_\alpha|/|S|$ and $P_\alpha(s) = 1/|S_\alpha|$, $s \in S_\alpha$, $\alpha \in A$, such that $a = |S|^{-1}$. In this case the procedure restarts with probability 0. We may consider Algorithm RAND-GRAPH in section 3.1 as such a degenerate case of RRANDELEM where

$$S = \bigcup_{\varphi \in \Phi} \{\varphi \times \text{Fix}(\varphi)\}, \qquad |S| = n! \cdot g_n, \qquad A = \Phi. \tag{3.2.1}$$

However, if we want to avoid evaluation of $|S|$ this case is not obtainable. What we can do instead is to evaluate the numbers $|S_\alpha|$ approximately as well as it is possible (with low effort). To be concrete, assume that we have found approximations $\sigma_\alpha \approx |S_\alpha|$, $\alpha \in A$. Put $\sigma = \sum \sigma_\alpha$ and define

$$P(\alpha) = \sigma_\alpha/\sigma. \qquad P_\alpha(s) = |S_\alpha|^{-1}, \alpha \in A.$$

Using these probability functions Algorithm RRANDELEM reduces the problem of generating u.a.r. an element from a 'large' set S to a series of generating tasks involving the 'small' sets S_α. This is advantageous at least in all those situations where the calculation of the numbers $|S_\alpha|$ is easy while the calculation of $|S|$ is difficult (for instance, because it is difficult to enumerate all sets S_α). Furthermore, since step (2) is again u.a.r. generation step we may use RRANDELEM recursively (after having specified a suitable partionning rule). Finally, we may use any (probabilistic) algorithm which produces elements s of S not necessarily u.a.r. but with Prob(s) that is computable at least a posteriori to build up a restarting procedure for u.a.r. generation, provided we can evaluate $a \leq \text{Min}_s \text{Prob}(s)$.

Based on the ideas outlined here Wormald has found generation procedures for several classes of unlabelled graphs [35]. In the case of arbitrary unlabelled graphs of size n he proposes a restarting procedure defined in the following way.

Let S and A be as in (3.2.1). To implement step (1) of RRANDELEM(S) Wormald uses a partition

$$\Phi = \Phi_1 \cup \Phi_2 \cup \cdots \cup \Phi_n$$

of Φ where

$\Phi_1 = \{id\}$ (id ... the identity permutation)

Φ_i = the set of all permutations $\varphi(\pi)$ where π has exactly $n - i$ fixed points

(cycles of length 1)

Furthermore he introduces upper bounds

$$\beta_1 = 2^N = |\Phi_1|,$$
$$\beta_i = 2^{N-H(n, i)}[n]_i \geq |\Phi_i|, \qquad 2 \leq i \leq n,$$
$$\beta = \sum \beta_i$$

where $N = \binom{n}{2}$, $[n]_i = n \cdot (n - 1) \cdot \ldots \cdot (n - i + 1)$ and $H(n, i) = i \cdot \left(\frac{n}{2} - \frac{i + 2}{4}\right)$, and replaces step (1) of RRANDELEM by

(1a) **Choose** a number i from $\{1, 2, \ldots, n\}$ with probability $\beta_i \beta^{-1}$;
(1b) **Choose** $\varphi \in \Phi_i$ with probability $|\Phi_i|^{-1}$;

In step (2) a graph G in Fix(φ) is generated u.a.r. as in RANDGRAPH (section 3.1). The parameter α used in determining the probability for acceptance P_{acc} has the value β^{-1}, hence

$$\text{Exp}(T) = \beta/g_n.$$

It can be shown that this value is bounded by a constant. The following theorem summarizes the results.

Theorem. *Wormald's restarting procedure generates unlabelled graphs of size n u.a.r., each graph being produced in expected time* $O(n^2)$. *No preprocessing for finding the number* g_n *is necessary.*

4. Labelled Graphs

In this section we deal with labelled graphs $G = (V, E)$ exclusively. In the first subsection we report on a general framework for the generation of a large variety of graphs of different types u.a.r. The next subsection is devoted to graphs with prescribed degree vector, and in the last part of this section we review some recent results concerning the generation of some types of planar graphs.

4.1. A Framework For Generating Labelled Graphs

Let Γ_n be the set of all labelled graphs of size n. Sampling from Γ_n u.a.r. can be done by constructing a list E of edges $\langle i,j \rangle$, $1 \le i < j \le n$, where the edges $\langle i,j \rangle$ are drawn independently, each with probability $1/2$. The resulting graph $G = (V, E)$ has probability 2^{-N}, where $N = \binom{n}{2}$.

Compared with the situation with unlabelled graphs (section 3.1), generating labelled graphs u.a.r. from Γ_n is a trivial problem. However, difficulties soon appear when Γ_n is replaced by a subset of graphs having some specified additional properties. To handle these cases successfully we introduce a particularly convenient representation for labelled graphs.

A widely used tool for representing graphs are adjacency lists. For $i \in V$ let N_i be the list of vertices adjacent to i. The combined list $(N_1, \ldots, N_n)$, however, is a redundant representation of G, since any $\langle i,j \rangle \in E$ is noted twice, due to $i \in N_j$ and $j \in N_i$. To avoid this redundancy we can use a list of *irredundant adjacency lists* $(A_1, \ldots, A_{n-1})$ defined in the following way.

$$A_j \subset V - \{x_1, \ldots, x_j\}, \qquad 1 \le j \le n - 1, \tag{4.1.1}$$

$$x_1 = 1, \qquad x_j = \mathrm{Min}\{i | i \in B_{j-1}\} \tag{4.1.2}$$

$$B_j = \begin{cases} \bigcup_{1 \le i \le j} A_i - \{x_1, \ldots, x_j\} & \text{if this set is non-empty} \\ V - \{x_1, \ldots, x_j\} & \text{otherwise} \end{cases}$$

$$1 \le j \le n - 1, \qquad B_0 = V$$

Now, again, $(A_1, \ldots, A_{n-1})$ represents a graph $G = (V, E)$ uniquely. A sublist $(A_1, \ldots, A_i)$, $i < n - 1$, may be considered as the subgraph containing $x_1, \ldots, x_i$, the edges $\langle x_k, j \rangle, j \in A_k, 1 \le k \le i$, and the isolated vertices

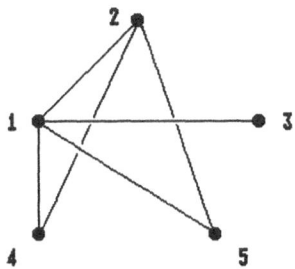

Figure 3

$$j \in V - \bigcup_{1 \le k \le i} A_k - \{x_1, \dots, x_i\}.$$

When using irredundant adjacency lists we shortly speak of the graph (or subgraph) $(A_1, \dots, A_i)$.

The representation of graphs by irredundant adjacency lists has been introduced in [30] in order to facilitate the solution of counting problems and to avoid stochastic dependencies.

For an example see the graph in Fig. 3 which is represented by $((2, 3, 4, 5), (4, 5), -, -)$ where '$-$' means the empty list.

We will introduce now an algorithm scheme, called PROTOGRAPH, which by specialization yields several efficient generation algorithms for labelled graphs of various kinds. Assume that we want to determine the sequence $x_1, \dots, x_{n-1}$ and the adjacency lists $A_1, \dots, A_{n-1}$ according to (4.1.1) and (4.1.2) for a random graph G. Assume Ψ_n is a subclass of Γ_n and, given $(\bar{A}_1, \dots, \bar{A}_i)$, $\psi(\bar{A}_1, \dots, \bar{A}_i)$ is the number of graphs $(A_1, \dots, A_{n-1}) \in \Psi_n$ with $A_j = \bar{A}_j$ for $1 \le j \le i$. Define

$$U_j = \bigcup_{1 \le k \le j} A_k - \{x_1, \dots, x_j\}, \quad V_j = V - \bigcup_{1 \le k \le j} A_k - \{x_1, \dots, x_j\},$$

$$u_j = |U_j|, \qquad v_j = |V_j|, \qquad 1 \le j \le n - 1. \tag{4.1.3}$$

$$U_0 = \{1\}, \qquad V_0 = \{2, 3, \dots, n\}, \qquad u_0 = 1, \qquad v_0 = n - 1.$$

The algorithm scheme is as follows.

Algorithm PROTOGRAPH

(1) $U_0 := \{1\}$; $V_0 := \{2, 3, \dots, n\}$; $j := 1$;

(2) **If** $U_{j-1} = 0$ **then** $x_j := \operatorname{Min}\{i | i \in V_{j-1}\}$ **else** $x_j := \operatorname{Min}\{i | i \in U_{j-1}\}$;

(3) **Select** $Y \subset U_{j-1} \cup V_{j-1} - \{x_j\}$ with probability

$$\operatorname{Prob}(Y) = \frac{\psi(A_1, \dots, A_{j-1}, Y)}{\psi(A_1, \dots, A_{j-1})}, \qquad (\psi(A_1, \dots, A_{j-1}) = |\Psi_n| \quad \text{for} \quad j = 1),$$

(4) $A_j := Y$;

(5) $U_j := U_{j-1} \cup A_j - \{x_j\};\ V_j := V_{j-1} - A_j - \{x_j\};\ j := j + 1;$
    If $j < n$ **then goto** (2);
(6) **Output** $(A_1, \ldots, A_{n-1})$;

It is evident that this algorithm scheme after specification of a subroutine for calculating $\psi(A_1, \ldots, A_j)$, $1 \le j \le n - 1$, yields an algorithm for generating the elements of Ψ_n u.a.r. By the way, $\psi(A_1, \ldots, A_j)$ counts the number of ways one can extend the subgraph $(A_1, \ldots, A_j)$ to a graph in Ψ_n. Again, the problem of constructing an efficient algorithm for generating u.a.r. graphs of a given type has been reduced to efficiently solving a corresponding counting problem. In general, it will not be known how to compute $\psi(A_1, \ldots, A_j)$ efficiently. However, there is some reasonable chance to find an efficient counting method if $\psi(A_1, \ldots, A_j)$ does not depend on the 'structure' of $(A_1, \ldots, A_j)$ but depends only on the parameters u_j and v_j defined in (4.1.3) and/or some additional easily handled parameters. We give a list of examples where algorithm PROTOGRAPH can be adapted successfully.

Example 1. $\Psi_n = \Gamma_n$(*the trivial case* mentioned at the begin of this subsection). Obviously, $\psi(A_1, \ldots, A_j) = 2^{N(j)}$ where $N(j) = \binom{n-j}{2}$.

Example 2. $\Psi_n = \Gamma_{n,m}$, *the set of all labelled graphs of size n with m edges*. Here we find

$$\psi(A_1, \ldots, A_j) = \binom{N(j)}{m_j}$$

where $N(j) = \binom{n-j}{2}$, $m_0 = m$ and $m_j = m - \sum_{i \le j} |A_j|$.

Example 3. $\Psi_n = C_n$, *the set of all connected labelled graphs of size n*. It is well-known that $c_n = |C_n|$ satisfies the recurrency equation

$$c_n = 2^N - \frac{1}{n} \cdot \sum_{1 \le k \le n-1} k \cdot \binom{n}{k} \cdot 2^{\binom{n-k}{2}} \cdot c_k \tag{4.1.4}$$

(see [10]). Furthermore, one can easily show that $\psi(A_1, \ldots, A_j)$ depends on u_j and v_j only. Write $c(u_j, v_j) = \psi(A_1, \ldots, A_j)$. These numbers satisfy the following generalization of (4.1.4) ($u = u_j, v = v_j$ for abreviation):

$$c(u, v) = 2^{\binom{u+v}{2}} - \sum_{1 \le t \le v} \binom{v}{t} \cdot 2^{\binom{t}{2}} c(u, v - t). \tag{4.1.5}$$

However, even (4.1.5) is still too monstruous for practical purposes. We better establish a table of the numbers $c(u, v)$, $1 \le u \le \bar{u}$, $1 \le v \le \bar{v}$ (once for ever) for some $\bar{u}, \bar{v}$ and use $2^{\binom{u+v}{2}}$ as an approximation for $c(u, v)$ in the range outside $[1, \bar{u}] \times [1, \bar{v}]$. (In practice $\bar{u} = \bar{v} = 12$ suffices). With these ideas behind it is possible to implement an appropriate version of PROTOGRAPH which generates connected labelled graphs u.a.r. within any $\varepsilon > 0$, which means that a graph $G \in C_n$ is produced with probability $\text{Prob}(G)$ satisfying

$$(1 - \varepsilon)c_n^{-1} \le \text{Prob}(G) \le (1 + \varepsilon)c_n^{-1}.$$

This result is of some theoretical interest. In practice, at least for vertex numbers $n \geq 12$, it is even better and easier to use a restarting procedure (see section 3.2) which generates u.a.r. graphs in Γ_n and accepts them if they are connected. Such a restarting procedure has been mentioned first in [30].

Example 4. $\Psi_n = C_{n,m}$, *the set of connected graphs of size* n *with* m *edges.* Here $\psi(A_1, \ldots, A_j)$ depends on u_j, v_j and m_j only, where m_j is defined in Example 2. An appropriate treatment is found by a combination of Example 2 and Example 3.

Example 5. $\Psi_n = T_n$, *the set of free labelled trees of size* n. This is a special case of Example 4 with $m = n - 1$ and $v_j = m_j, j \geq 1$. With $u_j = u$, $v_j = v$ and again putting $\psi(A_1, \ldots, A_j) = c(u, v)$ we get

$$c(u, v) = \sum_{0 \leq t \leq v} \binom{v}{t} (t + 1)^t \, {}^1 c(u - 1, v - t)$$

with solution $c(u, v) = u(u + v)^{v-1}$, $u \geq 0$, $v \geq 0$.

Example 6. $\Psi_n = \Gamma_{n,\text{even}}$, *the set of labeled graphs with all degrees* $d(i)$, $i \in V$, *even.* Here $\psi(A_1, \ldots, A_j)$ depends on u_j and v_j only, and with the same notation as above we have

$$c(u, v) = 2^{\binom{u+v-1}{2}}.$$

Example 7. $\Psi_n = \Gamma_{n,\text{eul}}$, *the set of Eulerian graphs of size* n (connected graphs with all vertex degrees even). As above $\psi(A_1, \ldots, A_j)$ depends only on u_j and v_j. Use again $c(u, v) = \psi(A_1, \ldots, A_j)$ to get

$$c(u, v) = 2^{\binom{u+v-1}{2}} - \sum_{1 \leq t \leq v} \binom{v}{t} 2^{\binom{t-2}{2}} c(u, v - t),$$

a formula which may be used in an analoguous way to Example 3.

Example 8. $\Psi_n = \Gamma_{n,bp}$, *the set of bipartite graphs.* Let Y_j, Z_j be a bipartition of $(A_1, \ldots, A_j)$. $\psi(A_1, \ldots, A_j)$ depends on $\alpha = |U_j \cap Y_j|$, $\beta = |U_j \cap Z_j|$ and $\gamma = |V_j|$. We find

$$\psi(A_1, \ldots, A_j) = 2^{\alpha(\beta+\gamma)} \sum_{0 \leq t \leq \gamma} 2^{t(\beta - \alpha + \gamma - t)} \binom{\gamma}{t}$$

Example 9. $\Psi_n = \Gamma_{n,d}$, *the set of graphs with degree vector* $d = (d_1, \ldots, d_n)$. Here $\psi(A_1, \ldots, A_j)$ depends highly on the structure of $(A_1, \ldots, A_j)$. [30] contains a discussion of this case.

Example 10. Digraphs

In the case of digraphs we have to replace the lists A_j in the representation of a graph by a pair of lists (A'_j, A''_j) containing predecessors and successors of x_j, respectively, being defined analoguously to (4.1.1) and (4.1.2). In [30] algorithms of type PROTOGRAPH are given for arbitrary digraphs, weakly connected digraphs, strongly connected digraphs, tournaments, and other types of digraphs.

All details concerning the particular graph classes addressed in Examples 1–10 are found in [29, 30].

4.2. Graphs With Prescribed Degree Vectors

The problem of generating random graphs with given degree vectors was already addressed in [30] (see Example 9 in section 4.1) where it was shown how to generate such graphs and how to compute a posteriori the probability that the particular output graph was produced. Although the probabilities associated with the different graphs may vary considerably this method can be used for constructing a restarting procedure (see section 3.2) which finally generates graphs of the prescribed type u.a.r. The expected time needed for generating a single graph depends on the minimum probability which may occur, therefore, since the algorithm presented in [30] is not very complicated, there is some chance for getting a complete analysis of the expected time consumed by the restarting procedure. However, as yet this subject has not been investigated sufficiently.

The special case in which the graphs to be generated are regular is of particular interest. Wormald [34] gives an efficient algorithm for generating labelled cubic graphs of size n. However, his method is based on a specific recurrence equation (see the remark at the end of section 2.3) for the associated counting problem, and it is not to see how to generalize it successfully to higher degrees. A simple method proposed by Wormald [34], which is based on an idea of Bollobas and Thomasson [4], generates regular graphs of arbitrary degree r u.a.r., but the algorithm fails to produce an output with some probability which remains bounded only for $r = 0((\log n)^{1/2})$.

More recently, the problem of this subsection has received a considerably amount of attention, caused partially by some very attractive results of Jerrum and Sinclair [15], [16], [27], [28], and others. By an indirect method based on approximate counting Jerrum and Sinclair [16] give an almost uniform generation algorithm which runs in polynomial time for regular graphs up to degree $\leq n/2$. The method is based on a Markov chain simulation technique which has turned out to be a powerful tool for the random generation of various combinatorial configurations [14, 15, 16]. To sketch the idea, let $\Gamma(d)$ denote the set of all labelled graphs with degree vector $d = (d_1, \ldots, d_n)$. With $\Gamma(d)$ a Markov chain MC(d) is associated whose states include the elements of $\Gamma(d)$ together with some auxiliary structures, and whose transitions correspond to simple random perturbations such as edge additions or deletions. This process turns out to converge asymptotically to a stationary distribution which is uniform over the states. Moreover, under some restrictions on the values of d, the convergence is fast in the sense that the distribution gets very close to uniform after a polynomial number of steps. This property of Markov chains is called rapidly mixing [1, 7. 14]. Thus one can generate elements of $\Gamma(d)$ almost u.a.r. by simulating the evolution of MC(d) for some small number of steps and outputting the final state.

Finally, two very recent papers [8, 20] deal with regular graphs of degree $o(n^{1/5})$ and $o(n^{1/3})$, respectively, using more direct methods.

4.3. *Two Classes of Planar Graphs*

4.3.1. Preliminaries

The counting problem for general planar graphs is still unresolved. However, there are several subclasses of planar graphs whose cardinalities depending on the size n are known or are efficiently computable (see [18]). In this section we present two recent results in this direction concerning 2-connected outerplanar graphs and maximal planar graphs and describe the corresponding generation algorithms.

As already mentioned at the end of section 2.3 the existence of a recurrence relation for the numbers of certain combinatorial configurations of variable size may lead to a generation algorithm for these configurations in a straightforward way. This was already pointed out in [22]. Here, we will pick up this idea and outline it briefly, giving a slightly modified presentation.

As before, let Ψ be a certain class of labelled graphs defined by some graph-property, $\Psi(n)$ the set of all such graphs of size n, and $\psi_n = |\Psi(n)|$, $n \in \mathbb{N}$. Let $l = \mathrm{Min}\{n | \Psi(n) \neq 0\}$. Assume that for $n \in \mathbb{N}$ the set $\Psi(n)$ is a union

$$\Psi(n) = \bigcup_{1 \leq k \leq k(n)} M_k$$

where each $M_k \subset \Psi(n)$ is an image

$$M_k = \theta_k(\Psi(j(k,1)) \times \Psi(j(k,2)) \times \cdots \times \Psi(j(k,i(n)))), \quad 1 \leq j(k,i) < n, \quad 1 \leq i \leq i(n)$$

under some injective mapping θ_k from some product of sets $\Psi(j(k,i))$ of graphs of smaller size into $\Psi(n)$. This means that each element $G \in M_k$ can be constructed in a unique way using elements $G_i \in \Psi(j(k,i))$ and applying some well-defined construction rule to them. It is not necessary that the sets M_k are disjoint. If not, however, assume in addition that there exist non-negative numbers $p_{n,1}, p_{n,2}, \ldots, p_{n,k(n)}$ such that

$$|\Psi(n)| = \sum_{1 \leq k \leq k(n)} p_{n,k} \cdot |M_k| = \sum_{1 \leq k \leq k(n)} p_{n,k} \cdot \prod_{1 \leq i \leq i(n)} |\Psi(j(k,i))|$$

and

$$\sum_{1 \leq k \leq k(n)} p_{n,k} = 1.$$

In the case where the $p_{n,k}$'s are known we get the recurrence equations

$$\psi_n = \sum_{1 \leq k \leq k(n)} p_{n,k} \cdot \psi_{j(k,1)} \cdot \psi_{j(k,2)} \cdots \psi_{j(k,i(n))}, \quad n > 1. \tag{4.3.1}$$

which like (2.2.1) after dividing both sides by ψ_n may be interpreted as a sum of probabilities

$$\sum_{1 \leq k \leq k(n)} p_{n,k} \frac{\psi_{j(k,1)} \cdots \psi_{j(k,i(n))}}{\psi_n} = 1. \tag{4.3.2}$$

Obviously, the formulas (2.2.1) and (2.3.1) are special cases of (4.3.1).

Assume that RANDΨL is an algorithm which generates elements of $\Psi(l)$ u.a.r. The following algorithm which is based on (4.3.2) and the 'start'-procedure RANDΨL generates elements from an arbitrary $\Psi(n)$ u.a.r.

Algorithm RAND$\Psi(n)$

(1) **Choose** k from the set $\{1, 2, \ldots, k(n)\}$ with probability

$$\text{Prob}(k) = \frac{p_{n,k}\psi_j(k, 1) \cdot \ldots \cdot \psi_j(k, i(n))}{\psi_n};$$

(2) **For** $1 \leq i \leq i(n)$ **do**
   if $j(k, i) > l$ **then** $G_i := \text{RAND}\Psi(j(k, i))$ **else** $G_i := \text{RAND}\Psi\text{L}$;
(3) **Output** $G := \theta_k(G_1, \ldots, G_{i(n)})$;

This algorithm requires preprocessing for calculating the sequence $\psi_1, \ldots, \psi_n$. With these numbers given, the time-complexity of RANDΨ is $0(n^2)$, provided step (3) can be done in time $0(n^2)$.

4.3.2. Two-connected Outerplanar Graphs and Maximal Planar Graphs

A planar graph is called *outerplanar* if it can be embedded in the plane so that all vertices lie on the same face. A 2-connected outerplanar graph possesses a unique Hamiltonian cycle C, and we may assume that this cycle is the boundary of the exterior face. All the remaining edges of such a graph are chords of C. Hence, two 2-connected outerplanar graphs are distinguished firstly by the arrangement of the vertices on the cycle C and, secondly, by the number and arrangement of their chords. Let $op(n)$ be the number of (labelled) 2-connected graphs on n vertices. Obviously, $op(3) = 1$. Furthermore, $op(n) = \frac{(n-1)!}{2} \cdot ch(n)$ where $ch(n)$ is the number of different ways one can draw pairwise non-crossing chords in the cycle $(1, 2, \ldots, n)$. In [11] the following recurrence relation for the sequence $ch(n)$ has been derived

$$ch(n) = 3 \cdot ch(n-1) + 2 \cdot \sum_{4 \leq j \leq n-1} ch(j-1) \cdot ch(n-j+2).$$

This implies

$$op(n) = 3(n-1) \cdot op(n-1)$$

$$+ 4 \sum_{4 \leq j \leq n-1} \binom{n-1}{j-2} \cdot op(j-1) \cdot op(n-j+2), \quad n \geq 4, \quad (4.3.3)$$

which, clearly, is a special case of (4.3.1).

A *maximal planar graph* is one to which no edge can be added without loosing planarity, or equivalently, for which every face is a triangle. The counting problem for such graphs has been solved in [31].

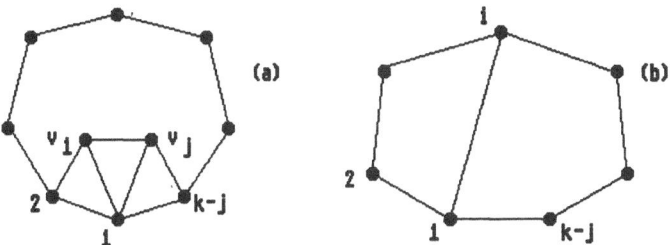

Figure 4

Let $mp(n)$ be the number of maximal planar graphs of size n. Based on the results in [31] in [11] a recurrence relation for the sequence $mp(n)$ is established. Any maximal planar graph G can be constructed by first choosing the exterior triangle and the triangulating then 'inner part' of this triangle by inserting $n - 3$ additional vertices and $3n - 6$ additional edges. If we choose the exterior triangle in the $\binom{n}{3}$ possible ways, since any triangle of G can be made the exterior one, by this method method we get any particular graph in exactly $2n - 4$ different ways. Therefore, we have

$$mp(n) = \frac{1}{2n - 4}\binom{n}{3}\cdot\mathrm{ins}(n, n - 3) \qquad (4.3.4)$$

where for arbitrary k and j $\mathrm{ins}(k, j)$ denotes the number of different ways one can triangulate the closed polygon $(1, 2, \ldots, k - j)$ by inserting the vertices $k - j + 1, \ldots,$ k and $2k - 6$ appropriate edges. There are two possible types of triangulations distinguished by the shape of the neighbourhood of the vertex 1 (see Fig. 4).

Fig. 4a shows the first type where no edge $\langle 1, i\rangle$, $3 \leq i \leq k - j - 1$, exists. Fig. 4b shows the second type where such an edge is drawn. Let $\mathrm{ins}'(k, j)$ and $\mathrm{ins}''(k, j)$ denote the numbers of triangulations of these two types, respectively. The following equations are valid:

$\mathrm{ins}(k, j) = \mathrm{ins}'(k, j) + \mathrm{ins}''(k, j)$

$$\mathrm{ins}'(k, j) = \sum_{0 \leq t \leq j}\binom{j}{t}\cdot t!\cdot\mathrm{ins}(k - 1, j - t) \qquad (4.3.5)$$

$$\mathrm{ins}''(k, j) = \sum_{3 \leq s \leq k - j - 1}\ \sum_{0 \leq t \leq j}\binom{j}{t}\cdot\mathrm{ins}'(s + t, t)\cdot\mathrm{ins}(k - s - t + 2, j - t)$$

with obvious initial conditions. This together with (4.3.4) yields a recurrence relation for $mp(n)$ which is of the form (4.3.1).

Implementations of RANDΨ for generating u.a.r. outerplanar graphs and maximal planar graphs, respectively, which are based on the results presented in this section can be found in [11].

References

[1] Aldous, D.: Random walks on finite groups and rapidly mixing Markov chains. Seminare de Probabilite XVII, 1981/82, Springer Lecture Notes in Mathematics 986, pp. 243–297.

[2] Baudon, O.: Generating random graphs, connected with the editor of graphs CABRI. Preprint, 1989.

[3] Berge, C.: Graphes et hypergraphes, Pairs: Dunod 1970.

[4] Bollobas, B.: The asymptotic number of unlabelled regular graphs. J. London Math. Soc. 26, 201–206 (1982).

[5] Cameron, R. D., Colbourn, C. J., Read, R. C., and Wormald, N. C.: Cataloguing the graphs on 10 vertices. Journal of Graph Theory 9, 551–562 (1985).

[6] Dixon, J. D., and Wilf, H. S.: The random selection of unlabelled graphs. Journal of Algorithms 4, 205–213 (1983).

[7] Dyer, M., Frieze, A., and Kannan, R.: A random polynomial time algorithm for approximating the volume of convex bodies. Preprint (1988)

[8] Frieze, A.: On random regular graphs with non-constant degree. Preprint (1988).

[9] Harary, F.: Graph Theory, Reading: Addison Wesley 1969.

[10] Harary, F., and Palmer, E.M.: Graphical Enumeration, New York: Academic Press 1973.

[11] Hofmann, Ch.: Über Erzeugungsverfahren für Graphen, insbesondere planare Graphen, Diplomarbeit an der Technischen Universität München, 1989.

[12] James, K. R., and Riha, W.: Algorithm for generating graphs of a given partition. Computing 16, 153–161 (1976).

[13] Jerrum, M. R., Valiant, L. G., and Vazirani, V. V.: Random generation of combinatorial structures from a uniform distribution. Theoretical Computer Science 43, 169–188 (1986).

[14] Jerrum, M. R., and Sinclair, A. J.: Conductance and the rapid mixing property for Markov chains. Proc. 20th ACM Symposium on Theory of Computing (1988).

[15] Jerrum, M. R., and Sinclair, A. J.: Approximating the permanent. Internal report CSR-275-88, Department of Computer Science, University of Edinburgh, 1988.

[16] Jerrum, M. R., and Sinclair, A. J.: Fast uniform generation of regular graphs. Preprint CSR-281-88, University of Edinburgh, 1988.

[17] Kerber, A., Laue, R., Hager, R., and Weber, W.: Cataloguing graphs by generating them uniformly at random, Preprint, 1989.

[18] Liskovets, V. A.: Ten steps to counting planar graphs. Congressus Numerantium 60, 269–277 (1987).

[19] McKay, B. D., and Wormald, N. C.: Asymptotic enumeration by degree sequence of graphs with degrees $o(n^{1/2})$. Preprint (1988).

[20] McKay, B. D., and Wormald, N. C.: Asymptotic enumeration by degree sequence of graphs of high degree. Preprint (1988).

[21] McKay, B. D., and Wormald, N. C.: Uniform generation of random regular graphs of moderate degree. Preprint (1988).

[22] Nijenhuis, A., and Wilf, H. S.: Combinatorial Algorithms (2nd edition). Orlando: Academic Press (1978).

[23] Prüfer, H.: Neuer Beweis eines Satzes über Permutationen. Arch. Math. Phys. 27, 742–744 (1918).

[24] Read, R. C.: A survey of graph generation techniques. Combinatorial Mathematics VIII (K. L. McAvaney ed.), Lecture Notes in Mathematics 884, p. 77–89, 1980.

[25] Read, R. C.: Everyone a winner or How to avoid isomorphism search when cataloguing combinatorial configurations. Annals of Discrete Mathematics 2, 107–120 (1978).

[26] Robinson, R. W.: Enumeration of Euler Graphs, in "Proof Techniques in Graph Theory" (F. Harary, ed.), pp. 147–153, New York: Academic Press, 1969.

[27] Sinclair, A. J.: Randomised algorithms for counting and generating combinatorial structures. PhD Thesis, University of Edinburgh, June 1988.

[28] Sinclair, A. J. and Jerrum, M. R.: Approximate counting, uniform generation and rapidly mixing Markov chains, Lecture Notes in Computer Science 246, 134–148 (1987).

[29] Tinhofer, G.: On the generation of random graphs with given properties and known distribution. Appl. Comp. Sci. 13, 265–297 (1979).

[30] Tinhofer, G.: Zufallsgraphen (in German). Appl. Comp. Sci. 17 (1980).

[31] Tutte, W. T.: A census of planar triangulations. Canad. J. Math. 14, 21–38 (1962).

[32] Valiant, L. G.: The complexity of computing the permanent. Theoretical Computer Science 8, 189–201 (1979).

[33] Valiant, L. G.: The complexity of enumeration and reliability problems. SIAM Journal on Computing 8, 410–421 (1979).

[34] Wormald. N. C.: Generating random regular graphs. Journal of Algorithms 5, 247–280 (1984).
[35] Wormald, N. C.: Generating random unlabelled graphs. SIAM Journal on Computing *16*, 717–727 (1987).
[36] Wilf, H. S.: The uniform selection of free trees. Journal of Algorithms 2, 204–207 (1981).

Prof. Dr. G. Tinhofer
Institut für Mathematik
Technische Universität München
Postfach 202420
D-8000 München 2
West-Germany

Computing Suppl. 7, 257–282 (1990)

Embedding one Interconnection Network in Another

B. Monien, Paderborn, and **H. Sudborough,** Dallas, Tex.

Abstract — Zusammenfassung

Embedding one Interconnection Network in Another. We review results on embedding network and program structures into popular parallel computer architectures. Such embeddings can be viewed as high level descriptions of efficient methods to simulate an algorithm designed for one type of parallel machine on a different network structure and/or techniques to distribute data/program variables to achieve optimum use of all available processors.

AMS Subject Classifications: 68Q10, 94-02, 94C15.

Key words: Embedding, simulation, interconnection network

Vergleich von Rechnerverbindungsnetzen. Wir rezensieren Ergebnisse über die Einbettung von Netzwerken und Programmstrukturen in populäre Rechnerarchitekturen. Solche Einbettungen können als hochsprachliche Beschreibungen effizienter Methoden angesehen werden mit deren Hilfe Algorithmen, die für eine parallele Maschine entwickelt wurden, auf einer anderen Netzwerkstruktur simuliert werden können. Auf der anderen Seite werden auf diese Weise Techniken zur Verteilung von Daten bzw. Programmvariablen beschrieben, die eine optimale Ausnutzung aller verfügbaren Prozessoren sicherstellen.

I. Common Network and Algorithm Structures

Various parallel computer architectures have gained favor and are in use today. Other structures included here are often used as program/data structures. The techniques we survey compare networks by considering the ability of one to simulate other network structures. Such simulations are studied by embeddings. A good simulation is said to exist when adjacent processors in the guest network are mapped to reasonably close processors in the host network, when the messages between processors in the guest can be routed in the host without incurring significantly larger delay, when the host network is not too much larger than the guest network, and, in the case of mapping guest networks into smaller hosts, when the processors of the host have been assigned a reasonably similar number of processes from the guest. The quality of a network as an interconnection structure is often discussed by other measurements. Typical measurements include a network's diameter, namely the maximum distance between any pair of nodes, and its maximum node degree, i.e. the maximum number of edges incident to a node. These properties are important, as (a) a network's diameter measures how much distance exists between processors and hence gives a lower bound on communication time and (b) a network's maximum node degree describes the largest number of connections made to an individual processor.

Binary Hypercubes   The binary hypercube of dimension n, denoted by $Q(n)$, is the graph whose nodes are all binary strings of length n and whose edges connect those binary strings which differ in exactly one position.

Clearly, a binary hypercube $Q(n)$ has 2^n nodes and, as each node is connected to n edges, a total of $n2^{n-1}$ edges. It is also easily seen that the diameter of the hypercube $Q(n)$ is n, which is the logarithm of the number of its nodes. An illustration of $Q(4)$ is shown in Figure 1.

Binary Trees   The complete binary tree of height n, denoted by $B(n)$, is the graph whose nodes are all binary strings of length at most n and whose edges connect each string x of length i ($0 \leq i \leq n$) with the strings xa, a in $\{0, 1\}$, of length $i + 1$. The node e, where e is the empty string, is the *root* of $B(n)$ and a node x is at level i, $i \geq 0$, in $B(n)$ if x is a string of length i. A binary tree is a connected subgraph of $B(n)$, for some $n \geq 0$. A variation of a complete binary tree allows for double roots, denoted by $DRB(n)$, i.e. its nodes are all binary strings of length at most n plus one new node e' (called the *alternate root*), where e represents the empty string, obtained from $B(n)$ by simply inserting e' into the edge connecting e with the node 1. (The new node e' thus has two neighbors: the root e and the node 1.) See Figure 2 for an illustration of $DRB(2)$.

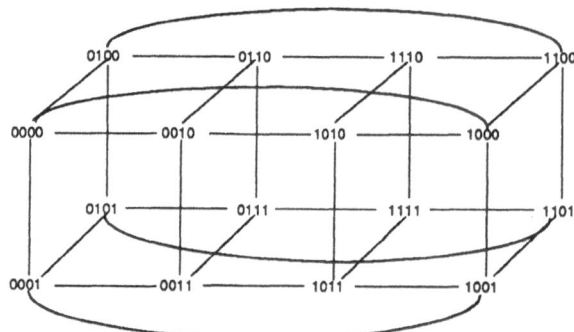

Figure 1

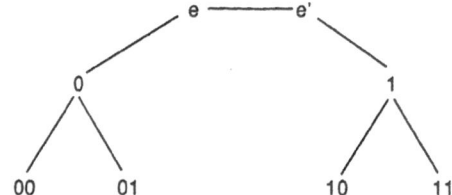

Figure 2

Clearly, $B(n)$ has $2^{n+1} - 1$ nodes and $2^{n+1} - 2$ edges. It is also easily seen that the diameter of $B(n)$ is $2n$, which is $O(\log N)$, where N is the number of its nodes, and the maximum node degree is 3.

Meshes The d-dimensional mesh of dimensions a_1, a_2, ..., a_d, denoted by $[a_1 \times a_2 \times \cdots \times a_d]$, is the graph whose nodes are all d-tuples of positive integers $(z_1, z_2, \ldots, z_d)$, where $1 \leq z_i \leq a_i$, for all i $(0 \leq i \leq d)$, and whose edges connect d-tuples which differ in exactly one coordinate by one.

Clearly, $[a_1 \times a_2 \times \cdots \times a_d]$ has $a_1 \times a_2 \times \cdots \times a_d$ nodes. Its diameter is $(a_1 - 1) + (a_2 - 1) + \cdots + (a_d - 1)$ and maximum node degree is $2d$, if each a_i is at least three.

Pyramids The pyramid of height n, denoted by $P(n)$, is the graph whose nodes are all triples of nonnegative integers (i, x, y), where $0 \leq i \leq n$ and $1 \leq x, y \leq 2^i$, and whose edges connect (i, x, y) with the vertices in $\{(i + 1, u, v)|u$ in $\{2x, 2x - 1\}$ and v in $\{2y, 2y + 1\}\}$ as well as with all vertices (i, u, v) such that (x, y) and (u, v) are adjacent nodes in the mesh $[2^i \times 2^i]$, for all i $(0 \leq i < n)$ and all x, y $(1 \leq x, y \leq 2^i)$.

$P(n)$ has $1 + 4 + 4^2 + \cdots + 4^n$ nodes. Its diameter is $2n - 1$ and it has maximum node degree 9. An illustration of $P(2)$ is shown in Figure 3.

X-trees The X-tree of height n, denoted by $X(n)$, is the graph whose nodes are all binary strings of length at most n and whose edges connect each string x of length i $(0 \leq i < n)$ with the strings xa, a in $\{0, 1\}$, of length $i + 1$ and, when $\text{binary}(x) < 2^i - 1$, connects x with $\text{successor}(x)$, where $\text{binary}(x)$ is the integer x represents in binary notation and $\text{successor}(x)$ denotes the unique binary string of length i such that $\text{binary}(\text{successor}(x)) = \text{binary}(x) + 1$. (For completeness let $\text{binary}(e) = 0$, where e is the empty string.)

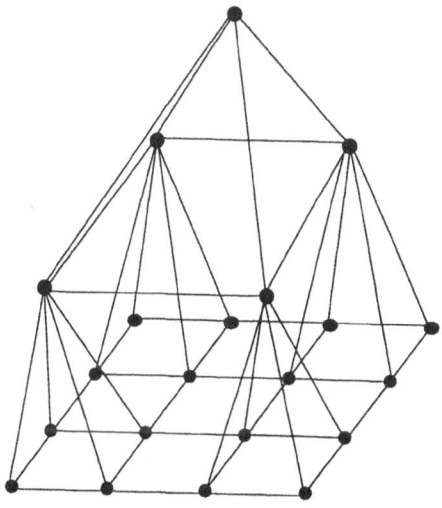

Figure 3

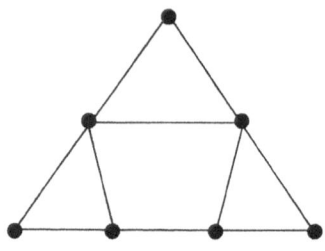

Figure 4

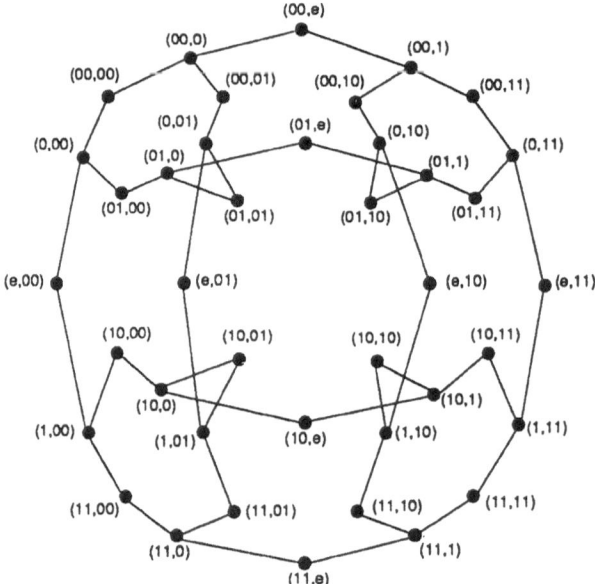

Figure 5

$X(n)$ has $2^{n+1} - 1$ nodes and $2^{n+2} - n - 4$ edges. Its diameter is $2n - 1$ and it has maximum node degree 5. An illustration of $X(2)$ is shown in Figure 4.

Mesh-of-Trees The mesh-of-trees of dimension n, denoted by $MT(n)$, is the graph whose nodes are all pairs (x, y), where x and y are binary strings of length at most n, with at least one of x, y of length exactly n, and whose edges connect, when x is of length less than n, (x, y) with (xa, y), and, when y is of length less than n, (x, y) with (x, ya), where a is in $\{0, 1\}$.

$MT(n)$ has $2^{n+1}(2^{n+1} - 2^{n-1} - 1)$ nodes and $2^{n+2}(2^n - 1)$ edges. Its diameter is $4n$ and it has maximum node degree 3. An illustration of $MT(2)$ is shown in Figure 5.

Butterflies The butterfly network of dimension n, denoted by $BF(n)$, is the graph whose nodes are all pairs (i, x), where i is a nonnegative integer $(0 \le i < n)$ and x is a binary string of length n and whose edges connect (i, x) with both $(i + 1 \pmod n, x)$

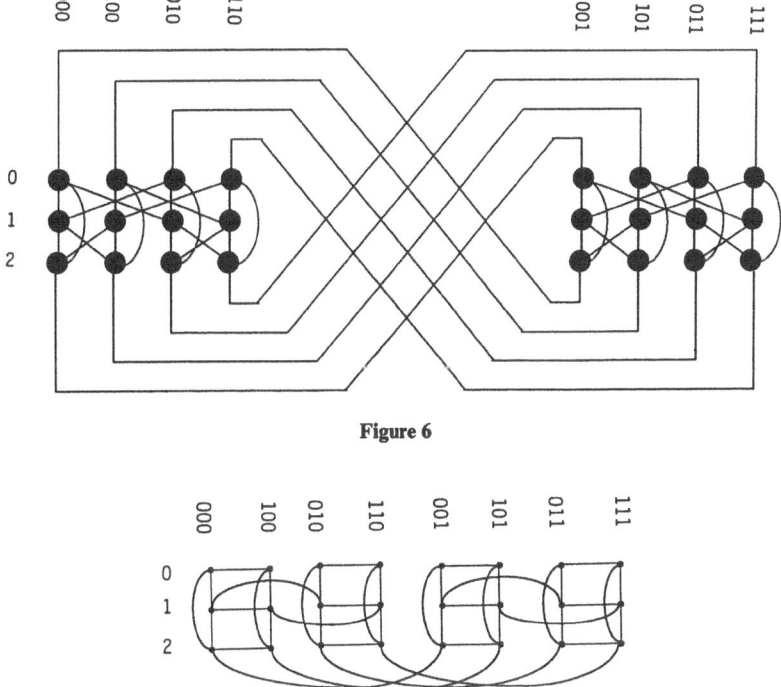

Figure 6

Figure 7

and with $(i + 1(\bmod n), x|i + 1)$, where $x|i + 1$ denotes the binary string which is identical to x except in the $((i + 1) \bmod n)$-th bit.

$BF(n)$ has $n2^n$ nodes and $n2^{n+1}$ edges, for all $n > 2$. Its diameter is $n + \text{floor}(n/2)$ and it has maximum node degree 4. An illustration of $BF(3)$ is shown in Figure 6.

Cube connected cycles   The cube connected cycle network of dimension n, denoted by $CCC(n)$, is the graph whose nodes are all pairs (i, x), where i is a nonnegative integer $(0 \le i < n)$ and x is a binary string of length n and whose edges connect (i, x) with both $(i + 1(\bmod n), x)$ and with $(i, x|i)$, where $x|i$ denotes the binary string which is identical to x except in the i-th bit.

$CCC(n)$ has $n2^n$ nodes and $3n2^{n-1}$ edges, for all $n > 2$. Its diameter is $2n + \text{floor}(n/2)$ and it has maximum node degree 3. An illustration of $CCC(3)$ is shown in Figure 7.

Shuffle-Exchange Networks   The shuffle-exchange network of dimension n, denoted by $SE(n)$, is the graph whose nodes are all binary strings of length n and whose edges connect each string xa, where x is a binary string of length $n - 1$ and a is in $\{0, 1\}$, with the string xb, where $b \ne a$ is a symbol in $\{0, 1\}$, and with the string ax. (An edge connecting xa with xb, $a \ne b$, is called an *exchange* edge and an edge connecting ax with xa is called a *shuffle* edge.)

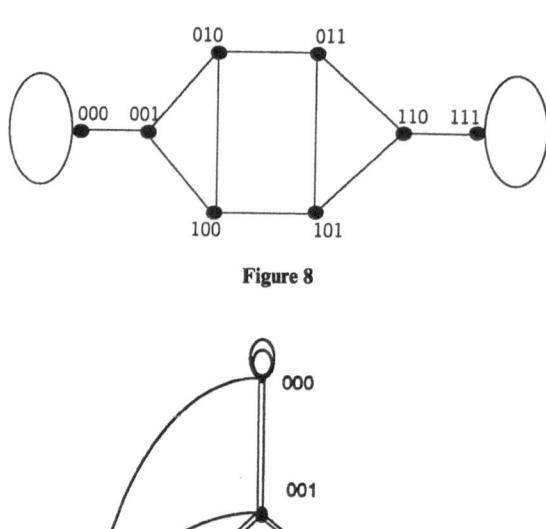

Figure 8

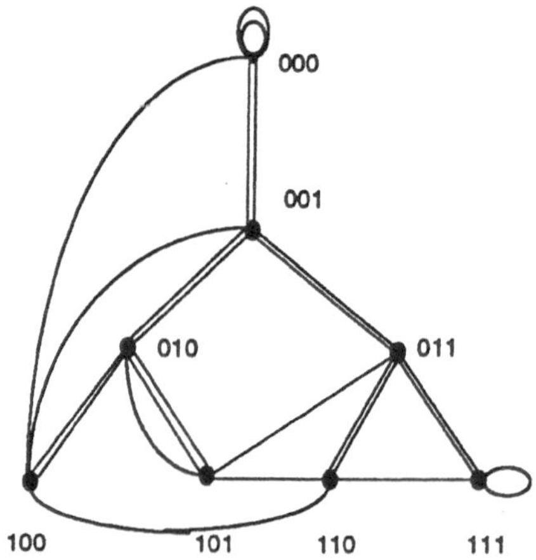

Figure 9

$SE(n)$ has 2^n nodes and 2^{n+1} edges. (Actually, the latter is a count of the directed edges, namely two from each node. The number of undirected edges will be smaller, as some of the connections are identical). Its diameter is $2n - 1$ and it has maximum node degree 3. An illustration of $SE(3)$ is shown in Figure 8.

DeBrujn Networks   The DeBrujn network of dimension n, denoted by $DB(n)$, is the graph whose nodes are all binary strings of length n and whose edges connect each string xa, where x is a binary string of length $n - 1$ and a is in $\{0, 1\}$, with the string bx, where $b \neq a$ is a symbol in $\{0, 1\}$, and with the string ax. (An edge connecting xa with bx, $a \neq b$, is called a *shufflexchange* edge and an edge connecting xa with ax is called a *shuffle* edge.)

$DB(n)$ has 2^n nodes and $2(2^n - 1)$ edges. Its diameter is n and it has maximum node degree 4. An illustration of $DB(3)$ is shown in Figure 9.

II. Measuring the Quality of Embeddings

Let G and H be finite undirected graphs. An *embedding* of G into H is a mapping f from the nodes of G to the nodes of H. G is called the *guest* graph and H is called the *host* graph of the embedding f. Most of the results we describe here are for one-to-one mappings, where each processor in the host network is assigned at most one process (represented by an assigned guest node). However, we shall also consider many-to-one embeddings, where each processor in the host can have many assigned processes. This has been done before in several places, for instance [FF], [BovL], [EMS], [GuH], [GuH2], [DS1], [DS2]. When considering many-to-one embeddings load factor is an important issue. That is, the *load factor* of an embedding f is the maximum, over all host graph nodes x, of the number of guest nodes assigned to x. Clearly it is advantageous to minimize the load factor in a simulation of one network by another, as the distinct processes assigned to the same processor will be run sequentially. Thus, the amount of time needed to simulate one step of the guest network is proportional to the maximum number of processes assigned to the same host processor. The *dilation* of an embedding f is the maximum distance in the host between the images of adjacent guest nodes, i.e. $\max\{\text{distance}_H(f(x), f(y)) | (x, y) \text{ is an edge in } G\}$, where $\text{distance}_H(a, b)$ denotes the length of the shortest path in H between the nodes a and b. Clearly one wishes to minimize dilation in a simulation of one network by another, as the amount of time to communicate between formerly adjacent processors is proportional to the distance between the host nodes to which they have been assigned. The *expansion* of the embedding f is the ratio of the number of nodes in the host graph to the number of nodes in the guest graph, i.e. $|\text{nodes}(H)|/|\text{nodes}(G)|$. When hosts are chosen from a collection C and no graph K in C satisfies $|\text{nodes}(G)| \leq |\text{nodes}(K)| < |\text{nodes}(H)|$, then H is called an *optimal* host in C for G. If there is a unique optimal host graph H in C for G, then H is called the *optimum* host in C for G. We also want to minimize expansion, as we want to use the smallest possible host network. (In fact, we may only have a fixed size host network and, consequently, we may have to consider many-to-one embeddings for large source structures.) We shall sometimes augment an embedding of G in H by a routing of G's edges, i.e. a mapping r of G's edges to paths in H. The *edge congestion* of such a routing r of G's edges, is the maximum, over all edges e in H, of the number of edges in G mapped to a path in H which includes e. That is, it is the maximum over all edges e in H of the number of edges of G routed through e. Clearly we also would like to minimize edge congestion, as it measures the amount of possible contention in the host for the same network link. If too many messages need to be passed through the same link, then some will need to be stored temporarily at the bottleneck and sent later. This will also add extra time to the communication between processors.

III. Embedding into Binary Hypercubes

As a binary hypercube has a regular structure and its diameter and number of connections at each node is logarithmic in its size, it is a popular architecture in the design of parallel computer networks. Several papers discuss the ability of binary hypercubes to simulate other network and algorithm structures. The following is a survey of some of this work:

A. Binary Trees

The complete binary tree $B(n)$, which has $2^{n+1} - 1$ nodes, can be embedded into the hypercube $Q(n + 1)$, which has 2^{n+1} nodes, with dilation 2. In fact, $B(n)$ can be embedded into $Q(n)$ in such a way that exactly one of its edges connects nodes assigned to positions at distance 2 in the hypercube and all others connect nodes at distance 1 [BhCLR], [BhI], [Hav], [Ne]. To see this observe that the double rooted binary tree $DRB(n)$ is a subgraph of $Q(n)$. This can be seen by a simple inductive argument. Observe that $DRB(1)$ is a subgraph of $Q(2)$. Now assume that $DRB(n)$ is embedded in $Q(n + 1)$ by a dilation 1 embedding f. Consider the positions assigned in the hypercube for the root e, the alternate root e', and the neighbors of these two nodes: 0 and 1. These four nodes form a chain of length 4, say 0, e, e', 1. As it is a dilation 1 embedding the successive positions they are mapped to must differ in exactly one bit position, say the first differ in the i-th bit, the next differ in the j-th bit, and the last differ in the k-th bit, where $1 \le i, j, k \le n + 1$. Then consider the embedding f' illustrated in Figure 10(a), where $f'(e) = f(0)$, $f'(e') = f(e)$, and $f'(1) = f(e')$. A dilation 1 embedding g of $DRB(n + 1)$ into $Q(n + 2)$ is obtained from the embeddings f and f'. That is, one views f as embedding the left subtree of $DRB(n + 1)$, which is a copy of $DRB(n)$, into the left half of $Q(n + 2)$, i.e. the copy of $Q(n + 1)$ which consists of all nodes whose bit string begins with 0, and, similarly, views f' as embedding the right subtree of $DRB(n + 1)$ into the right half of $Q(n + 2)$. The embedding g is illustrated in Figure 10(b).

Note that an inorder numbering of the nodes of a complete binary tree of height n also describes a dilation 2 embedding [BhCLR]. This is illustrated in Figure 11. Dilation 1 is not possible, as it is known that the complete binary tree $B(n)$ is not

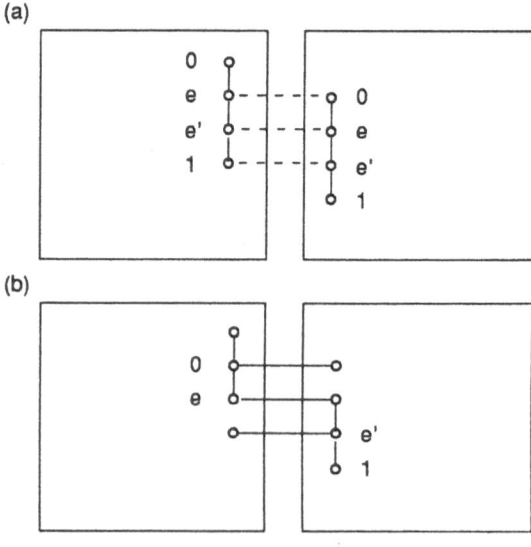

Figure 10

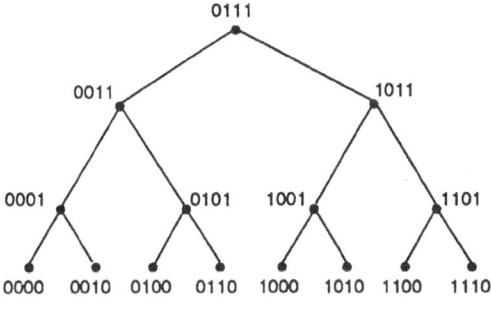

Figure 11

a subgraph of $Q(n + 1)$, for all $n > 1$. The argument is straightforward. Both binary trees and hypercubes are bipartite graphs, i.e. their nodes can be assigned two colors, say black and white, so that adjacent nodes do not receive the same color. Such a two coloring of $B(n)$, for $n > 1$, must result in $2^n + 2^{n-2} + \cdots > 2^n$ nodes receiving the same color, as all nodes at the same level must receive the same color and so must all nodes at odd (even) levels. Similarly, $Q(n + 1)$ is bipartite and any two coloration of its nodes results in all nodes with an even number of occurrences of the bit 1 getting the same color and similarly with those with an odd number of occurrences of the bit 1. Thus any two coloration of $Q(n + 1)$ has exactly 2^n nodes in each color class. So, $B(n)$ cannot be a subgraph of $Q(n + 1)$, as it has too many nodes in the same color class. Note that dilation 2 embeddings are possible, as we've seen, as they allow nodes in the same color class of $B(n)$ to change color classes in $Q(n + 1)$.

Embeddings of arbitrary binary trees into hypercubes with small dilation have also been described. The principal technique is the use of an appropriate bisection theorem, i.e. a result describing a set of edges in the tree whose deletion results in two collections of subtrees, each having half of the total number of nodes. Bhatt, Chung, Leighton, and Rosenberg [BhCLR] described a dilation 10 embedding with small expansion (small here means roughly 4). An alternative construction was described by Monien and Sudborough [MoSu2], giving a dilation 5 embedding without expansion and a dilation three embedding with constant expansion.

Other embeddings of trees into hypercubes include results about caterpillars and refinements of caterpillars. A *caterpillar* is a tree in which there is a simple path P such that every vertex is either included in P or is adjacent to a node in P. (The edges connecting nodes in P to nodes not in P are called *legs*.) A tree T is a refinement of a caterpillar if it is possible to obtain T from some caterpillar by the addition of degree two nodes into some number of the caterpillar's legs. For example, caterpillars and certain refinements of caterpillars are known to be subgraphs of hypercubes [MoSpUW], [HavL].

Embeddings of binary trees into hypercubes by many-to-one maps have not received much attention. Presumably the dilation can be lowered when the embeddings are many-to-one (not one-to-one), even when small load factor is required.

It is known that the complete binary tree of height k, for all $k > 0$, can be embedded into the hypercube with half as many points, namely $Q(k)$, with dilation 1 and load factor 2 [Mo3]. The embedding to achieve this is a straightforward application of the one-to-one embedding described earlier for double rooted binary trees. That is, let $DRB(k)$ be embedded into $Q(k + 1)$ by a dilation 1 embedding f. Without loss of generality, let the root be assigned to 0^{k+1} and the alternate root be assigned to $0^j 10^{k-j}$, i.e. the two nodes are mapped to hypercube nodes that differ in the $(j + 1)$-th position. Then, consider the mapping f' of the nodes of $DRB(k)$ to the nodes of $Q(k)$ which maps each node x to the string obtained from $f(x)$ by deleting the $(j + 1)$-th position in the string. Now remove the alternate root and view f' as a two-to-one mapping from $B(k)$ to $Q(k)$. Clearly, as the mapping has not increased the distance between the images of any nodes and it has decreased by one the distance between the root and the image of its child that was at distance 2, the new embedding has dilation 1. Furthermore, every node of $Q(k)$ hosts two guest processes except the one that hosts the root (it has only one assigned process). Thus, this embedding is optimum.

B. Meshes

Any mesh whose dimensions are a power of 2 is a subgraph of its optimum hypercube. That is, for all $n > 0$, if $n = n_1 + n_2 + \cdots n_k$, then $[n_1 \times n_2 \times \cdots \times n_k]$ is a subgraph of $Q(n)$. This is easily seen by induction on n. For example, this means that $Q(4)$ contains as a subgraph the meshes $[2 \times 8]$, $[4 \times 4]$, $[4 \times 2 \times 2]$ and $Q(4)$ is, of course, identical to the mesh $[2 \times 2 \times 2 \times 2]$. It follows from this that many meshes whose dimensions are not all a power of two are also subgraphs of their optimum hypercubes. For example, the mesh $[7 \times 7]$ with 49 points is a subgraph of the mesh $[8 \times 8]$ and, therefore, of its optimum hypercube $Q(6)$. The general statement is that a d-dimensional mesh $[a_1 \times a_2 \times \cdots a_d]$ is a subgraph of its optimum hypercube if and only if $\text{ceiling}(\log_2 a_1) + \text{ceiling}(\log_2 a_2) + \cdots + \text{ceiling}(\log_2 a_d) = \text{ceiling}(\log_2 a_1 + \log_2 a_2 + \cdots + \log_2 a_d)$ [BrS], [ChC], [Gr].

That this condition is necessary is easily seen. For example, suppose we have a dilation 1 embedding f of a 2-dimensional mesh $[m \times n]$ in its optimum hypercube, i.e. the hypercube $Q(t)$, for $t = \text{ceiling}(\log_2 (m \times n))$. Call, for any s, nodes (i, s) and $(i + 1, s)$ *column-adjacent nodes in the i-th row* and (s, i) and $(s, i + 1)$ *row-adjacent nodes in the i-th column*. First, observe that any dilation 1 embedding must map all column-adjacent nodes in the same row and all row-adjacent nodes in the same column to hypercube nodes that differ in the same bit position. For instance, let f map (i, s) to 0^t and $(i + 1, s)$ to $0^{k-1} 10^{t-k}$, i.e. hypercube nodes that differ in just the k-th bit. Let f map $(i, s + 1)$ to $0^{p-1} 10^{t-p}$, for some p, which (without any loss of generality) we assume is greater than k. Then, the mesh node $(i + 1, s + 1)$, which is a neighbor of both $(i + 1, s)$ and $(i, s + 1)$ must map to the hypercube node $0^{k-1} 10^{p-k-1} 10^{t-p}$, as f is a dilation 1 embedding. Therefore, $(i, s + 1)$ and $(i + 1, s + 1)$ also map to hypercube nodes that differ in just the k-th bit. The general statement follows. Secondly, observe that row-adjacent nodes in the same column and column-adjacent nodes in the same row cannot be mapped to hypercube nodes that differ in the same position, as each row and column intersect and this would result in mesh nodes being mapped to the same hypercube node. So, if f is a dilation 1 embedding of the mesh $[m \times n]$ into its optimum hypercube $Q(t)$, then there must be (a) at least ceiling($\log_2 m$) bits in the binary strings denoting hypercube positions that are altered for column-adjacent nodes in the m rows and (b) at least ceiling($\log_2 n$) bits, distinct from those described in (a), for the row-adjacent nodes in the n columns. This is only possible, if $t \geq$ ceiling($\log_2 m$) + ceiling($\log_2 n$).

It is known that every 2-dimensional mesh can be embedded into its optimum hypercube with dilation 2 [Ch]. This is optimum, as the preceding paragraph shows many 2-dimensional meshes are not subgraphs of their optimum hypercubes. An earlier technique [BeMS] for embedding a $[m \times n]$ mesh into its optimum hypercube actually did so by embedding with dilation 2 it into the mesh $[2^{m'} \times p]$, where $m' =$ ceiling($\log_2 m$) and p is determined by the technique. (As the latter mesh has a power of two rows, it is a subgraph of a hypercube.) For example, the mesh $[5 \times 50]$ by this technique is embedded with dilation 2 in the mesh $[8 \times 32]$. The latter mesh is a subgraph of the optimum hypercube $Q(8)$ for the $[5 \times 50]$ mesh. The embedding of a $[m \times n]$ mesh into a mesh $[2^{m'} \times p]$, where $m' =$ ceiling($\log_2 m$) is done via the construction of tiles. A $(m, 2^i)$-tile, for any $i > 0$ and any m ($2^{i-1} \leq m \leq 2^i$), is an embedding of a $[m, 2^i]$ mesh into a $[2^i, m]$ mesh such that rows of the original mesh are embedded as horizontal chains, i.e. the nodes in the first (last) column of the original $[m, 2^i]$ mesh are embedded into the first (last, respectively) column of the host $[2^i, m]$ mesh. (In particular, the embedding that simply rotates the original mesh and maps rows to columns with dilation 1 is not satisfactory.) A recursive construction of $(m, 2^i)$ tiles, for all $i > 0$, is described and it is shown that each constructed tile describes a dilation 2 embedding. (An example of the (5, 8)-tile constructed is shown in Figure 12.)

For a mesh $[m \times n]$ one performs the embedding into $[2^{m'} \times p]$, for some p, by taking the $(m, 2^{m'})$-tile T and chaining it together in the form $T\text{-}T^R - T\text{-}T^R$-formed by a vertical reflection of T [BeMS]. Although this technique falls short of embedding every 2-dimensional mesh into its optimum hypercube, it does describe a dilation 2 embedding for a large number. As indicated, moreover, it is now known that all 2-dimensional meshes can be embedded with dilation 2 into their optimum hypercubes [Ch]. The techniques is similar, but does not embed meshes into meshes. Instead it embeds a mesh into the optimum hypercube directly (using

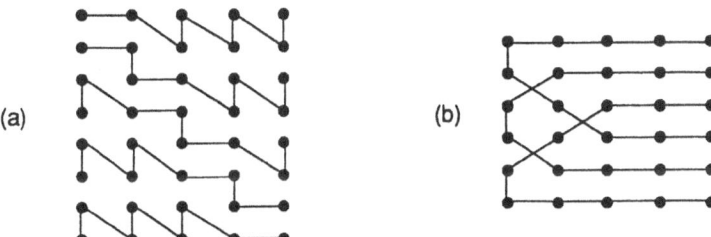

(a)   (b)

Figure 12

binary reflected gray codes) and thereby uses the additional edges available in a hypercube.

These techniques have also been investigated for their ability to embed multi-dimensional meshes into their optimum hypercubes. The technique of embedding meshes into meshes (using explicitly constructed dilation 2 tiles) results in a method to embed d-dimensional tiles with dilation at most d into hypercubes. Under certain conditions (described in [BeMS]) the technique is guaranteed to embed a d-dimensional mesh into its optimum hypercube. As this condition is not satisfied by a large number of d-dimensional meshes, the general question of embedding multi-dimensional meshes into their optimum hypercubes is still open. The technique used to show that all 2-dimensional meshes can be embedded with dilation 2 into their optimum hypercube has also been extended to the multi-dimensional case [Ch].

Results are also known about many-to-one embeddings of meshes into hypercubes [EMS]. For an arbitrary mesh M and positive integer i, let M's $1/2^i$-size hypercube be the hypercube with $1/2^i$ as many processors as M's optimum size hypercube. For example, if M is a [5×5] mesh (it has 25 nodes), then M's optimum size hypercube is $Q(5)$ with 32 nodes, its 1/2-size hypercube is $Q(4)$ with 16 nodes, its 1/4-size hypercube is $Q(3)$ with 8 nodes, etc. In [EMS] it is shown that, for all i, every 2-dimensional mesh can be embedded into its $1/2^i$-size hypercube with dilation 1 and load factor $1 + 2^i$. For example, each $2D$ mesh can be embedded into its 1/2-size hypercube with load factor 3 and dilation 1 and into its 1/4-size hypercube with load factor 5 and dilation 1. In many cases better results are known. For example, it is known that every mesh which has a number of rows which can be expressed as either 2^i or $2^i(1 + 2^j)$, for some nonnegative integers i and j, can be embedded into its $1/2^i$-size hypercube with dilation 1 and load factor 2^i. For example, every mesh with 2, 3, 4, 5, 6, 8, 9, 10, 12, 16, 17, 18, 20, 24, 32, or 33 rows can be embedded with dilation 1 and load factor 2 into its 1/2 size hypercube. We illustrate an embedding of the [5×5] mesh into the 16 point binary hypercube in Figure 13.

The technique used to obtain these embeddings of meshes into smaller hypercubes is called *braiding*. To illustrate, we describe in Figure 14 a braiding of 5 rows of a mesh to yield load factor 2 and dilation 1, and in Figure 15 a braiding of 9 rows of a mesh to yield load factor 2 and dilation 1. It should be noted that we say that r rows can be braided on 2^s *rods* with load factor f and dilation 1, when there is an embedding of the points of an r row mesh onto the points of a 2^s row mesh type structure with extra column edges (to be described) that assigns the points column by column from left to right across the guest and host mesh structures and assigns *uniformly* f points of the r row mesh to each point in a given column of the 2^s row mesh structure before assigning points to the next column. (The extra column edges of the 2^s row mesh structure are the same as in a hypercube. In particular, let the 2^s rows be labeled by binary strings of length s and in the order given by a binary reflected Gray code. There are edges between points in the same column whose row labels differ in one position, as well as between corresponding row positions in adjacent columns.) Clearly, such a 2^s mesh structure is a subgraph of a hypercube. In fact, it is a subgraph of the hypercube formed by increasing, if necessary, the

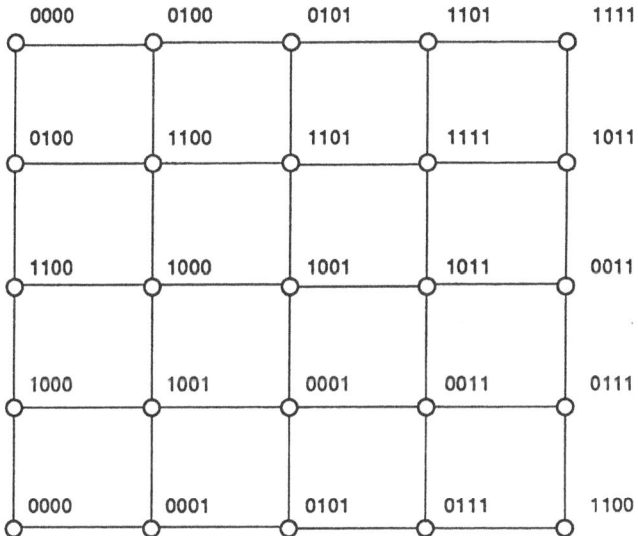

Figure 13

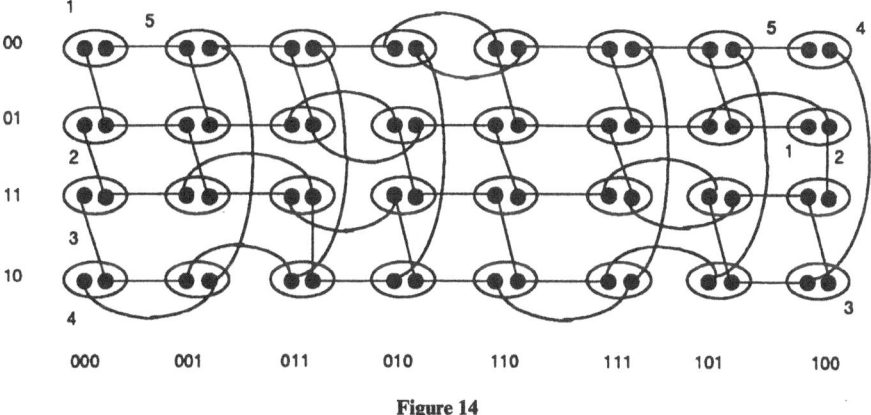

Figure 14

number of columns to the next power of 2 and then, after labeling, say the resulting 2^t, columns with successive strings in a binary reflected Gray code of all
binary strings of length t, adding edges between columns whose labels differ in one
position.

There are various theorems about braidings that help to find efficient embeddings
of meshes into smaller hypercubes. For instance, in [EMS] a product theorem
states: If r_1 rows can be braided on s_1 rods with uniform load factor f_1 and dilation
1 and r_2 rows can be braided on s_2 rods with uniform load factor f_2 and dilation 1,
then $r_1 \times r_2$ rows can be braided on $s_1 \times s_2$ rods with uniform load factor $f_1 \times f_2$
and dilation 1. For example, as 5 rows can be braided onto 4 rods with uniform

B. Monien and H. Sudborough

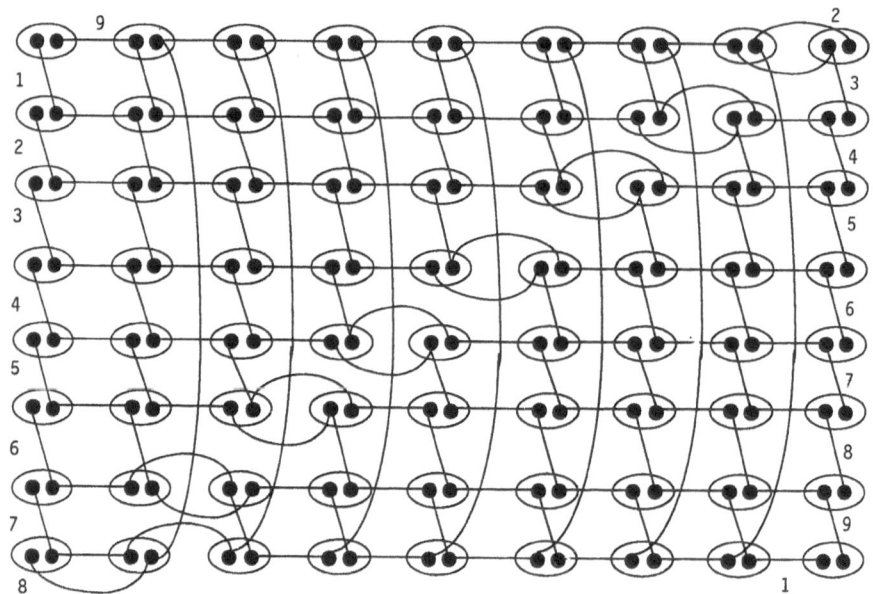

Figure 15

load factor 2 and dilation 1, it follows that 25 rows can be braided on 16 rods with uniform load factor 4 and dilation 1.

These techniques and other similar results are used in [EMS] to embed multidimensional meshes into smaller hypercubes with small dilation and optimum or nearly optimum load factor. For example, a $3D$ [$5 \times 5 \times 9$] mesh can be embedded into its 1/2-size hypercube, $Q(7)$, with dilation 2 and load factor 2.

C. Pyramids

The pyramid $P(k)$, for all $k > 0$, can be embedded into its optimum hypercube, $Q(2k + 1)$, with dilation 2 and edge congestion 2 [St]. (Stout did not consider edge congestion, but it can be seen that the embedding he describes does indeed have edge congestion 2.) We describe a different embedding here with the same bounds on dilation and edge congestion. Our embedding is described recursively. To begin with, a dilation 2, edge congestion 2 embedding of $P(1)$ into $Q(3)$ is shown in Figure 16. Define the following invariant property, for the sake of induction: $P(k)$ can be embedded into $Q(2k + 1)$ by an embedding f_k which:

(a) has dilation 2 and edge congestion 2,
(b) maps the apex of the pyramid $P(k)$ to a hypercube node, called the *standard apex position*, which has an unassigned neighbor, called the *alternate apex position* such that at most one edge is routed through the edge connecting the standard and alternate apex positions, and

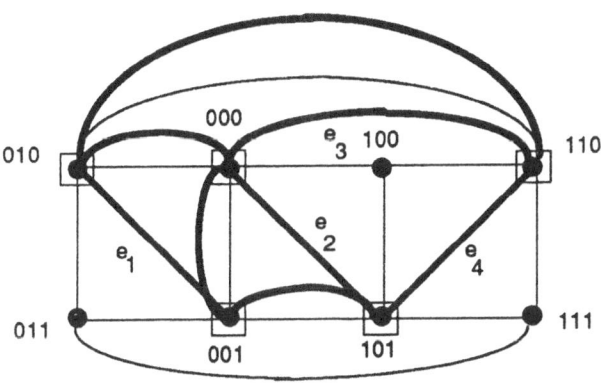

Figure 16

(c) the embedding g_k that agrees with f_k on every node of $P(k)$ except the apex and maps the apex to the alternate instead of the standard apex position also satisfies conditions (a)–(b), where the role of the standard and alternate apex positions are reversed.

Embed the nodes and edges of $P(k + 1)$ into $Q(2k + 3)$ by the embedding f_{k+1} which is defined as follows:

(1) View $Q(2k + 3)$ as partitioned into four copies of $Q(2k + 1)$, which we refer to as the quadrants of $Q(2k + 3)$. The four quadrants are defined by the sets of nodes in $Q(2k + 3)$ that begin with the prefixes 00, 01, 10, and 11, respectively,

(2) Embed a copy of $P(k)$ into each of the four quadrants, where the copies embedded in the 10 and 11 quadrants are mapped by f_k and the copies embedded in the 00 and 01 quadrants are mapped by g_k, i.e. the apexes of the copies of $P(k)$ embedded in the 00 and 01 quadrants are placed at the alternate apex positions, and

(3) Place the apex of $P(k + 1)$ in the apex position of the 00 quadrant and then route the edges of $P(k + 1)$ as shown in Figure 17.

It is easily seen that conditions (a)–(c) are satisfied by f_{k+1}. Note that the edges connecting corresponding nodes in the four copies of the pyramid $P(k)$, while not explicitly shown in Figure 17, connect nodes assigned to corresponding hypercube positions (hence are neighbors in the embedding). Furthermore these edges connect nodes assigned to distinct quadrants and are not used for other edges in the embedding. Thus f_{k+1} is a dilation 2, edge congestion 2 embedding of the pyramid $P(k + 1)$ into its optimum hypercube.

Dilation 3, edge congestion 2 and dilation 2, edge congestion 3 embeddings of pyramids into their optimum hypercubes have also been described [LaW], [LaW2]. These authors conjectured earlier that no embedding could achieve dilation 2 and edge congestion 2 simultaneously, apparently unaware of the earlier result of Stout [St]. In fact, this conjecture has also been disproved in another direction.

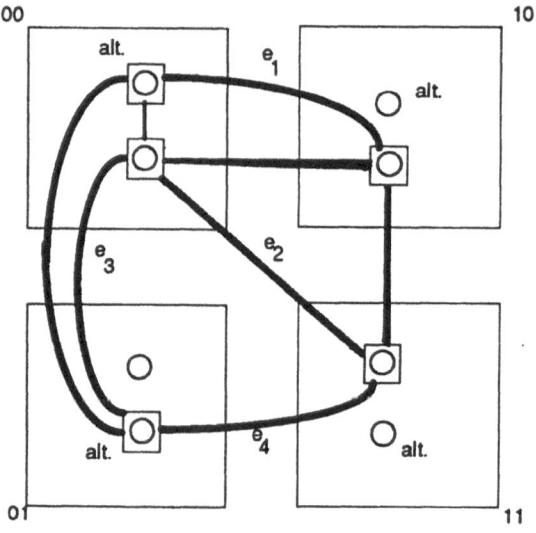

Figure 17

In [DS2] a dilation 3, edge congestion 1 embedding of $P(k)$ into $Q(2k + 5)$ has been described. This embedding can again be described by induction on k. To begin consider the following dilation 3 and edge congestion 1 embedding of $P(1)$ and $Q(7)$.

Map the apex of $P(1)$ to 0^7 and the four base nodes to 10^6, $10^5 1$, $10^2 10^3$, and 11010^3. Then route the edges as follows.

(1) edges from the apex of $P(1)$ to the base nodes:
    (a) $0^7 \rightarrow 10^6$
    (b) $0^7 \rightarrow 0^6 1 \rightarrow 10^5 1$
    (c) $0^7 \rightarrow 0^3 10^3 \rightarrow 10^2 10^3$
    (d) $0^7 \rightarrow 010^5 \rightarrow 01010^3 \rightarrow 11010^3$

(2) edges between the base nodes:
    (a) $10^6 \rightarrow 10^2 10^3$
    (b) $10^2 10^3 \rightarrow 11010^3$
    (c) $11010^3 \rightarrow 110^5 \rightarrow 110^4 1 \rightarrow 10^5 1$
    (d) $10^5 1 \rightarrow 10^6$

It is then easily verified that the edges of the pyramid are routed through unique hypercube edges (i.e. the embedding has edge congestion 1) and that the dilation is 3. Furthermore, for the sake of our induction step, we observe that there is a copy of $Q(5)$ as a subgraph of $Q(7)$ in which only one node of the pyramid is assigned. This is, the set of nodes $\{0x_1 x_2 0x_3 x_4 x_5 |$for each i $(1 \le i \le 5)$, x_i is in $\{0, 1\}\}$ is a set of nodes that appropriately induces a copy of $Q(5)$ with only the node 0^7 used as a host. In addition, only the nodes $0^6 1$ and 010^5 in this set are used for routing pyramid edges in the embedding of $P(1)$ into $Q(7)$. The inductive step is then accomplished by assuming a dilation 3, edge congestion 1 embedding of $P(k)$ into

$Q(2k + 5)$ in which there is a copy of $Q(5)$ as a subgraph with only one of its points assigned (namely for the apex of $P(k)$) and only two other points (as indicated) used for routing edges. Then, $P(k + 1)$ is embedded in $Q(2k + 7)$ with dilation 3 and edge congestion 1 by viewing $P(k + 1)$ as four copies of $P(k)$ with an additional apex, using the points in the indicated copies of $Q(5)$ in the hosts of the inductively given embeddings of each of the four $P(k)$'s to host the additional apex node of $P(k + 1)$ and to route the edges, and finally to obtain a new copy of $Q(5)$ as indicated for the inductive step. The details can be found in [DS2].

We note that edge congestion 1 is not possible into the smallest possible hypercubes, at least for small pyramids. For example, the pyramids $P(1)$, $P(2)$, and $P(3)$ have maximum node degrees of 5, 7, and 9, respectively, while their optimum hypercubes, namely $Q(3)$, $Q(5)$, and $Q(7)$, have maximum node degree 3, 5, and 7, respectively. Therefore, edge congestion 2 is necessary for any embedding of these pyramids into their optimum hypercubes. It is unknown, as yet, whether there exists a dilation 2, edge congestion 1 embedding of $P(k)$ into $Q(2k + 1)$, for $k > 3$. It is known that any embedding of a pyramid into a hypercube must have dilation at least 2, as pyramids have odd length cycles. Other work on embeddings of pyramids into hypercubes has been described recently by [HoJ2].

D. X-trees

There is a dilation 2, edge congestion 2 embedding of $X(k)$ into $Q(k + 1)$, for all $k > 0$. The embedding strategy is similar to that used for embedding complete binary trees and pyramids and is easily defined recursively. In particular, we assume for an inductive hypothesis that there is a dilation 2, edge congestion 2 embedding of $X(k)$ into $Q(k + 1)$ such that the root of the X-tree is assigned to a hypercube position that has an unassigned neighbor and that there is one edge routed through the edge connecting the position of the root and this neighbor. Such an embedding of $X(1)$ into $Q(2)$ is shown in Figure 18(a). Let f_k denote such a dilation 2, edge congestion 2 embedding of $X(k)$ into $Q(k + 1)$. An appropriate embedding f_{k+1} of $X(k + 1)$ into $Q(k + 2)$ is defined by the following:

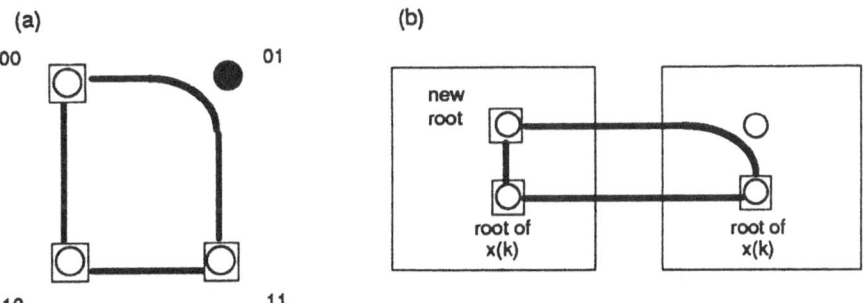

Figure 18

(1) Embed a copy of $X(k)$ into each of two copies of $Q(k+1)$ by f_k,
(2) Place a new node, the root of $X(k+1)$, into the unassigned position adjacent to the position of the root in the embedding f_k of one of the two copies, and
(3) Route the edges as shown in Figure 18(b).

Note that edges connecting nodes in copies of $X(k)$ will connect nodes assigned to corresponding positions in each of the two copies of $Q(k+1)$. So, these edges connect nodes placed at distance 1 in the hypercube and they also have edge congestion 1.

E. Mesh of Trees

A dilation 2 embedding of $MT(n)$ into $Q(2n+2)$ is easily described using a dilation 2 embedding of the complete binary tree $B(n)$ into $Q(n+1)$ and the observation that $MT(n)$ is a product of two such trees. For example, take the dilation 2 embedding f_n of $B(n)$ into $Q(n+1)$ given by the inorder numbering of nodes (using binary notation). Then define the embedding g_n of the mesh of trees $MT(n)$ into $Q(2n+2)$ by $g_n(x,y) = f_n(x)f_n(y)$. The embedding g_2 of $MT(2)$ into $Q(6)$ is shown in Figure 19.

(e,00)	→ 011000	(e,01)	→ 011010	(e,10)	→ 011100	(e,11)	→ 011110
(0,0)	→ 001000	(0,01)	→ 001010	(0,10)	→ 001100	(0,11)	→ 001110
(1,00)	→ 101000	(1,01)	→ 101010	(1,10)	→ 101100	(1,11)	→ 101110
(00,00)	→ 000000	(00,01)	→ 000000	(00,10)	→ 000100	(00,11)	→ 000110
(01,00)	→ 010000	(01,01)	→ 010010	(01,10)	→ 010100	(01,11)	→ 010110
(10,00)	→ 100000	(10,01)	→ 100010	(10,10)	→ 100100	(10,11)	→ 100110
(11,00)	→ 110000	(11,01)	→ 110010	(11,10)	→ 110100	(11,11)	→ 110110
(00,e)	→ 000011	(01,e)	→ 010011	(10,e)	→ 100011	(11,e)	→ 110011
(00,0)	→ 000001	(01,0)	→ 010001	(10,0)	→ 100001	(11,0)	→ 110001
(00,1)	→ 000101	(01,1)	→ 010101	(10,1)	→ 100101	(11,1)	→ 110101

Figure 19

F. Butterflies and Cube Connected Cycles

The butterfly $BF(n)$, for all even integers $n > 0$, can be embedded into its optimum hypercube $Q(n + \text{ceiling}(\log_2 n))$ with dilation 1 [Stoe]. That is, $BF(n)$ is a subgraph of its optimum size hypercube, for even integers $n > 0$. As there is a dilation 1 embedding of $CCC(n)$ into $BF(n)$ [FU], i.e. $CCC(n)$ is a subgraph of $BF(n)$, it follows that $CCC(n)$ is, of course, also a subgraph of its optimum size hypercube for even integers $n > 0$.

G. Shuffle-Exchanges and DeBrujn Networks

It remains open whether either of these networks can be embedded into a hypercube with $O(1)$ dilation and $O(1)$ expansion. As the Shuffle-Exchange network can be embedded with dilation 1 into the DeBrujn network [Fu], a positive resolution of both questions can be obtained by an appropriate embedding of the DeBrujn graph.

H. Complexity Issues

Hypercube Embedding Problem
Instance: A finite undirected graph G and positive integers k and n.
Qestion: Does there exist a dilation k embedding of G into $Q(n)$?

This problem is known [KrVC] to be NP-complete even when $k = 1$ (by a reduction from the 3-partition problem [GaJ]). In fact, it is known to be NP-complete even to decide if a tree can be embedded with dilation 1 into its optimum size hypercube [CW].

IV. Embeddings into Binary Trees

A simple path can be embedded into its optimum complete binary tree with dilation 3 [Se]. An outerplanar graph with maximum vertex degree d can be embedded into a binary tree with dilation ceiling($\log_2 2d$) + ceiling($\log_2 \log_2 2d$) + 5 [Mo]. For any $n > 0$, the X-tree $X(n)$ can be embedded in the complete binary tree $B(n)$ with dilation $O(\log n) = O(\log \log N)$, where N is the number of vertices in the X-tree [BhCHLR].

It is known to be NP-complete to decide, given a graph G and a positive integer k, whether G can be embedded into a binary tree with dilation k [Mo]. In fact, it is NP-complete even for trees. On the other hand, for each fixed k, there is a poly-nomial algorithm (using dynamic programming), which when given a graph G, decides if G can be embedded with dilation at most k in a binary tree [MaSS]. It is easily established, using the respective diameters of a guest graph and the intended binary tree host, that many graphs cannot be embedded into a complete binary tree with $O(1)$ dilation. For example, complete ternary trees require $c(\log \log n)$ dilation [HoMR], for some $c > 0$, and a dilation $O(\log \log n)$ embedding exists [Ell1].

V. Embeddings into Meshes

Embedding meshes into meshes is an interesting issue. Every 2-dimensional mesh can be embedded into either its optimum square 2-D mesh or the next-larger-size square 2-D mesh with dilation at most 3 [Ell2]. The technique uses *squeezing* and *folding*, which were described in [AlR]. Examples of these operations are shown in Figure 20. In fact, a similar squeezing operation was used, via the recursive construc-tion of tiles, as described earlier, to embed a mesh M into a mesh M' in which M' is a subgraph of M's optimum hypercube [BeMS]. A recent paper [MeH], in fact, shows that every rectangular grid can be embedded into a small square grid with dilation 2. Other related work was done in [FS].

Embeddings into meshes of complete binary trees, meshes of trees, planar graphs, shuffle-exchange networks, and other network structures have also been described in work on VLSI [Ull]. For example, complete binary trees are embedded into meshes by the well known H-tree construction and with better dilation by a modified H-tree construction, as described in [Ull]. Embeddings of general network structures into meshes are described by separator theorems, such as the $O(n^{1/2})$ planar separator theorem, and the recursive construction of a network layout based on separator results [Ull].

The problem of, given a finite undirected graph G and positive integers k and d, deciding whether G can be embedded with dilation at most k into a d-dimensional mesh is known [BhCo] to be NP-complete (by a reduction from 1-in-3 3SAT) even when the graph G is a binary tree, $k = 1$, and $d = 2$. There are also interesting upper and lower bound results on embeddings of meshes into meshes of a different number of dimensions, the routing of messages between the processors in such a simulation [KoA], [KRT], and results about embeddings of meshes with "wrap around edges" (called *toruses*) [MaT].

Results are also known about many-to-one embeddings of meshes into meshes [SS]. This work describes how to embed meshes into smaller meshes with optimum load factor and dilation 1. These techniques do not work for all possible host meshes though; it is easily seen that for some host meshes large dilation and/or large load factor is required. For example, if the host mesh is a $1 \times k$ mesh, for some k, i.e. a linear chain, then it is straightforward to show that large dilation and/or large load factor is necessary. Other work has been described in [EMS] to embed multidimensional meshes into smaller meshes with small dilation and optimum or nearly optimum load factor. Similarly results are known about many-to-one embeddings of torus networks into smaller torus networks [PS]

VI. Embeddings into Butterfly and Cube Connected Cycle Networks

The complete binary tree $B(n + \text{floor}(\log_2 n))$ can be embedded into the butterfly network $BF(n + 3)$ with dilation 4 [BhCHLR]. This shows that an n-vertex X-tree, for example, can be embedded into a butterfly network with dilation $O(\log \log n)$ and $O(1)$ expansion, as the paper also describes an embedding of X-trees into complete binary trees, as described earlier. The paper also shows that, there is a constant $c > 0$, such that for any nontree planar graph G whose smallest $1/3:2/3$ separator is of size $S(n)$ and in which $F(G)$ is the largest number of vertices in any internal face (of a planar embedding), any embedding of G into a butterfly must have dilation at least $[c \cdot \log S(n)]/F(G)$. In particular, as a 2-dimensional mesh is planar, has an $n^{1/2}$ separator, and has 4 nodes per face, any embedding of 2-D meshes into a butterfly must have dilation at least $c \cdot \log n$, for some constant $c > 0$. This is proportional to the butterfly's diameter and hence a random placement of the mesh nodes achieves this order of magnitude dilation.

VII. Embeddings into Pyramids

In [DS1] it is shown that, for any $k \geq 0$, the complete binary tree of height $2k + 1$ is a subgraph of its optimum size pyramid, namely $P(k + 1)$. This dilation 1 embedding of $B(2k + 1)$ into $P(k + 1)$ can perhaps best be described by induction on $k \geq 0$. For the basis step, we map $B(1)$ into $P(1)$ by the embedding f, where $f(e) = (1, 0, 0), f(0) = (1, 0, 1)$, and $f(1) = (1, 1, 0)$. For the inductive step, assume that one is given already an embedding of $B(2k + 1)$ into $P(k + 1)$ with the leaves occupying all of the points in the odd diagonal positions of level $k + 1$ (the bottom level) of $P(k + 1)$. Then, the embedding can be extended to a dilation 1 embedding of $B(2k + 3)$ into $P(k + 2)$ by: (a) assigning the points in level $2k + 2$ of the complete binary tree to points in even diagonal positions of level $k + 2$ of the host pyramid that are adjacent in the pyramid to where their parent (in the tree) has been assigned, (b) assigning the points in level $2k + 3$ of the complete binary tree to points in odd diagonal points of level $k + 2$ of the host pyramid that are adjacent in the host to where their parent (in the tree) has been assigned. Details can be found in [DS1]. Such a dilation 1 embedding of $B(3)$ into $P(2)$ is shown in Figure 20.

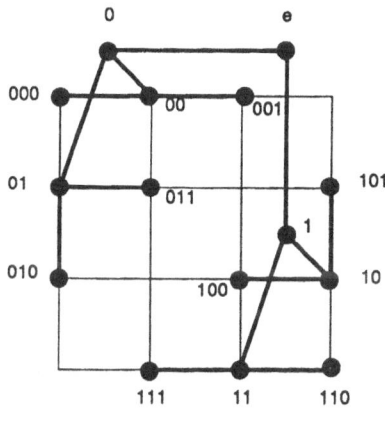

Figure 20

Furthermore, for any $k \geq 0$, the X-tree of height $2k + 1$ can be embedded into its optimum size pyramid, $P(k + 1)$, with dilation 2 and edge congestion 2 [DS1]. The embedding is again described by induction on k. The embedding has greater depth technically than the one just described, however, as the X-tree has edges than the complete binary tree does not (namely, those connecting points in the same level) and additional care must be taken to assign points of the X-tree and route its edges to achieve dilation 2 and edge congestion 2. However, such an efficient embedding of X-trees yields an important corollary. Namely, using the efficient embedding of arbitrary binary trees into X-trees [Mo2], it shows that arbitrary binary trees can be efficiently embedded into pyramids. That is, one first embeds an arbitrary binary tree into a X-tree and then embeds the host X-tree into an appropriate pyramid.

Many-to-one embeddings into pyramids have also been considered. In [DS2] it is shown that pyramids can be embedded into smaller pyramids with dilation 1 and optimum load factor. For example, it is shown that, for all $k > 0$, a pyramid of height $k + 1$ can be embedded into a pyramid of height k with dilation 1 and load factor 5 in such a way that only one node of the host pyramid receives 5 guests, namely the apex of the host. As $P(k + 1)$ has one more than 4 times the nodes of $P(k)$, this is an optimum embedding. Furthermore, this has been extended to yield a dilation 1, optimum load factor embedding of $P(k + j)$ into $P(k)$, for all k, $j > 0$. For example, [DS2] describes, for all $k > 0$, an embedding of $P(k + 2)$ into $P(k)$ with dilation 1 and load factor 17 in which 5 nodes host 17 guests, namely those at levels 0 and 1, and all other nodes host 16 guests. This is optimum, as $P(k + 2)$ has 5 more than 16 times the nodes of $P(k)$. These compression embeddings mean that one can compute rather efficient embeddings of large structures into small pyramids by first embedding into the structure's optimum size pyramid and then compressing the optimum size pyramid into a smaller size pyramid. In [DS2] optimum dilation and load factor embeddings of binary trees and X-trees into small pyramids have explicitly been described. (These embeddings of complete binary trees and X-trees into small pyramids are better than what one would obtain by the process mentioned of (1) embedding into the optimum size pyramid and then (2) compressing the optimum size pyramid into a smaller one.

VIII. Embeddings into X-trees

Using bisection lemmas for arbitrary binary trees, as described in [MoSu2], Monien [Mo2] has described techniques for embedding arbitrary binary trees into X-trees with dilation 10 and $O(1)$ expansion. As indicated earlier, this result enables one to describe efficient embeddings of arbitrary binary trees into other networks by simply describing an embedding of an appropriate X-tree.

IX. Concluding Remarks

Not much is known about embeddings into shuffle-exchange or DeBrujn networks. For example, can arbitrary binary trees be embedded with $O(1)$ dilation and $O(1)$ expansion in shuffle-exchange graphs? Clearly, the complete binary tree $B(n - 1)$ is a subgraph of the DeBrujn network $DB(n)$, as the DeBrujn graph $DB(n)$ can be viewed as a complete binary tree (with an added node adjacent to the root) together with edges forming another complete binary tree added on. (See Figure 9.) The DeBrujn graph $DB(n)$ can also be embedded with dilation 2 in the shuffle-exchange $SE(n)$, as the shufflexchange edge of the DedBrujn can be simulated by a shuffle edge followed by an exchange edge of the shuffle-exchange graph. Thus, the completebinary tree $B(n - 1)$ can be embedded with dilation 2 in the shuffle-exchange network $SE(n)$.

There is a wealth of results about embedding graphs into simple paths, i.e. linear layouts. There dilation is customarily called *bandwidth* and edge congestion is usually called *cutwidth*. The interested reader should consult some of the literature sources [ChiCDG], [Chu], [ChuLR], [ChuMST], [ElST], [FeL], [GuS], [MaPS], [MaS], [MaS2], [Mi], [MiS], [MoSu1], [MoSu3], [MoSu4], [Si], [Su], [Ya], [Ya2]. In particular, many early papers on the subject of embedding graphs with small dilation or small average dilation were written by A. Rosenberg, and co-authors, for example, in [Ro], [Ro2], [Ro3], [RS]. We are guilty of a possibly unavoidable (certainly unintentional) sin of not including all relevant references about embedding problems. Hopefully, some of these omissions will be forgiven by referring interested readers to the following valuable sources for additional work: [AnBR], [BeS], [Bi], [ChS], [CyKVC], [Gr], [HeR], [HasLN], [HoJ], [HoJ3], [Ne], [SaS], [Wu].

References

[AlR]	R. Aleliunas, A. L. Rosenberg, "On Embedding Rectangular Grids in Square Grids", *IEEE Trans. on Computers, C-31, 9* (1982), pp. 907–913.
[AnBR]	F. N. Annexstein, M. Baumslag, A. L. Rosenberg, "Group Action Graphs and Parallel Architectures", manuscript, Computer and Info. Sci., University of Massachusetts, Amherst, Massachussetts 01003, U.S.A., 1987.
[BeS]	B. Becker, H. U. Simon, "How Robust is the *n*-Cube?" *Information and Computation, 77,* (1988), pp. 162–178.
[BeMS]	S. Bettayeb, Z. Miller, I. H. Sudborough, "Embedding Grids into Hypercubes", *Proc. of Aegean Workshop on Computing*, Springer Verlag's *Lecture Notes in Computer Science*, Vol. 319 (1988), pp. 201–211.
[BhCLR]	S. Bhatt, F. Chung, T. Leighton, A. Rosenberg, "Optimal Simulation of Tree Machines", *Proc. 27th Annual IEEE Symp. Foundations of Computer Sci.*, Oct. 1986, pp. 274–282.
[BhCHLR]	S. Bhatt, F. Chung, J.-W. Hong, T. Leighton, A. Rosenberg, "Optimal Simulations by Butterfly Networks", *Proc. 20th Annual ACM Theory of Computing Symp.*, 1988, pp. 192–204.
[BhCo]	S. Bhatt, S. S. Cosmadakis, "The Complexity of Minimizing Wire Lengths for VLSI Layouts", *Info. Processing Letters 25* (1987).
[BhI]	S. N. Bhatt, I. C. F. Ipsen, "How to Embed Trees in Hypercubes", Research Report YALEU/DCS/RR-443, Yale University, Dept. of Computer Science, 1985.
[Bi]	D. Bienstock, "On Embedding Graphs in Trees", manuscript, Bell Communications Research, Morristown, New Jersey 07060, U.S.A., 1988.
[BovL]	H. L. Bodlaender and J. van Leeuwen, "Simulation of Large Networks on Smaller Networks", *Information and Control 71* (1986), pp. 143–180.
[BrS]	J. E. Brandenburg, D. S. Scott, "Embeddings of Communication Trees and Grids into Hypercubes", *Intel Scientific Computers Report*, #280182-001, 1985.
[Ch]	M. Y. Chan, "Dilation 2 Embedding of Grids into Hypercubes", Tech. Report, Computer Science Program, Univ. Texas at Dallas, 1988.
[ChC]	M. Y. Chan, F. Y. L. Chin, "On Embedding Rectangular Grids in Hypercubes", *IEEE Trans. on Computers, 37* (1988), pp. 1285–1288.
[ChS]	T. F. Chan, Y. Saad, "Multigrid Algorithms on the Hypercube Multiprocessor", *IEEE Trans. on Comp.*, Vol c-35, No. 11, Nov. 1986, pp. 969–977.
[ChiCDG]	P. Z. Chinn, J. Chvatalova, A. K. Dewdney, N. E. Gibbs, "The Bandwidth Problem for Graphs and Matrices—A Survey", *J. Graph Theory, 6* (1982), pp. 223–254.
[Chu]	F. P. K. Chung, "Labelings of Graphs", A chapter in *Selected Topics in Graph Theory*, III, (eds. L. Beinike and R. Wilson).
[ChuLR]	F. R. K. Chung, F. T. Leighton, A. L. Rosenberg, "A Graph Layout Problem with Applications to VLSI Design", manuscript, 1985.
[ChuMST]	M.-J. Chung, F. Makedon, I. H. Sudborough, J. Turner, "Polynomial Algorithms for the Min-Cut Linear Arrangement Problem on Degree Restricted Trees", *SIAM J. Computing 14, 1* (1985), pp. 158–177.
[CW]	D. Corneil and A. Wagner, manuscript, Department of Computer Science, University of British Columbia, Vancouver, B.C., Canada, 1988.
[CyKVC]	G. Cybenko, D. W. Krumme, K. N. Venkataraman, A. Couch, "Heterogeneous Processes on Homogeneous Processors", manuscript, Dept. of Computer Sci., Tufts University, Medford, Massachusetts 02155 U.S.A., 1986.
[DS1]	A. Dingle, I. H. Sudborough, "Simulation of Binary Trees and *X*-trees on Pyramid Networks", *Proc. of the 1st Annual IEEE Symp. on Parallel and Distributed Processing*, 1989, pp. 210–219.

[DS2]   A. Dingle, I. H. Sudborough, "Efficient Uses of Pyramid Networks", *Proc. of the 1st Annual IEEE Symp. on Parallel and Distributed Processing*, 1989, pp. 220–229.

[Ell1]   J. A. Ellis, "Embedding Graphs in Lines, Trees, and Grids", Ph.D. Thesis, Northwestern Univ., Evanston, Illinois, U.S.A. (1984).

[Ell2]   J. A. Ellis, "Embedding Rectangular Grids into Square Grids", *Proc. of Aegean Workshop on Computing*, Springer Verlag's *Lecture Notes in Computer Science*, Vol. 319 (1988), pp. 181–190.

[ElST]   J. A. Ellis, I. H. Sudborough, J. Turner, "Graph Separation and Searching", manuscript, Computer Science Program, University of Victoria, P. O. Box 1700, Victoria, B.C. V8W 2Y2, Canada (1987).

[EMS]   J. A. Ellis, Z. Miller, I. H. Sudborough, "Embedding Large Meshes into Small Hypercubes", manuscript, Computer Science Program, M.P. 31, University of Texas at Dallas, Richardson, Texas, 75083-0688 (1989).

[FeL]   M. R. Fellows, M. A. Langston, "Layout Permutation Problems and Well-Partially-Ordered Sets", manuscript, Department of Computer Science, Washington State University, Pullman, Washington, 99164-1210 U.S.A., 1988.

[FF]   J. P. Fishburn and R. A. Finkel, "Quotient Networks", *IEEE TC*, *31*, 1982, pp. 288–295.

[FS]   A. Fiat, A. Shamir, "Polymorphic Arrays: A Novel VLSI Layout for Systolic Computation", *Proc. of IEEE Foundations of Computer Sci. Conf.*, 1984, pp. 37–45.

[FU]   R. Feldmann, W. Unger, "The Cube Connected Cycle network is a subgraph of the Butterfly network", submitted for publication

[GaJ]   M. R. Garey, D. S. Johnson, *Computers and Intractability: A Guide to the Theory of NP-Completeness*, W. H. Freeman and Co., San Francisco, 1979.

[Gr]   D. S. Greenberg, "Optimum Expansion Embeddings of Meshes in Hypercube", Technical Report YALEU/CSD/RR-535, Yale University, Dept. of Computer Science.

[GuH1]   A. K. Gupta, S. E. Hambrusch, "A Lower Bound on Embedding Tree Machines with Balanced Processor Utilization", manuscript, Dept. of Computer Sciences, Purdue University, West Lafayette, IN 47907, 1988.

[GuH2]   A. K. Gupta, S. E. Hambrusch, "New Cost Measures in the Embedding of Tree Machines", manuscript, Dept. of Computer Sciences, Purdue University, West Lafayette, IN 47907, 1988.

[GuS]   E. Gurari, I. H. Sudborough, "Improved dynamic programming algorithms for bandwidth minimization and the min cut linear arrangement problem" *J. Algorithms*, 5 (1984), pp. 531–546.

[HasLN]   J. Hastad, T. Leighton, M. Newman, "Reconfiguring a Hypercube in the Presence of Faults", *Proc. 19th Annual ACM Symp. Theory of Computing*, May 25–27, 1987.

[Hav]   I. Havel, "On Hamiltonian Circuits and Spanning Trees of Hypercubes", *Cas. Pest. Mat.* (in Czech.), *109* (1984), pp. 135–152.

[HavL]   I. Havel, P. Liebl, "One Legged Caterpillars Span Hypercubes", *J. Graph Theory*, 10 (1986), pp. 69–76.

[HeR]   L. S. Heath, A. L. Rosenberg, "An Optimal Mapping of the FFT Algorithm onto the Hypercube Architecture", COINS Tech. Report 87-19, Computer and Info. Sci., University of Massachusetts, Amherst, Massachussets 01003, U.S.A., 1987.

[HoJ]   C.-T. Ho, S. L. Johnson, "On the Embedding of Arbitrary Meshes in Boolean Cubes with Expansion Two Dilation Two", *Proc. 1987 International Conference on Parallel Processing*, pp. 188–191.

[HoJ2]   C.-T. Ho, S. L. Johnson, "Embedding Hyper-Pyramids into Hypercubes" Dept. of Computer Science, Yale University, New Haven, CT 06520 (1988).

[HoJ3]   C.-T. Ho, S. L. Johnson, "Embedding Meshes in Boolean Cubes by Graph Decomposition", Dept. of Computer Science, Yale University, New Haven, CT 06520 (1989).

[HoR]   J. W. Hong, A. L. Rosenberg, "Graphs that are Almost Binary Trees", *SIAM J. Computing* 11, 2 (1982), pp. 227–242.

[HoMR]   J. W. Hong, K. Mehlhorn, A. Rosenberg, "Cost Trade-offs in Graph Embeddings with Applications", *J. ACM*, *30*, 4 (1983), pp. 709–728.

[KoA]   S. R. Kosaraju, M. J. Atallah, "Optimal Simulations Between Mesh-Connected Arrays of Processors", *Proc. 1986 ACM Theory of Computing Symp.*, pp. 264–272.

[KRT]   D. Krisanc, S. Rajasekaran, T. Tsantilas, "Optimal Routing Algorithms for Mesh-Connected Processor Arrays", *Proc. of Aegean Workshop on Computing*, Springers *Lecture Notes in Computer Science*, Vol. 319 (1988), pp. 411–422.

[KrVC]   D. W. Krumme, K. N. Venkataraman, G. Cybenko, "Hypercube Embedding is *NP*-complete", *Proc. of Hypercube Conf.*, SIAM, Knoxville, Tennessee, Sept., 1985.

[LaW]   T.-H. Lai, W. White, "Embedding Pyramids into Hypercubes", OSU-CISRC-11/87-TR41, Dept. of Computer and Info. Sci., The Ohio State Univ., Columbus, Ohio, 43210, U.S.A., 1988.

[LaW2]   T.-H. Lai, W. White, "Mapping Multiple Pyramids into Hypercubes Using Unit Expansion", manuscript, Dept. of Computer and Info. Sci., The Ohio State Univ., Columbus, Ohio, 43210, U.S.A., 1988.

[MaT]   Y. E. Ma, L. Tao, "Embeddings among Toruses and Meshes", *Proc. of the 1987 Int. Conf. on Parallel Processing*, August, 1987, pp. 178–187.

[MaPS]   F. Makedon, C. H. Papadimitriou, I. H. Sudborough, "Topological Bandwidth", *SIAM J. Alg. and Discrete Meth. 6* (1985), pp. 418–444.

[MaS]   F. Makedon, I. H. Sudborough, "Minimizing Width in Linear Layouts", *Discrete Applied Math., 23* (1989), pp. 243–265.

[MaS2]   F. Makedon, I. H. Sudborough, "Graph Layout Problems", *Surveys in Computer Science* (ed. H. Maurer), Bibliographisches Insitut, Zurich, 1984, pp. 145–192

[MaSS]   F. Makedon, C. G. Simonson, I. H. Sudborough, "On the complexity of tree embedding problems", manuscript, Computer Science Program, University of Texas at Dallas, Richardson, Texas, 75083-0688, U.S.A., 1988.

[MeH]   R. Melhem, G.-Y. Hwang, "Embedding Rectangular Grids into Square Grids with Dilation 2", manuscript, Dept. of Computer Science, Univ. of Pittsburgh, Pittsburgh, PA, 15260, 1989.

[Mi]   Z. Miller, "A Linear Algorithm for Topological Bandwidth in Degree 3 Trees", *SIAM J. Computing, 17* (1988), pp. 1018–1035.

[MiS]   Z. Miller, I. H. Sudborough, "A Polynomial Algorithm for Recognizing Small Cutwidth in Hypergraphs", *Proc. of Aegean Workshop On Computing*, Springer Verlag's *Lecture Notes in Computer Science*, vol. 227 (1986), pp. 252–260.

[Mo]   B. Monien, "The Problem of Embedding Trees into Binary Trees is NP-Complete", *Proc. FCT '85*, Springers *Lecture Notes in Computer Science*, vol. 199, pp. 300–309.

[Mo2]   B. Monien, "Simulating Binary Trees on X-Trees", manuscript, 1988

[Mo3]   B. Monien, personal communication, 1989

[MoSpUW]   B. Monien, G. Spenner, W. Unger, and G. Wechsung, "On the Edge Length of Embedding Caterpillars into Various Networks", manuscript, Dept. of Math. and Computer Science, Univ. Paderborn, Paderborn, W. Germany, 1988.

[MoSu1]   B. Monien, I. H. Sudborough, "Min Cut is NP-complete for Edge Weighted Trees", *Theoretical Computer Science, 58* (1988), pp. 209–229.

[MoSu2]   B. Monien, I. H. Sudborough, "Simulating Binary Trees on Hypercubes", *Proc. of Aegean Workshop on Computing*, Springer Verlag's *Lecture Notes in Computer Science*, Vol. 319 (1988), pp. 170–180.

[MoSu3]   B. Monien, I. H. Sudborough, "Bandwidth constrained NP complete problems", *Theoretical Computer Science 41* (1985), pp. 141–167.

[MoSu4]   B. Monien, I. H. Sudborough, "On eliminating nondeterminism from Turing machines that use less than logarithm worktape space", *Theoretical Computer Science 21* (1982), pp. 237–253.

[Ne]   L. Nebesky, "On Cubes and Dichotomic Trees", *Cas. Pest. Mat.* (in Czech.), *99* (1974), pp. 164–167.

[PS]   T. Peng, I. H. Sudborough, "Embedding Large Torus Networks into Small Meshes, Toruses, and Hypercubes", manuscript, Computer Science Program, M.P. 31, University of Texas at Dallas, Richardson, Texas, 75083-0688 (1989).

[Ro]   A. L. Rosenberg, "Preserving Proximity in Arrays", *SIAM J. Computing* 1979, pp. 443–460.

[Ro2]   A. L. Rosenberg, "Data Graphs and Addressing Schemes", *J.C.S.S., 5*, 1971, pp. 193–238.

[Ro3]   A. L. Rosenberg, "An extrinsic characterization of addressable data graphs", *Discrete Math., 9*, 1974, pp. 61–70.

[RS]   A. L. Rosenberg, L. Snyder, "Bounds on the Costs of Data Encodings", *Math. Systems Theory, 12*, 1978, pp. 9–39.

[SaS]   Y. Saad and M. H. Schultz, "Data Communication in Hypercubes", Yale University Research Report RR-428, October 1985.

[Se]   M. Sekanina, "On an Ordering of the Set of Vertices of a Connected Graph", Publications Faculty Science, Univ. Brno 412 (1960), pp. 137–142.

[Si]   C. G. Simonson, "A Variation on the Min Cut Linear Arrangement Problem", *Math. Systems Theory, 20* (1987), pp. 235–252.

[SS]   D. Sáng, I. H. Sudborough, "Embedding Large Meshes into Small Ones", manuscript, Computer Science Program, M.P. 31, University of Texas at Dallas, Richardon, Texas, 75083-0688 (1989).

[St]   Q. Stout, "Hypercubes and Pyramids", *Pyramidal Systems for Computer Vision*, V. Cantoni and S. Levialdi, eds., Springer, 1986, pp. 75–89.

[Stoe]   E. Stoehr, "An optimum embedding of the butterfly network in the hypercube", manuscript, Akadamie der Wissenschaften der DDR, Karl Weierstrass-Institut-fuer-Mathematik, Mohrenstr. 39, Berlin, DDR-1086, German Democratic Republic (1989).

[Su]   I. H. Sudborough, "Bandwidth constraints on problems complete for polynomial time", *Theoretical Computer Science*, 26 (1983), pp. 25–52.

[Ull]   J. D. Ullman, *Computational Aspects of VLSI*, Computer Science Press, 11 Taft Court, Rockville, Maryland 20850, U.S.A., 1984.

[Wu]   A. Y. Wu, "Embedding of Tree Networks into Hypercubes", *J. of Parallel and Distributed Computing*, 2, 3 (1985), pp. 238–249.

[Ya]   M. Yannakakis, "A Polynomial Algorithm for the Min Cut Linear Arrangement of Trees", *J. ACM*, 32, 4 (1985), pp. 950–959.

[Ya2]   M. A. Yannakakis, "Linear and Book Embeddings of Graphs", *Proc. of Aegean Workshop On Computing*, Springers *Lecture Notes in Computer Science*, vol. 227 (1986), pp. 226–235.

B. Monien
Fachbereich Mathematik/Informatik
Universität Paderborn
D-4790 Paderborn
Federal Republic of Germany

H. Sudborough
Computer Science Program, MP 31
University of Texas at Dallas
Richardson, TX 75083-0688
U.S.A.

Scientific Computation with Automatic Result Verification

Edited by **U. Kulisch** and **H. J. Stetter**

1988. 22 figures. VIII, 244 pages.
Soft cover DM 128,–, öS 900,–
Reduced price for subscribers to "Computing":
Soft cover DM 115,20, öS 810,–
ISBN 3-211-82063-9

(Computing, Supplementum 6)

This Computing Supplementum collects a number of original contributions which are all aiming to compute rigorous and reliable error bounds for the solution of numerical problems. An introductory article of the editors about the meaning and diverse methods of automatic result verification is followed by 16 original contributions. The first chapter deals with automatic result verification for standard mathematical problems like enclosing the solution of ordinary boundary value problems, linear programming problems, linear systems of equations and eigenvalue problems. The second chapter deals with applications of result verification methods to problems of the technical sciences. The contributions consider critical bending vibrations stability tests for periodic differential equations, geometric algorithms in the plane, and the periodic solution of the oregonator, a mathematical model in chemical kinetics. The contributions of the third chapter are concerned with extending and developing the tools required in scientific computation with automatic result verification: evaluation of arithmetic expressions of polynomials in several variables and of standard functions for real and complex point and interval arguments with dynamic accuracy. As an appendix, a short account of the Fortran-SC language was added which permits the programming of algorithms with result verification in a natural manner.

Springer-Verlag Wien New York

Moelkerbastei 5, A-1010 Wien · Heidelberger Platz 3, D-1000 Berlin 33 ·
175 Fifth Avenue, New York, NY 10010, USA ·
37-3, Hongo 3-chome, Bunkyo-ku, Tokyo 113, Japan

Computer Algebra

Symbolic and Algebraic Computation

Edited by
B. Buchberger, G. E. Collins, and **R. Loos,**
in cooperation with **R. Albrecht**

Second Edition
1983. 5 figures. VII, 283 pages.
Soft cover DM 64,–, öS 448,–
ISBN 3-211-81776-X

Contents: Loos, R.: Introduction. – Buchberger, B., Loos, R.: Algebraic Simplification. – Neubüser, J.: Computing with Groups and Their Character Tables. – Norman, A. C.: Integration in Finite Terms. – Lafon, J. C.: Summation in Finite Terms. – Collins, G. E.: Quantifier Elimination for Real Closed Fields: A Guide to the Literature. – Collins, G. E., Loos, R.: Real Zeros of Polynomials. – Kaltofen, E.: Factorization of Polynomials. – Loos, R.: Generalized Polynomial Remainder Sequences. – Lauer, M.: Computing by Homomorphic Images. – Norman, A. C.: Computing in Transcendental Extensions. – Loos, R.: Computing in Algebraic Extensions. – Collins, G. E., Mignotte, M., Winkler, F.: Arithmetic in Basic Algebraic Domains. – van Hulzen, J. A., Calmet, J.: Computer Algebra Systems. – Calmet, J., van Hulzen, J. A.: Computer Algebra Applications. – Mignotte, M.: Some Useful Bounds. – Author Index. – Subject Index.

Computer algebra is an alternative and complement to numerical mathematics. Its importance is steadily increasing. This volume is the first systematic and complete treatment of computer algebra. It presents the basic problems of computer algebra and the best algorithms now known for their solution with their mathematical foundations, and complete references to the original literature. The volume follows a top-down structure proceeding from very high-level problems which will be well-motivated for most readers to problems whose solution is needed for solving the problems at the higher level. The volume is written as a supplementary text for a traditional algebra course or for a general algorithms course. It also provides the basis for an independent computer algebra course.

Springer-Verlag Wien New York

Moelkerbastei 5, A-1010 Wien · Heidelberger Platz 3, D-1000 Berlin 33 ·
175 Fifth Avenue, New York, NY 10010, USA ·
37-3, Hongo 3-chome, Bunkyo-ku, Tokyo 113, Japan

Defect Correction Methods

Theory and Applications

Edited by **K. Böhmer** and **H. J. Stetter**

1984. 32 figures. IX, 243 pages.
Soft cover DM 72,–, öS 504,–
Reduced price for subscribers to "Computing":
Soft cover DM 64,80, öS 453,60
ISBN 3-211-81832-4

(Computing, Supplementum 5)

Defect Correction Methods comprise an important class of constructive mathematical methods, many of which have been developed within the past 10 years. This volume contains a collection of papers from the major areas where defect correction methods have been devised and applied, and an introductory survey. It originated from an Oberwolfach meeting in July 1983; the articles were written by an international group of scientists who are active in this field. The volume contains the first comprehensive presentation of this important area of numerical and applied mathematics, an area whose results have so far only been published in journals and reports.

The reader will get acquainted with the ideas of defect correction through major theoretical results and through a variety of applications. The articles relate defect correction with discretization methods of many kinds (e. g. the novel multigrid technique), with algorithms for the computation of guaranteed high-accuracy results, and with design techniques in numerical software. The lists of references of the individual articles provide an easy access to the current literature on the subject.

Springer-Verlag Wien New York

Moelkerbastei 5, A-1010 Wien · Heidelberger Platz 3, D-1000 Berlin 33 ·
175 Fifth Avenue, New York, NY 10010, USA ·
37-3, Hongo 3-chome, Bunkyo-ku, Tokyo 113, Japan